MW01102403

Thermal Vibrational Convection

Thermal Vibrational Convection

G. Z. Gershuni and **D.V. Lyubimov**
Perm State University
Perm, Russia

JOHN WILEY & SONS
Chichester · New York · Weinheim · Brisbane · Singapore · Toronto

Baffins Lane, Chichester,
West Sussex PO19 1UD, England

National 01243 779777
International (+44) 1243 779777
e-mail (for orders and customer service enquiries): cs-books@wiley.co.uk
Visit our Home Page on http://www.wiley.co.uk
or http://www.wiley.com

Other Wiley Editorial Offices

John Wiley & Sons, Inc., 605 Third Avenue,
New York, NY 10158–0012, USA

VCH Verlagsgesellschaft mbH, Pappelallee 3,
D-69469 Weinheim, Germany

Jacaranda Wiley Ltd, 33 Park Road, Milton,
Queensland 4064, Australia

John Wiley & Sons (Asia) Pte Ltd, Clementi Loop #02–01,
Jin Xing Distripark, Singapore 129809

John Wiley & Sons (Canada) Ltd, 22 Worcester Road,
Rexdale, Ontario M9W 1L1, Canada

Library of Congress Cataloging–in–Publication Data

Gershuni, G.Z. (Grigoriĭ Zinov'evich)
Thermal vibrational convection / G.Z. Gershuni and D.V. Lyubimov.
p. cm.
Includes bibliographical references and index.
ISBN 0-471-97385-8 (alk. paper)
1. Fluid dynamics—Mathematics. 2. Heat—Convection—Mathematics.
3. Mass transfer—Mathematics. I. Li͡ubimov, D.V. (Dmitriĭ
Viktorovich) II. Title
QC151. G47 1997 97–14117
536′. 25—dc21 CIP

British Library Cataloguing in Publication Data

A catalogue record for this book is available from the British Library

ISBN 0 471 97385 8

Typeset in 10/12 pt Times by Thomson Press (India) Ltd., New Delhi
Printed and bound in Great Britain by Biddles Ltd, Guildford, Surrey
This book is printed on acid-free paper responsibly manufactured from sustainable forestry, in which at least two trees are planted for each one used for paper production.

To the memory of

Professor E.M. Zhukhovitsky,

one of the founders of this scientific field

Contents

Preface

The term 'thermal vibrational convection' is used by the authors in two ways. The first one is connected with the study of vibrational effects on convection, heat and mass transfer caused by a static gravity field. The second one is proposed to refer to the specific regular flows that appear under vibration in a cavity filled with a fluid. The thermovibrational flow of the second kind can be observed even in the case of pure weightlessness, because it is not connected with the static gravity.

Most of the material in this book is devoted to consideration of the case of high-frequency vibrations, i.e. when the period of vibration is much smaller than the reference hydrodynamic time and the displacement amplitude is small. In this limit, the method of averaging can be effectively applied to study the phenomena.

The first part of the book (written by G.Z. Gershuni) considers the situation where a cavity is completely filled with a fluid. In this case the fluid is at rest in the proper reference frame and its motion is uniform in the laboratory frame. The specific convective flow appears only in the presence of temperature non-uniformities.

In Chapter 1 the governing set of equations describing the vibrational convection phenomenon is written down, including the set of equations for the average fields of velocity, temperature and pressure obtained by means of the averaging method. The necessary conditions of quasi-equilibrium are derived (i.e. the state in which the average velocity is absent but the pulsation component is, in general, non-zero). Numerous equilibrium situations are considered for both weightlessness and a superposition of vibrations and static gravity. The problem of the quasi-equilibrium stability is formulated for the general case and the most interesting problems are solved. Stability borders are found as well as the characteristics of the most dangerous perturbations. Here the double diffusive situation is also considered for the case where a horizontal layer of a binary mixture with the Soret effect is subject to the static gravity and high-frequency vibration fields.

In Chapter 2 plane-parallel flows and their stability are studied. The cases of vertical and horizontal infinite layers are discussed, the main profiles are determined and their linear stability is investigated. The case of the plane-parallel vibrational convective flow under weightlessness is also considered, including its stability.

Chapter 3 is concerned with the solution (mainly, numerical) of some non-linear problems which arise when a fluid fills closed cavities of various geometries. In these cases equilibrium is not possible, and thus the goal is to study the intensity and structure of the flows. Bifurcation analysis is carried out for situations close to equilibrium. The problem of the vibrational convective boundary layer is also addressed.

The subject of Chapter 4 is the consideration of a situation when the temperature non-uniformity is caused by internal heat generation. The geometry is a plane horizontal layer under static gravity and a high-frequency vibration. Internal heat sources are distributed uniformly or the fluid is exothermal, obeying the Arrenius law. In both cases the linear stability of mechanical quasi-equilibrium is investigated and non-linear regimes bifurcating from the equilibrium are studied numerically.

Chapter 5 deals with the situation where the vibration frequency is finite. The new feature is the resonance phenomena which were excluded in the case of high frequencies. Parametric excitation is considered for a horizontal layer heated either from below or from above in the presence of modulation of the gravity field or longitudinal vibrations of finite frequency. The structure of secondary flows is also considered.

The main objective of the second part of the book (written by D.V. Lyubimov) is to investigate the situations when the structure of the pulsational field is non-uniform. A good example is a fluid-filled cavity where some part of the boundary is free. There the pulsational field exists in the proper reference frame, even when the temperature gradient is absent. The flow is generated by the Schlichting boundary conditions on the rigid part of the boundary and by the wave transport on its free part. Here the convective flow is not purely natural due to the component of the flow which is not connected with the temperature non-uniformity. That is why it is necessary to revise the conventional Boussinesq approximations. The same situation is encountered when some part of the boundary is vibrating while the rest of it is quiescent. A similar effect is also observed in a more complicated case—swing vibrations, for example.

Chapters 6 and 7 are devoted to the statement of the problem in the new situation. In Chapter 6 a method of averaging is developed and the equations for the average flow are obtained. Chapter 7 deals with the analysis of the boundary conditions.

In Chapter 8 the asymptotic transition to the limiting case of the conventional description is considered, which is possible when (a) the cavity is closed, (b) swing vibrations are applied and (c) surface tension on the free boundary is strong.

In Chapter 9 some particular situations are analyzed: a heated rigid body oscillating in the surrounding cold fluid which is at rest far from the body; linear vibrations of a liquid zone with a free surface; swing vibrations; thermovibrational convection in a layer between two coaxial cylinders when the inner one is heated and vibrates with circular polarization.

It is a pleasure for us to express our gratitude to our colleagues and coworkers, R.V. Birikh, L.M. Braverman, G.I. Burde, V.I. Chernatynsky, V.A. Demin, A.Yu. Gelfgat, I.O. Keller, A.K. Kolesnikov, V.G. Kozlov, J.-C. Legros, T.P. Lyubimova, A. Mojtabi, R. Monti, B.I. Myznikova, G.F. Putin, B. Roux, A.N. Sharifulin, V.M. Shikhov, R.R. Siraev, B.L. Smorodin, A.A. Tcherepanov, D.N. Volfson, Yu.S. Yurkov, M.A. Zaks and E.M. Zhukhovitsky, for their invaluable help and numerous useful discussions.

G.Z. Gershuni and D.V. Lyubimov

1 Basic Equations: Mechanical 'Quasi-equilibrium' and Its Stability

In this chapter the basic set of equations for thermovibrational convection is derived. We consider the limiting case of vibrations of very high frequency, which means that the period of vibrations is small in comparison with all the reference hydrodynamic times. In this situation the averaging technique may be effectively applied and the closed set of equations for averaged fields may be obtained. Further, we formulate the necessary conditions of a mechanical 'quasi-equilibrium', i.e. of the state with a zero average velocity component but, in general, a non-zero pulsation one. Linear stability of various 'quasi-equilibrium' configurations is studied by the method of small perturbations. First, the stability of equilibrium in weightlessness is investigated. Then we address the situations where both mechanisms of convection excitation—thermogravitational and thermovibrational—are superimposed. The last section of this chapter is devoted to the problem of stability of a binary mixture with the Soret effect.

1 Basic Equations of Thermovibrational Convection

Let us begin with the derivation of the equations describing natural thermal convection in the presence of vibrations. The closed cavity is assumed to be filled with a fluid. Let V be the volume of the cavity and F its boundary. Some temperature distribution is maintained on F, so that inside the fluid there exists some spatial non-uniformity of temperature. The cavity together with the confined fluid oscillates linearly and harmonically in some direction that may be characterized by the unit vector $\mathbf{n}$. Let b be the displacement amplitude and Ω the angular frequency of vibrations. The Boussinesq conditions are supposed to be valid, which means that a 'weak' convection exists in a cavity on the laboratory scale (see [1, 2]). Therefore, it is assumed that mechanical compressibility of the fluid is not considerable and the density variations caused by thermal expansion are relatively small: $\beta\Theta \ll 1$, where β is the thermal expansion coefficient and Θ is the reference temperature difference. Then the density non-uniformity may be taken into account only in the equation of motion where it produces the convective buoyancy force.

We introduce the proper coordinate system of the cavity. In this non-inertial system, the equations describing natural convection differ from the standard Boussinesq equations only by the complete acceleration, which now includes both the static gravity acceleration and the vibrational part:

$$\mathbf{g} \rightarrow \mathbf{g} + (b\Omega^2 \cos \Omega t)\mathbf{n} \tag{1.1}$$

Thus, the complete set of equations of natural convection under static gravity and vibration in the framework of Boussinesq approximation has the form

$$\frac{\partial \mathbf{v}}{\partial t} + (\mathbf{v}\nabla)\mathbf{v} = -\frac{1}{\rho}\nabla p + \nu\Delta\mathbf{v} + g\beta T\boldsymbol{\gamma} + \beta T b\Omega^2 \cos \Omega t\mathbf{n} \tag{1.2}$$

$$\frac{\partial T}{\partial t} + \mathbf{v}\nabla T = \chi\Delta T \tag{1.3}$$

$$\operatorname{div} \mathbf{v} = 0 \tag{1.4}$$

Here the notations are as usual: $\mathbf{v}$ is the velocity, T is the temperature counted from relative zero, ρ is the constant density corresponding to $T = 0$, p is the pressure deviation from the hydrostatic one at the constant density ρ, $\boldsymbol{\gamma}$ is the unit vector directed vertically upward and ν, χ, β are, respectively, the kinematic viscosity, heat conductivity and thermal expansion coefficients.

Evidently, in the closed cavity in the absence of temperature non-uniformities the fluid moves together with the cavity as a solid body. The appearance of a relative flow is due to the density non-uniformity, i.e. the non-uniformity of temperature.[†]

The existence of a vibrational convective force in the equation of motion (1.2) leads to the development of the convective flow component oscillating with time. Thus, the velocity, temperature and pressure fields may be presented as superpositions of a mean (average) and pulsational parts (the average

[†] The situation differs from the described one, e.g. in the case of a cavity with a partly free boundary. The relevant approach to that case may be found in Chapter 6 and further on.

contributions are marked with overbars):

$$\mathbf{v} = \overline{\mathbf{v}} + \mathbf{v}', \qquad T = \overline{T} + T', \qquad p = \overline{p} + p' \tag{1.5}$$

It is important to emphasize that, generally speaking, the average velocity component differs from that of a vibration-free flow. The first reason for that is the non-linearity of the system: its average response to a periodic excitation, even oscillating around zero, may be non-zero. The second reason is the structure of the vibrational convective force itself. Its expression includes the product of two oscillating factors, viz. vibrational acceleration and temperature. The average value of the product varies proportionally to $\overline{\cos\Omega t \cdot T}$ and depends on the phase ratio of acceleration and temperature oscillations which, in general, is non-zero. The mean value of the vibrational convective force induces the average component of the flow. This fact is most pronounced in the limiting case where the static gravity is absent, i.e. under pure weightlessness conditions, where harmonic oscillations of a cavity filled with a non-isothermal fluid are responsible for the appearance of the regular average flow (vibrational thermal convection in weightlessness [3]).

Now we focus on the average flow component. It may be described by the fields ($\overline{\mathbf{v}}$, $\overline{T}$ and $\overline{p}$). To obtain the closed set of equations for average fields, the averaging method is that proposed by P. L. Kapitsa when solving the problem of a pendulum with a vibrating pole [4]. The applicability of this technique is to be justified using some physical assumptions. It is necessary to suppose that the vibration frequency is high enough for the period to be small with respect to all the reference hydrodynamic and thermal times. Besides, the amplitude of vibration must be in some sense small. Then, in accordance with the main idea of the averaging method, the corrections ($\mathbf{v}'$, T', p') in equations (1.5) are assumed to oscillate quickly, with reference times of the order of the vibration period, whereas the average parts ($\overline{\mathbf{v}}$, $\overline{T}$, $\overline{p}$) are 'slow' functions of time, responding, for example, to some non-stationary external conditions or reflecting some transient process in the system.

Substituting (1.5) into the basic system (1.2) to (1.4) and selecting 'fast' terms, one can obtain the following equations:

$$\begin{aligned}\frac{\partial \mathbf{v}'}{\partial t} + (\overline{\mathbf{v}}\nabla)\mathbf{v}' + (\mathbf{v}'\nabla)\overline{\mathbf{v}} + (\mathbf{v}'\nabla)\mathbf{v}' = {} & -\frac{1}{\rho}\nabla p' + \nu\Delta\mathbf{v}' \\ & + g\beta T'\gamma + \beta(\overline{T} + T')b\Omega^2 \cos\Omega t\mathbf{n}\end{aligned} \tag{1.6}$$

$$\frac{\partial T'}{\partial t} + \overline{\mathbf{v}}\nabla T' + \mathbf{v}'\nabla\overline{T} + \mathbf{v}'\nabla T' = \chi\Delta T' \tag{1.7}$$

$$\operatorname{div}\mathbf{v}' = 0 \tag{1.8}$$

Let us simplify these equations. As assumed, the frequency is high and the amplitude is small. Due to that, in equation (1.6) all the terms containing $\mathbf{v}'$ except for the leading term $\partial \mathbf{v}'/\partial t$ are neglected. In a similar way, in equation (1.7) the only term $\partial T'/\partial t$ is retained of all the ones incorporating T'. Among the terms describing convective oscillatory forces the main one includes $\overline{T}$. Thus, retaining only main terms the simplified version of equations for the oscillatory components may be written as

$$\frac{\partial \mathbf{v}'}{\partial t} = -\frac{1}{\rho}\nabla p' + \beta \overline{T} b \Omega^2 \cos \Omega t \mathbf{n} \tag{1.9}$$

$$\frac{\partial T'}{\partial t} + \mathbf{v}' \nabla \overline{T} = 0 \tag{1.10}$$

$$\operatorname{div} \mathbf{v}' = 0 \tag{1.11}$$

These equations enable the closure problem to be solved, i.e. to express the pulsation components in terms of the average ones.

Let us consider equation (1.9). The vector field $\overline{T}\mathbf{n}$ may be presented as the sum of solenoidal ($\mathbf{w}$) and irrotational ($\nabla\varphi$) parts:

$$\overline{T}\mathbf{n} = \mathbf{w} + \nabla\varphi, \qquad \operatorname{div}\ \mathbf{w} = 0 \tag{1.12}$$

Here φ is some scalar. Substituting equations (1.12) into (1.9) and equalizing the solenoidal and irrotational parts separately, one gets

$$\frac{\partial \mathbf{v}'}{\partial t} = \beta b \Omega^2 \cos \Omega t \cdot \mathbf{w} \tag{1.13}$$

$$\nabla\left(-\frac{p'}{\rho} + \beta b \Omega^2 \cos \Omega t \cdot \varphi\right) = 0 \tag{1.14}$$

Function $\mathbf{w}$ on the right-hand side of equation (1.13) varies slowly with time. Hence, integration of (1.13) with respect to the 'fast' time yields

$$\mathbf{v}' = \beta b \Omega \sin \Omega t \cdot \mathbf{w} \tag{1.15}$$

Substituting $\mathbf{v}'$ into equation (1.10) and integrating over the 'fast' time, the oscillatory part of the temperature field takes the form

$$T' = \beta b \cos \Omega t (\mathbf{w} \nabla T) \tag{1.16}$$

as a result of substituting $\mathbf{v}'$ from equation (1.15) and T' from equation (1.16) into the complete set of equations and averaging them with respect to the 'fast' time, the governing system for averaged fields takes the form (hereafter we omit overbars)

$$\frac{\partial \mathbf{v}}{\partial t} + (\mathbf{v}\nabla)\mathbf{v} = -\frac{1}{\rho}\nabla p + \nu\Delta\mathbf{v} + g\beta T\boldsymbol{\gamma} + \varepsilon(\mathbf{w}\nabla)(T\,\mathbf{n} - \mathbf{w}) \tag{1.17}$$

$$\frac{\partial T}{\partial t} + \mathbf{v}\nabla T = \chi\Delta T \tag{1.18}$$

$$\operatorname{div}\mathbf{v} = 0 \tag{1.19}$$

Here $\varepsilon = 0.5(\beta b\Omega)^2$ is a single parameter determining the effect of vibrations on the system in the limiting case of high frequencies. As may be seen from equations (1.2) to (1.4), if the frequency has a finite value, the amplitude b and the frequency Ω are independent.

The last term in the right-hand side of equation (1.17) is apparently the average convective force of vibrational nature. It is expressed through the additional 'slow' variable—the solenoidal part $\mathbf{w}$ of the vector $\overline{T}\,\mathbf{n}$. As may be inferred from equation (1.12), $\mathbf{w}$ satisfies the equations

$$\operatorname{div}\mathbf{w} = 0, \qquad \operatorname{curl}\mathbf{w} = \nabla T \times \mathbf{n} \tag{1.20}$$

It follows from equation (1.15) that $\mathbf{w}$ is the proper amplitude of the pulsation velocity component, slowly varying with time.[†]

Let us now formulate the boundary conditions. Since the cavity is entirely filled with a fluid, the non-slip velocity condition takes place on the boundary. The temperature conditions are determined by the heating regime. For example, the temperature on the boundary may be fixed. In this case

$$\mathbf{v}|_F = 0, \qquad T|_F = T_{\mathrm{w}} \tag{1.21}$$

Now we consider the boundary conditions for $\mathbf{w}$. The viscous force driving the oscillatory flow has been neglected when deriving equation (1.9). This means that the existence of the Stokes boundary layer for the pulsation flow is not taken into account. Similarly, the temperature skin layer is not accounted for in equation (1.10). Thus, on the pulsation velocity component only the non-permeability condition may be imposed rather than the non-slip one:

$$w_{\mathrm{n}}|_F = 0 \tag{1.22}$$

Here w_{n} is the normal component of vector $\mathbf{w}$.

[†] A useful form of equation (1.17) may be derived by extracting the potential part from the averaged thermovibrational force and taking into account (1.20):

$$\frac{\partial \mathbf{v}}{\partial t} + (\mathbf{v}\nabla)\mathbf{v} = -\nabla\left(\frac{p}{\rho} + \varepsilon\frac{w^2}{2}\right) + \nu\Delta\mathbf{v} + g\beta T\boldsymbol{\gamma} + \varepsilon(\mathbf{w}\mathbf{n})\nabla T \tag{1.17a}$$

The set (1.17) to (1.20) with the boundary conditions (1.21) and (1.22) form the boundary problem to determine the average fields. The oscillatory components may then be found with the aid of the closure relations (1.15) and (1.16).

As already mentioned, the averaging technique is valid only in the high-frequency range. Simultaneously, there exists one more restriction owing to the fact that we use the Boussinesq model of an incompressible fluid. This condition requires that the acoustic wavelength must be greater than the reference spatial scale. Thus the vibration period τ is to satisfy the inequalities:

$$\frac{L}{c} \ll \tau \ll \min\left(\frac{L^2}{\nu}, \frac{L^2}{\chi}\right) \tag{1.23}$$

where L is the reference scale (the linear size of the cavity, boundary layer thickness, etc.) and c is the sound velocity in the fluid. At $L \sim 1$ cm, inequalities (1.23) yield the applicability ranges which are: for water $7 \times 10^{-1}\,\mathrm{s} \ll \tau \ll 100\,\mathrm{s}$ and for air $3 \times 10^{-5}\,\mathrm{s} \ll \tau \ll 5\,\mathrm{s}$. Therefore, there exists a very wide frequency range within which implementation of the averaging method is physically justified without the risk of falling outside the framework of the incompressible fluid model.

It is necessary to point out one more requirement ensuring feasibility of the averaging technique. It is connected with the amplitude b. When deriving the simplified equation (1.9), the inertial terms have been neglected. In fact, equation (1.9) is the linearized Euler equation. The term $(\mathbf{v}'\,\nabla)\mathbf{v}'$ must be much smaller than $\partial\mathbf{v}'/\partial t$. Taking into account the fact that in accordance with (1.15) $|\mathbf{v}'| \sim \beta b \Omega \Theta$, the following condition is given:

$$b \ll \frac{L}{\beta\Theta} \tag{1.24}$$

The Boussinesq parameter $\beta\Theta$ is small, so the value of b used in the experiment may well be of the order of the reference scale L or even greater.

On writing down equation (1.9) we have neglected the buoyancy force in the static gravity field $g\beta T'\gamma$. Comparing it with $\partial\mathbf{v}'/\partial t$ and using relations (1.15) and (1.16), one concludes that this neglect is justified under the following relationship between the gravity acceleration and the vibrational one:

$$\frac{g}{L\Omega^2}\beta\Theta \ll 1 \tag{1.24$'$}$$

Let us write down the boundary problem for the average flow in a dimensionless form. As the scales for length, time, velocity, temperature, $\mathbf{w}$ field and pressure we choose, respectively, L (the reference size), L^2/ν, χ/L, Θ (the reference temperature difference) and $\rho\nu\chi/L^2$. Retaining for the dimensionless variables the same notations as in equations (1.17) to (1.22), one obtains

$$\begin{aligned}
\frac{\partial \mathbf{v}}{\partial t}+\frac{1}{Pr}(\mathbf{v}\nabla)\mathbf{v} &= -\nabla p+\Delta \mathbf{v}+Ra\,T\boldsymbol{\gamma}+Ra_{\mathrm{v}}(\mathbf{w}\nabla)(T\,\mathbf{n}-\mathbf{w}) \\
Pr\frac{\partial T}{\partial t}+\mathbf{v}\nabla T &= \Delta T \\
\operatorname{div}\mathbf{v} &= 0
\end{aligned} \tag{1.25}$$

$$\begin{aligned}
\operatorname{div}\mathbf{w} &= 0, \qquad \operatorname{curl}\mathbf{w} = \nabla T\times\mathbf{n} \\
\mathbf{v}|_F &= 0, \qquad T|_F = T_{\mathrm{w}}, \qquad w_{\mathrm{n}}|_F = 0
\end{aligned} \tag{1.26}$$

The boundary problem (1.25) and (1.26) includes three dimensionless parameters determining the similarity of the averaged convective fields, namely the Rayleigh number Ra, its vibrational analog Ra_{v} and the Prandtl number Pr:

$$Ra=\frac{g\beta\Theta l^3}{\nu\chi}, \qquad Ra_{\mathrm{v}}=\frac{\varepsilon\Theta^2L^2}{\nu\chi}=\frac{(\beta b\Omega\Theta L)^2}{2\nu\chi} \qquad Pr=\frac{\nu}{\chi}. \tag{1.27}$$

The very first papers on the application of the averaging technique to hydromechanics of viscous flows were published by Schlichting [5] and Lin [6] (see [7]). These works deal with the problem of an oscillatory boundary layer. In the theory of convection, the method was used first by Zen'kovskaya and Simonenko [8, 9] to study the effect of high-frequency vibrations on the equilibrium stability of a plane horizontal layer in a static gravity field. Mathematical justification of the method was given by Simonenko [10, 11].

2 'Quasi-equilibrium' and the Stability Problem

The boundary problem formulated in the previous section allows average convective flows in cavities under various heating conditions and at various vibration parameters to be studied. Naturally, the questions begin with whether a mechanical quasi-equilibrium is possible in a non-uniformly heated fluid in a vibration field as well as questions about its stability. Here the term 'quasi-equilibrium' designates a state in which the average fluid velocity equals zero whereas the pulsation velocity component does not necessarily vanish. In the present and the two following sections, the quasi-equilibrium states are considered under conditions of pure weightlessness, i.e. the static gravity field is absent. In this case one should omit the term $g\beta T\boldsymbol{\gamma}$ in equation (1.17), rendering the convective buoyancy force, or the term $Ra\,T\boldsymbol{\gamma}$ in the set (1.25).

By setting $\mathbf{v}=0$ in the quasi-equilibrium state and assuming the temperature T_0, pressure p_0 and pulsation velocity $\mathbf{w}_0$ fields to be time independent, from the equation of motion one gets the following necessary condition:

$$-\nabla p_0 + Ra_{\mathrm{v}}(\mathbf{w}_0\nabla)(T_0\mathbf{n} - \mathbf{w}_0) = 0 \tag{1.28}$$

This relationship means that the average thermovibrational 'buoyancy' force is balanced (compensated) by the pressure gradient.

Applying the curl operation to both sides of equation (1.28), one has

$$\nabla(\mathbf{w}_0\mathbf{n}) \times \nabla T_0 = 0 \tag{1.29}$$

To the last equation we add the Laplace equation for temperature:

$$\Delta T_0 = 0 \tag{1.30}$$

the equations for the pulsation velocity:

$$\operatorname{div} \mathbf{w}_0 = 0, \qquad \operatorname{curl} \mathbf{w}_0 = \nabla T_0 \times \mathbf{n} \tag{1.31}$$

and the boundary conditions:

$$T_0|_F = T_{\mathrm{w}}, \qquad w_{0n}|_F = 0 \tag{1.32}$$

Then the set (1.29) to (1.32) makes a closed boundary problem to find the quasi-equilibrium fields T_0, p_0 and $\mathbf{w}_0$. The temperature is determined by a solution of a harmonic problem with the boundary conditions of the Dirichlet type. We remark that constancy of the temperature gradient in a fluid is not any longer the necessary condition of equilibrium, as it is for a static gravity field (see [2]). It may be seen that vector $\mathbf{w}_0$ obeys the Poisson equation with uniform boundary conditions of the Neumann type. However, the fields T_0 and $\mathbf{w}_0$ found from elliptic boundary value problems do not in general satisfy the 'vibrational hydrostatic condition' (1.29). That is why an equilibrium is possible only for a special geometry, special heating conditions and specially directed vibration axis.

As soon as the equilibrium takes place, the problem of its stability emerges. According to a conventional scheme of the linear stability theory, small perturbations T', p' and $\mathbf{w}'$ are introduced of the equilibrium fields T_0, p_0, $\mathbf{w}_0$, as well as a weak average flow of the velocity $\mathbf{v}$. Substituting the perturbed fields $T_0 + T'$, $p_0 + p'$, $\mathbf{w}_0 + \mathbf{w}'$ and $\mathbf{v}$ into the basic set (1.25) and linearizing it, one obtains a set of equations for the perturbations

$$\frac{\partial \mathbf{v}}{\partial t} = -\nabla p' + \Delta \mathbf{v} + Ra_{\mathrm{v}}[(\mathbf{w}_0\nabla)(T'\mathbf{n} - \mathbf{w}') + (\mathbf{w}'\nabla)(T_0\mathbf{n} - \mathbf{w}_0)]$$

$$Pr\frac{\partial T'}{\partial t} + \mathbf{v}\nabla T_0 = \Delta T' \tag{1.33}$$

$$\operatorname{div} \mathbf{v} = 0, \qquad \operatorname{div} \mathbf{w}' = 0, \qquad \operatorname{curl} \mathbf{w}' = \nabla T' \times \mathbf{n}$$

The cavity boundaries are assumed to be rigid and of high heat conductivity. That leads to the boundary conditions

$$\mathbf{v}|_F = 0, \qquad T'|_F = 0, \qquad \mathbf{w}'_{\mathrm{n}}|_F = 0 \tag{1.34}$$

For the case of 'normal' modes exponentially depending on time as $\exp(-\lambda t)$, where λ is called a decrement, the set (1.33) and (1.34) transforms into the amplitude problem. With the primes omitted this becomes

$$-\lambda \mathbf{v} = -\nabla p + \Delta \mathbf{v} + Ra_{\mathrm{v}}[(\mathbf{w}_0\nabla)(T\,\mathbf{n} - \mathbf{w}) + (\mathbf{w}\nabla)(T_0\,\mathbf{n} - \mathbf{w}_0)]$$

$$-\lambda\, Pr\, T + \mathbf{v}\nabla T_0 = \Delta T \tag{1.35}$$

$$\operatorname{div} \mathbf{v} = 0, \qquad \operatorname{div} \mathbf{w} = 0, \qquad \operatorname{curl} \mathbf{w} = \nabla T \times \mathbf{n}$$

$$\mathbf{v}|_F = 0, \qquad T|_F = 0, \qquad w_{\mathrm{n}}|_F = 0$$

The spectral amplitude problem (1.35) is not self-conjugated. Thus, its eigenvalues λ are complex. In what follows we adopt the notation $\lambda = \lambda_{\mathrm{r}} + i\lambda_{\mathrm{i}}$, where λ_{r} and λ_{i} are the real and imaginary parts, respectively.

The eigenvalue λ depends on the parameters Ra_{v} and Pr. The stability border of the mechanical equilibrium may be found from the condition $\lambda_{\mathrm{r}} = 0$. The dispersion relation corresponding to the problem (1.35) determines the neutral (critical) vibrational Rayleigh number Ra_{v} and the neutral frequency λ_{i} of oscillatory perturbations. If the instability is due to development of a monotonic perturbation ($\lambda_{\mathrm{i}} = 0$), then it is easy to see that at the stability border ($\lambda_{\mathrm{r}} = 0$) the term containing the Prandtl number falls out of the boundary-value problem (1.35). A similar situation occurs for the thermogravitational convective instability of an equilibrium under static gravity (see [2]). Thus, for the case of a monotonic instability the critical Rayleigh number does not depend on Pr.

First, let us consider one important particular case [12]. Let F at the cavity boundary be maintained with such a temperature distribution that it creates inside the fluid a constant temperature gradient that is parallel to the vibration axis: $\nabla T_0 \parallel \mathbf{n}$. Then the Laplace equation for T_0 is satisfied and the set (1.31) becomes homogeneous: $\operatorname{div} \mathbf{w}_0 = 0$, $\operatorname{curl} \mathbf{w}_0 = 0$. Taking into account the boundary condition $w_{0\mathrm{n}}|_F = 0$, the solution for the pulsation velocity becomes trivial ($\mathbf{w}_0 = 0$) and the 'hydrostatic' equation (1.29) holds identically. It should be emphasized that here we deal with a 'true' equilibrium, not quasi-equilibrium. Indeed, both average and pulsation velocities are zero in an appropriate coordinate system. Such an equilibrium under an appropriate temperature distribution on the boundary is possible in a cavity of an arbitrary form.

When $\nabla T_0 \parallel \mathbf{n}$ and $\mathbf{w}_0 = 0$, the first equation in (1.35) may be rewritten in the form

$$-\lambda \mathbf{v} = -\nabla p + \Delta \mathbf{v} + Ra_{\mathrm{v}}\, A w_1\, \mathbf{n} \tag{1.36}$$

Here A is the dimensionless temperature gradient and $w_l = \mathbf{wn}$ is the projection of $\mathbf{w}$ on to the axis of vibration. Further, both sides of the second equation of (1.35) are multiplied by the unit vector $\mathbf{n}$ and formula (1.12), $T\mathbf{n} = \mathbf{w} + \nabla\varphi$, is used:

$$-\lambda Pr(\mathbf{w} + \nabla\varphi) + Av_l\,\mathbf{n} = \Delta\mathbf{w} + \nabla(\Delta\varphi) \tag{1.37}$$

where $v_l = \mathbf{vn}$. Then both sides of equation (1.36) are multiplied by $\mathbf{v}^*$ and both sides of equation (1.37) by $\mathbf{w}^*$ (here the asterisk denotes complex conjugation). The resulting equations are integrated over the volume of the cavity. With allowance for the rest of the equations and boundary conditions of the problem (1.35) as well as some standard vector analysis formulas, one arrives at the relationships

$$-\lambda\int|\mathbf{v}|^2\mathrm{d}V = -\int|\mathrm{curl}\,\mathbf{v}|^2\mathrm{d}V + Ra_\mathrm{v}\,A\int w_l v_l^*\,\mathrm{d}V \tag{1.38}$$

$$-\lambda\,Pr\int|\mathbf{w}|^2\mathrm{d}V + A\int v_l w_l^*\,\mathrm{d}V = -\int|\mathrm{curl}\,\mathbf{w}|^2\,\mathrm{d}V \tag{1.39}$$

The complex-conjugate expressions are

$$-\lambda^*\int|\mathbf{v}|^2\,\mathrm{d}V = -\int|\mathrm{curl}\,\mathbf{v}|^2\,\mathrm{d}V + Ra_\mathrm{v}\,A\int w_l^* v_l\,\mathrm{d}V \tag{1.40}$$

$$-\lambda^*\,Pr\int|\mathbf{w}|^2\,\mathrm{d}V + A\int v_l^* w_l\,\mathrm{d}V = -\int|\mathrm{curl}\,\mathbf{w}|^2\,\mathrm{d}V \tag{1.41}$$

Finally, from equations (1.38) to (1.41) it follows that

$$(\lambda + \lambda^*)\int\left(|\mathbf{v}|^2 + Ra_\mathrm{v}\,Pr\,|\mathbf{w}|^2\right)\mathrm{d}V = 2\int\left(|\mathrm{curl}\,\mathbf{v}|^2 + Ra_\mathrm{v}|\mathrm{curl}\,\mathbf{w}|^2\right)\mathrm{d}V \tag{1.42}$$

$$(\lambda - \lambda^*)\int\left(|\mathbf{v}|^2 - Ra_\mathrm{v}\,Pr|\mathbf{w}|^2\right)\mathrm{d}V = 0 \tag{1.43}$$

From the integral relations (1.42) and (1.43) one can draw certain conclusions concerning the real and imaginary parts of the characteristic decrement. Indeed, since Ra_v by its definition (1.27) is positive, both integrals in equation (1.42) are positive definite. Hence $\lambda + \lambda^* = 2\lambda_\mathrm{r} > 0$, i.e. the real part of the decrement is positive and all the perturbations decay with time. Thus, the equilibrium configuration is stable with respect to the perturbations of the considered type. As for equation (1.43), the entering integral is not of a definite sign, and nothing may be inferred on the imaginary part $\lambda - \lambda^* = 2\lambda_\mathrm{i}$ of the decrement. At $Ra_\mathrm{v} = 0$, apparently $\lambda_\mathrm{i} = 0$, and perturbations evolve (decay) monotonically. As Ra_v increases, the appearance of oscillatory perturbations in the spectrum is quite possible. The situation is similar to that for a static gravity field when the system is heated from above.

Thus we conclude that even in the case when the equilibrium is unfeasible, high-frequency vibrations with the axis parallel to the temperature gradient have a damping effect on any non-isothermal flow.

3 Plane Layer in the Presence of a Transversal Temperature Gradient

3.1 Equilibrium

In this section the problem of a mechanical quasi-equilibrium stability in weightlessness is studied for a cavity of the simplest geometry [3, 12]. Let us consider an infinite plane layer bounded by two parallel rigid plates $z = 0$ and $z = h$ at which constant temperatures are maintained, being, respectively, $T = \Theta$ and $T = 0$. The Cartesian coordinate system is introduced as it is shown in Fig. 1. The vibration axis is situated in the xz plane at an angle α to the x axis. So the unit vector $\mathbf{n}$ has the components $(\cos\alpha,\ 0,\ \sin\alpha)$.

It is easy to see that under the adopted conditions the mechanical quasi-equilibrium exists and has the following structure:

$$T_0 = T_0(z), \qquad \mathbf{w}_0(w_0(z), 0, 0) \tag{1.44}$$

The 'hydrostatic' relation (1.29) is satisfied automatically for $\nabla(\mathbf{w}_0\mathbf{n}) \parallel \nabla T_0$. Taking into account the boundary conditions, the linear temperature profile is obtained:

$$T_0 = \Theta\left(1 - \frac{z}{h}\right) \tag{1.45}$$

For w_0 one has the equation

$$\frac{\mathrm{d}w_0}{\mathrm{d}z} = -\frac{\Theta}{h}\cos\alpha$$

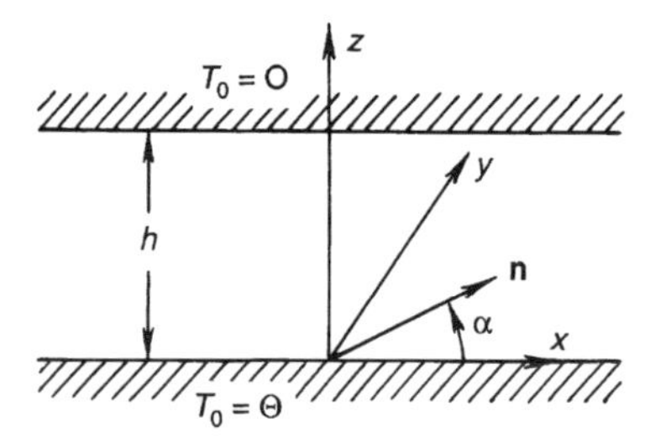

Fig. 1 Geometry of the problem and the coordinate system

which upon integration yields

$$w_0 = -\frac{\Theta}{h}\cos\alpha(z + C)$$

The integration constant C may be found from the condition that in a quasi-equilibrium the oscillatory flow must be closed-looped:

$$\int_0^h w_0 \, \mathrm{d}z = 0$$

Therefore $C = -h/2$ and the w_0 profile is

$$w_0 = -\frac{\Theta}{h}\cos\alpha\left(z - \frac{h}{2}\right) \tag{1.46}$$

Choosing h as a length scale and Θ as that for temperature and $\mathbf{w}_0$, the profiles are rewritten in a dimensionless form:

$$T_0 = 1 - z, \qquad w_0 = -\cos\alpha\left(z - \tfrac{1}{2}\right) \tag{1.47}$$

3.2 Spectral Amplitude Problem

In what follows, the equilibrium stability is studied. From the set (1.35) one may eliminate pressure and the longitudinal components of the average and pulsation velocities. Then the spectral problem for T and the transversal velocities v_z and w_z is obtained. From the heat transfer equation taking into account (1.47), one finds that

$$-\lambda \, Pr \, T - v_z = \Delta T \tag{1.48}$$

The equations

$$\operatorname{div}\mathbf{w} = 0, \qquad \operatorname{curl}\mathbf{w} = \nabla T \times \mathbf{n}$$

yield the Poisson equation

$$\Delta\mathbf{w} = -\operatorname{curl}\left(\nabla T \times \mathbf{n}\right)$$

which may be transformed into

$$\Delta w_z = -\cos\alpha\frac{\partial^2 T}{\partial x \, \partial z} + \sin\alpha\Delta_2 T \tag{1.49}$$

where Δ_2 is the two-dimensional Laplace operator in the plane of the layer. Applying the curl–curl operation to the equation of motion and taking into account equations (1.47) and (1.49) one arrives at

$$-\lambda\Delta v_z = \Delta\Delta v_z + Ra_{\mathrm{v}}\left(\cos^2\alpha\frac{\partial^2 T}{\partial x^2} + \cos\alpha\frac{\partial^2 w_z}{\partial x\,\partial z} - \sin\alpha\Delta_2 w_z\right) \quad (1.50)$$

Here we have introduced perturbations with a harmonic dependences on x and y:

$$(v_z, w_z, T) = [v(z), w(z), \theta(z)]\exp\left[\mathrm{i}(k_1 x + k_2 y)\right] \quad (1.51)$$

where v, w and θ are the amplitudes depending on the transversal coordinate z and k_1, k_2 are the wavenumbers along the x and y axes.

Substitution of equation (1.51) into equations (1.48) to (1.50) gives the set of amplitude equations

$$-\lambda\,\mathrm{D}v = \mathrm{D}^2 v + Ra_{\mathrm{v}}\left(-k_1^2\cos^2\alpha\cdot\theta + \mathrm{i}k_1\cos\alpha\cdot w' + k^2\sin\alpha\cdot w\right)$$

$$-\lambda\,Pr\,\theta - v = \mathrm{D}\theta \quad (1.52)$$

$$\mathrm{D}w = -\left(\mathrm{i}k_1\cos\alpha\cdot\theta' + k^2\sin\alpha\cdot\theta\right)$$

Here the prime denotes differentiation with respect to the transversal coordinate z, $k^2 = k_1^2 + k_2^2$ and $\mathrm{D} = \mathrm{d}^2/\mathrm{d}z^2 - k^2$.

On the rigid boundaries with a high thermoconductance, the conditions are

$$z = 0 \text{ and } z = 1: \qquad v = v' = 0, \qquad \theta = 0, \qquad w = 0 \quad (1.53)$$

The additional formula $v' = 0$ emerges as a consequence of the non-slip condition and the continuity equation.

The dispersion relation corresponding to the spectral amplitude problem (1.52) and (1.53) determines the eigenvalue, i.e. the decrement as a function of all the parameters: $\lambda = \lambda(Ra_{\mathrm{v}}, Pr, k_1, k_2, \alpha)$.

3.3 Plane Perturbations

We consider two-dimensional perturbations with the structure $v_y = 0$, $w_y = 0$, $\partial/\partial y = 0$. This type of perturbation is of special interest. As will be shown in what follows, they are the most dangerous in the situation under discussion. Besides, had the problem for the plane perturbations been solved, all the necessary information on the behavior of the normal three-dimensional ones becomes available.

To obtain the description of the plane perturbations, in the spectral problem (1.52) and (1.53) one must set $k_2 = 0$ and $k^2 = k_1^2$. The emerging set of the linear ordinary homogeneous differential equations with constant coefficients and homogeneous boundary conditions has been integrated numerically by the Runge–Kutta method using the shooting procedure. In this problem 'the

Table 1 Critical characteristics depending on the inclination angle of the vibration axis

α	Ra_{vm}	k_{m}
0°	2129	3.23
10°	2237	3.21
20°	2614	3.15
30°	3502	3.01
40°	5757	2.66
50°	1.289×10^4	1.98
60°	4.166×10^4	1.31
70°	2.13×10^5	0.78
80°	3.46×10^6	0.37
85°	5.55×10^7	0.18
90°	∞	0

principle of monotonicity of perturbations' does not apply, and the oscillatory perturbations are free to appear in the spectrum. However, the calculations show that the equilibrium loses its stability due to development of monotonic perturbations for which $\lambda_{\mathrm{i}} = 0$. On the stability border $\lambda_{\mathrm{r}} = 0$, i.e. $\lambda = 0$ entirely. Thus the neutral perturbation problem does not involve the Prandtl number. The eigenvalue of this problem is the 'neutral' vibrational Rayleigh number Ra_{v} depending on the wavenumber k and the angle α of the vibration axis inclination. At a given α, the neutral curve $Ra_{\mathrm{v}}(k)$ assumes a minimum at some value k_{m} that determines the wavelength of the most dangerous perturbation. The corresponding value Ra_{vm} yields the equilibrium stability border. The critical parameters as the functions of the inclination angle α are given in Table 1. It can be seen that the stability is minimal for longitudinal vibrations, where the value $Ra_{\mathrm{vm}} = 2129$ corresponds to the critical point $k_{\mathrm{m}} = 3.23$. The critical value of the vibrational Rayleigh number increases monotonically with the angle of inclination, and $Ra_{\mathrm{v}} \to \infty$ when $\alpha \to 90°$. The result is quite natural since according to the previous section the equilibrium is absolutely stable if the vibration axis is parallel to the temperature gradient. The critical wavenumber decreases as α grows, the most dangerous harmonics being shifted into the long-wave part of the spectrum. Close to the range of absolute stabilization ($\alpha \approx 90°$) the following asymptotics takes place:

$$Ra_{\mathrm{vm}} = \frac{3.46 \times 10^{10}}{(90° - \alpha)^4}, \qquad k_{\mathrm{m}} = 0.037(90° - \alpha)$$

Here the angle α is measured in degrees.

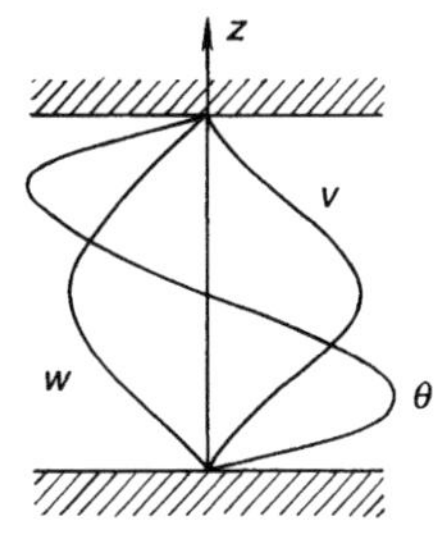

Fig. 2 Distribution of the critical perturbation amplitudes across the layer ($\alpha = 0$)

The above-considered plane perturbations have the form of rolls with their axes parallel to Oy, i.e. perpendicular to ∇T_0 and $\mathbf{n}$. The periodicity of rolls is determined by k_{m}. At $\alpha = 0$ the size of a convective cell in the x direction is close to unity (in terms of h). The distribution of the perturbation amplitude over the layer cross-section in the critical point is shown in Fig. 2; the normalizing is arbitrary. Thus the instability situation is close to the well-known Rayleigh–Benard instability which takes place in a horizontal layer heated from below under static gravity.

We remark that we are considering the main level of vibrational convective instability in a plane layer. The second instability level sets out with the advent of two-vortice perturbations in the transversal direction. The critical parameter values for the second level are: $Ra_{\mathrm{vm}} = 29.5 \times 10^3$ and $k_{\mathrm{m}} = 6.5$ at $\alpha = 0°$.

Let us estimate the critical temperature difference Θ_{c} necessary to excite the thermovibrational convection by breaking the equilibrium through instability. In the case of longitudinal vibration ($\alpha = 0°$) the critical vibrational Rayleigh number is $Ra_{\mathrm{vm}} = 2129$. From the Rayleigh number definition (1.27) it follows that

$$\Theta_{\mathrm{c}} = \frac{\sqrt{2 \times 2129 \nu \chi}}{\beta h b \Omega} \tag{1.54}$$

The critical temperature difference is in inverse proportion to the vibration intensity $b\Omega$. As the reference values we choose the frequency 100 Hz, displacement amplitude 1 mm and layer thickness 1 cm. Then for the main instability level one gets $\Theta_{\mathrm{c}} \approx 20°$ for water and $\Theta_{\mathrm{c}} \approx 50°$ for air.

3.4 Three-dimensional Perturbations

As soon as the problem for two-dimensional perturbations has been solved, the three-dimensional problem solution might be found by a simple recalculation [12].

Let us write down the problem for neutral three-dimensional perturbations setting $\lambda = 0$ in (1.52):

$$
\begin{aligned}
&\left(v^{\mathrm{IV}} - 2k^2 v'' + k^4 v\right) + Ra_{\mathrm{v}}\left(-k_1^2 \cos^2\alpha \cdot \theta + \mathrm{i}k_1 \cos\alpha \cdot w' + k^2 \sin\alpha \cdot w\right) = 0 \\
&\theta'' - k^2\theta = -v \\
&w'' - k^2 w = -\left(\mathrm{i}k_1 \cos\alpha \cdot \theta' + k^2 \sin\alpha \cdot \theta\right) \\
&z = 0, \quad z = 1: \qquad v = v' = 0, \qquad \theta = 0, \qquad w = 0
\end{aligned}
\tag{1.55}
$$

The problem for two-dimensional neutral perturbations may be written as follows:

$$
\begin{aligned}
&\left(\overline{v}^{\mathrm{IV}} - 2\overline{k}^2 \overline{v}'' + \overline{k}^4 \overline{v}\right) + \overline{Ra}_{\mathrm{v}}\left(-\overline{k}_1^2 \cos^2\overline{\alpha} \cdot \overline{\theta} + \mathrm{i}\overline{k}_1 \cos\overline{\alpha} \cdot \overline{w}' + \overline{k}^2 \sin\overline{\alpha} \cdot \overline{w}\right) = 0 \\
&\overline{\theta}'' - \overline{k}^2\overline{\theta} = -\overline{v} \\
&\overline{w}'' - \overline{k}^2 \overline{w} = -\left(\mathrm{i}\overline{k}_1 \cos\overline{\alpha} \cdot \overline{\theta}' + \overline{k}^2 \sin\overline{\alpha} \cdot \overline{\theta}\right) \\
&z = 0, \quad z = 1: \qquad \overline{v} = \overline{v}' = 0, \qquad \overline{\theta} = 0, \qquad \overline{w} = 0
\end{aligned}
\tag{1.56}
$$

Hereafter we use the overbar to denote all the amplitudes and parameters of the plane problem.

It is easy to see that the three-dimensional problem (1.55) may be reduced to the two-dimensional one (1.56), transforming the amplitudes according to

$$
v = \overline{v}, \qquad \theta = \overline{\theta}, \qquad \sqrt{Ra_{\mathrm{v}}}\, w = \sqrt{\overline{Ra}_{\mathrm{v}}}\, \overline{w} \tag{1.57}
$$

and simultaneously transforming the other parameters, including the inclination angle of the vibration axis, as

$$
k^2 = \overline{k}^2, \qquad Ra_{\mathrm{v}} \sin^2\alpha = \overline{Ra}_{\mathrm{v}} \sin^2\overline{\alpha}, \qquad Ra_{\mathrm{v}}\, k_1^2 \cos^2\alpha = \overline{Ra}_{\mathrm{v}}\, \overline{k}^2 \cos^2\overline{\alpha} \tag{1.58}
$$

From (1.58) follow the formulas linking the parameters of both problems:

$$
k^2 = \overline{k}^2, \qquad a \,\mathrm{cotan}\,\alpha = \mathrm{cotan}\,\overline{\alpha}, \qquad Ra_{\mathrm{v}} = \overline{Ra}_{\mathrm{v}}\left(\sin^2\overline{\alpha} + \frac{1}{a^2}\cos^2\overline{\alpha}\right) \tag{1.59}
$$

Here we introduced the notation for the parameter of three-dimensional perturbations:

$$
a = \frac{k_1}{k} = \frac{k_1}{\sqrt{k_1^2 + k_2^2}} \tag{1.60}
$$

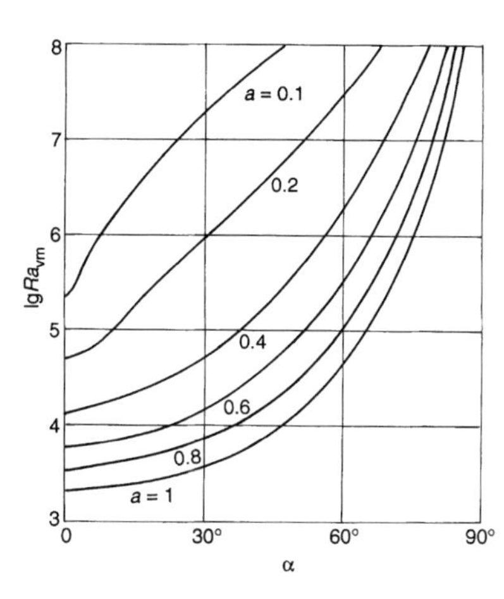

Fig. 3 Critical value of the vibrational Rayleigh number as a function of inclination angle for different values of a three-dimensional perturbation parameter

The limiting case $a = 1$ ($k_2 = 0$) corresponds to plane perturbations in the form of two-dimensional rolls whose axes are parallel to the y axis. In the opposite limiting case $a = 0$ ($k_1 = 0$) there arise three-dimensional spiral perturbations in the form of rolls parallel to the x axis.

By means of formulas (1.59) and (1.60) the critical Rayleigh number may be determined for three-dimensional perturbations with the wavenumbers k_1 and k_2 at some inclination angle α if the critical value $\overline{Ra}_v$ is known for a two-dimensional perturbation with the wavenumber $\bar{k}$ at the angle of inclination $\bar{\alpha}$ which may not coincide with α. An illustration is presented in Fig. 3 where the dependence $Ra_v(\alpha)$ is plotted for different values of the three-dimensional parameter a. It is seen that for all α values the plane perturbations are the most dangerous ones. In some sense, we have come out with an analog of the Squire theorem [13] which exists in the stability theory of plane-parallel isothermal flows.

It is known that in the theory of convective stability of a plane horizontal fluid layer heated from below in static gravity the degeneracy of two- and three-dimensional perturbations takes place. This fact reflects the isotropy in the plane of layer. In the problem under consideration, such an isotropy is absent since there exists some singled-out direction determined by the vibration axis. That is why the plane and three-dimensional perturbations are not equivalent.

3.5 Effect of the Thermal Properties of the Boundaries

The problem of convective instability of equilibrium has been solved for the layer whose boundaries $z = 0$ and $z = 1$ are isothermal planes where the temperature perturbations vanish. Physically it corresponds to the limiting situation where the heat conductivity of the boundary material is much higher than that of the fluid, e.g. a water layer between metal plates. If the heat conductivities of the fluid and boundaries are comparable, it is necessary to take into account penetration of temperature perturbations into the rigid boundaries, i.e. the stability problem must be considered as a conjugated one. It may be formulated as follows.

As before, let us consider a plane fluid layer of thickness h and heat conductivity κ. However, unlike the previous case, the layer separates two semi-infinite solid masses with finite heat conductivities different from that of the fluid. The region $z < 0$ is occupied by the solid mass number 1 with the heat conductivity κ_1 and the region $z > h$—by the solid mass number 2 with the heat conductivity κ_2. Thus, we have to add the equations for temperature perturbations within the solid masses:

$$\Delta T_1 = 0 \qquad \text{and} \qquad \Delta T_2 = 0$$

Here we mean neutral perturbations which determine the stability border. At the boundaries between the solid mass and the fluid, the continuity conditions for temperatures and heat fluxes are satisfied:

$$z = 0: \qquad T_1 = T, \qquad \kappa_1 \frac{\partial T_1}{\partial z} = \kappa \frac{\partial T}{\partial z}$$

$$z = h: \qquad T = T_2, \qquad \kappa \frac{\partial T}{\partial z} = \kappa_2 \frac{\partial T_2}{\partial z}$$

Besides, when going far away from the layer, the temperature perturbations decay:

$$z \to -\infty: \qquad T_1 \to 0 \qquad \text{and} \qquad z \to \infty: \qquad T_2 \to 0$$

Then the problem is customarily brought to a dimensionless form and the normal modes are assumed to be periodic in the plane of the layer. The spectral amplitude problem is obtained for the neutral perturbation amplitudes in the case of longitudinal vibration ($\alpha = 0$):

$$\begin{gathered} \mathrm{D}^2 v + Ra_{\mathrm{v}}\left(-k^2\theta + \mathrm{i}kw'\right) = 0 \\ \mathrm{D}\theta = -v \\ \mathrm{D}w = -\mathrm{i}k\theta' \\ \mathrm{D}\theta_{\mathrm{m1}} = 0 \\ \mathrm{D}\theta_{\mathrm{m2}} = 0 \end{gathered} \tag{1.61}$$

$$z = 0: \qquad v = v' = 0, \qquad w = 0, \qquad \theta_{\mathrm{m1}} = \theta, \qquad \theta'_{\mathrm{m1}} = \widetilde{\kappa}_1 \theta'$$

$$z = 1: \qquad v = v' = 0, \qquad w = 0, \qquad \theta = \theta_{\mathrm{m2}}, \qquad \widetilde{\kappa}_2 \theta' = \theta'_{\mathrm{m2}}$$

$$z \to -\infty: \qquad \theta_{\mathrm{m1}} \to 0$$

$$z \to \infty: \qquad \theta_{\mathrm{m2}} \to 0$$

Here $D = d^2/dz^2 - k^2$, θ_{m1} and θ_{m2} are the amplitudes of the temperature perturbations in both solid masses, $\widetilde{\kappa}_1 = \kappa/\kappa_1$ and $\widetilde{\kappa}_2 = \kappa/\kappa_2$ are the ratios of the heat conductivity coefficients.

The solutions for the auxiliary problems within the solid masses are

$$\theta_{m1} = A\,e^{kz}, \qquad \theta_{m2} = B\,e^{-k(z-1)} \tag{1.62}$$

Therefore, it is evident that the spectral problem (1.61) may be reformulated by eliminating the additional variables θ_{m1} and θ_{m2} and imposing the matching conditions

$$z = 0: \qquad \widetilde{\kappa}_1\theta' = k\theta; \qquad z = 1: \qquad \widetilde{\kappa}_2\theta' = -k\theta \tag{1.63}$$

The problem was solved numerically in [14]. The critical Ra_v value depends on k, $\widetilde{\kappa}_1$ and $\widetilde{\kappa}_2$. If $\widetilde{\kappa}_1$ and $\widetilde{\kappa}_2$ are fixed, the neutral curve $Ra_v(k)$ has a minimum at the point k_m. The corresponding value of Ra_{vm} determines the threshold of the vibrational convective instability.

In Fig. 4 the instability characteristics Ra_{vm} and k_m are presented for the symmetrical case $\widetilde{\kappa}_1 = \widetilde{\kappa}_2 = \widetilde{\kappa}$ as the functions of $\widetilde{\kappa}$. At $\widetilde{\kappa} = 0$ we recover the stability problem for a fluid layer between two parallel plates of high conductivities. As $\widetilde{\kappa}$ grows, the critical values Ra_{vm} and k_m monotonically decrease. This is due to the fact that penetration of the temperature perturbations into masses increases the effective size of the region available for perturbations. This additional

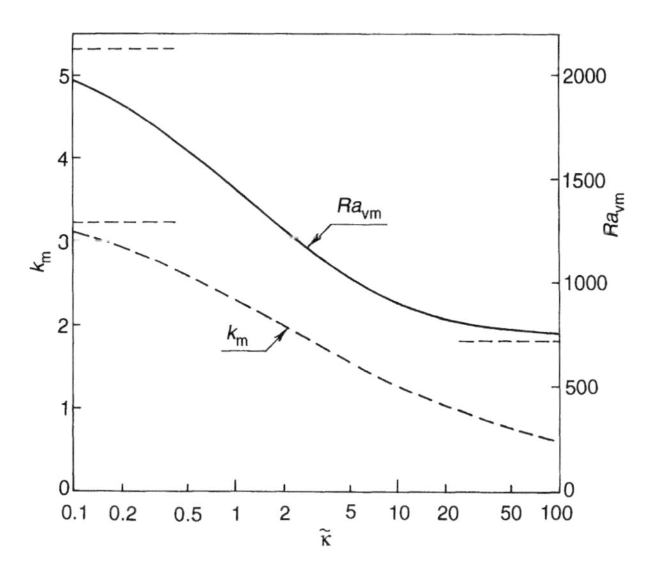

Fig. 4 Critical instability parameters Ra_{vm} and k_m depending on κ for the symmetrical case $\widetilde{\kappa}_1 = \widetilde{\kappa}_2 = \kappa$

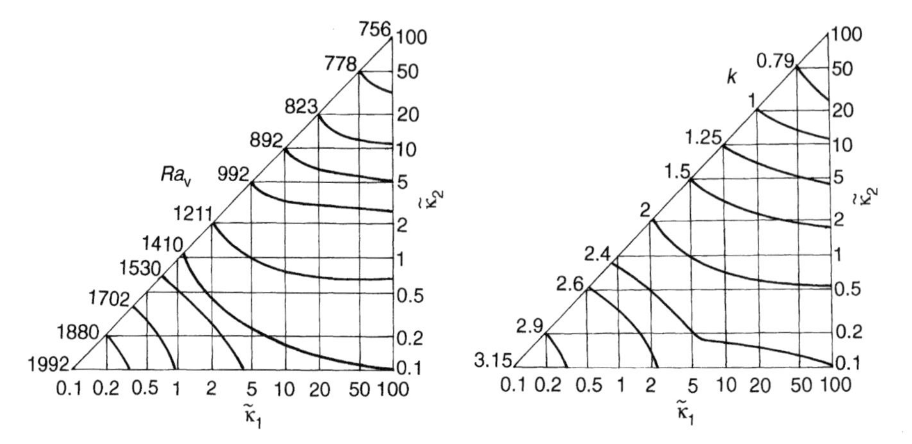

Fig. 5 Isolines of Ra_{vm} and k_{m} in the plane $(\widetilde{\kappa}_1, \widetilde{\kappa}_2)$ in the asymmetrical case

freedom favors the perturbation development, thus bringing the stability down and enlarging the critical perturbation wavelength.

Let us consider the limiting case of heat-insulated boundaries. Setting $\widetilde{\kappa} \to \infty$, we assume that both masses have relatively low heat conductivity. As an example, one may take a layer of a molten metal confined between two glass plates. In this situation the critical Ra_{vm} value has the finite limit $Ra_{\mathrm{vm}} \to 720$ and the critical wavenumber tends to zero. As in the case of thermogravitational stability of a horizontal fluid layer heated from below, the instability of the layer with the heat-insulated boundaries is caused by long-wave perturbations.

In an asymmetric case of different conductivities $(\widetilde{\kappa}_1 \neq \widetilde{\kappa}_2)$ it is obvious that $Ra_{\mathrm{vm}}(\widetilde{\kappa}_1, \widetilde{\kappa}_2) = Ra_{\mathrm{vm}}(\widetilde{\kappa}_2, \widetilde{\kappa}_1)$. Therefore the stability border does not change in response to repositioning of the masses. In Fig. 5 the maps of Ra_{vm} and k_{m} isolines are shown in the plane $(\widetilde{\kappa}_1, \widetilde{\kappa}_2)$. One may see that in a general asymmetric case the tendency to lowering the stability boundary holds, and the critical wavelength grows if the heat conductivity, even of one of solid mass, decreases. In the limiting asymmetric situation when one of the masses is heat-insulated $(\widetilde{\kappa}_1 \to \infty)$ and the other has a high conductivity $(\widetilde{\kappa}_2 = 0)$, one has $Ra_{\mathrm{vm}} = 1469$ and $k_{\mathrm{m}} = 2.7$. The maps of isolines in Fig. 5 are quite analogous to those for thermogravitational instability in a horizontal fluid layer heated from below (see [2]).

3.6 Oscillatory Instability

The analysis of the spectrum of equilibrium perturbations in a high-frequency vibration field shows that, in contrast to the case of thermogravitational convection, the perturbations of the oscillatory type are quite possible. At the same time, the instability in all the considered situations is caused by monotonic

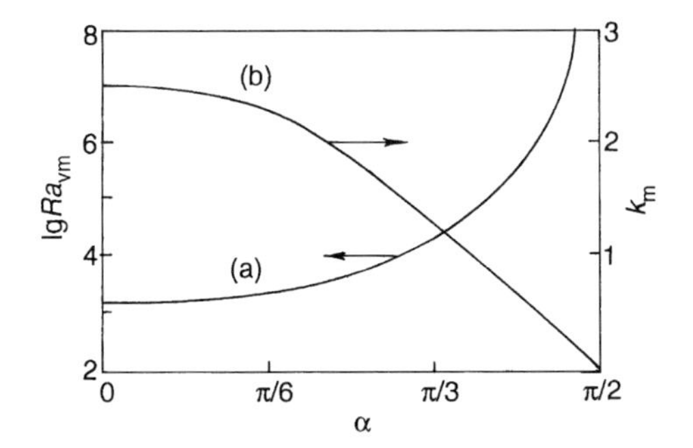

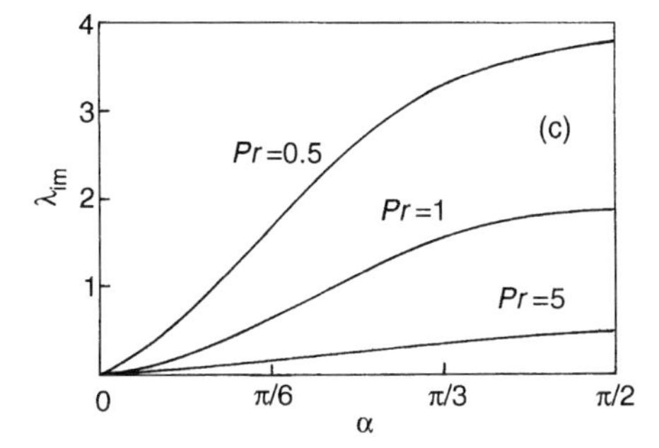

Fig. 6 Characteristics of the oscillatory instability as functions of the angle [15]. (a) Critical vibrational Rayleigh number Ra_{vm}; (b) critical wavenumber k_m; (c) critical phase velocity λ_{im} for three Pr values. Ra_{vm} and k_m do not depend on Pr in practice

perturbations. This fact is, of course, related to the symmetry of the problem. Braverman and Oron [15] have found an example of an asymmetric problem where the oscillatory instability exists. They studied the vibrational convective instability in a plane layer with rigid boundaries, one of which was thermally insulated whereas the other one was isothermal. If the vibration axis is inclined to the temperature gradient by some angle, the equilibrium of such a system becomes oscillatory unstable. The spectral problem for plane perturbations may be derived from (1.52) under the boundary conditions:

$$z = 0: \quad v = v' = 0, \quad \theta = 0, \quad w = 0$$
$$z = 1: \quad v = v' = 0, \quad \theta' = 0, \quad w = 0$$

The characteristics of the oscillatory instability are given in Fig. 6. The amplitude problem yields that if $\lambda \neq 0$, the solution of the eigenvalue problem is a function of the Prandtl number. However, computations show that the critical Rayleigh number Ra_{vm} and the critical wavenumber k_m practically do not depend on Pr. Conversely, the critical value of the phase velocity λ_{im} does depend on Pr decreasing as Pr grows. In the problem under consideration, the phase velocity of critical perturbations is positive ($\lambda_{\rm im} > 0$). At the stability border the perturbation assumes the form of a periodic system of convective cells drifting in the positive direction of the x axis.

4 Other Quasi-equilibrium Configurations in Weightlessness

4.1 Plane Layer

In the previous section the convective instability of equilibrium has been considered in a plane fluid layer when the equilibrium temperature gradient is transversal and the axis of vibration is arbitrarily directed with respect to the layer. This is not the only case where the equilibrium in a horizontal layer is possible. Some other situations of such a kind are discussed in this section, together with the results of their stability analysis. As well we discuss the examples of equilibrium configurations for domains of various geometric arrangements.

The equilibrium state in a plane layer and its stability were studied by Braverman [16–19]. Let us begin with the case where the temperature gradient and the vibration axis are parallel to the plane of the layer and inclined to each other by the angle α. The y axis is chosen along the temperature gradient (see Fig. 7).

It may easily be proved that in this case the mechanical equilibrium exists for the following geometry: $\nabla T_0(0,1,0)$; $\mathbf{w}_0(w_0,0,0)$ where $w_0 = \sin\alpha \cdot y$. Here h and Ah are taken as the length and temperature scales, A being the equilibrium temperature gradient and h the thickness of the layer. The scalar function φ_0 is introduced to describe the potential part of the equilibrium field $T_0\mathbf{n} = \mathbf{w}_0 + \nabla\varphi$; $\varphi_0 = (y^2/2)\cos\alpha$.

Now the spectral amplitude problem is written down for neutral monotonic three-dimensional perturbations proportional to $\exp[\,\mathrm{i}(k_1x + k_2y)]$. An additional variable is introduced to describe the perturbation amplitude of a scalar function φ. The resulting set of equations for the amplitudes v_x, v_y, v_z, θ and φ

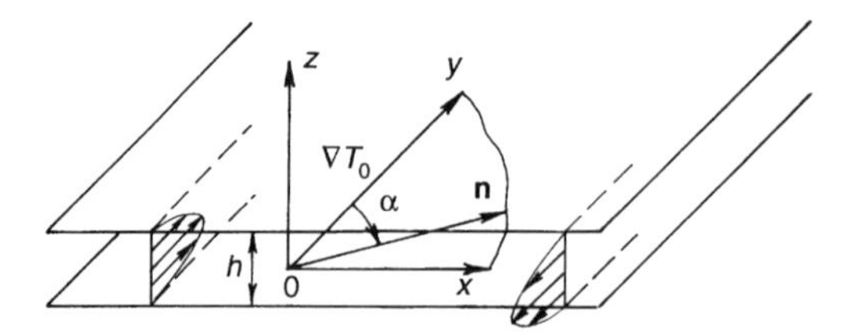

Fig. 7 The system of coordinates and the relative positions of the vectors ∇T_0 and $\mathbf{n}$. The velocity profiles are sketched for the perturbation (1.66)

has the form

$$-\mathrm{i}k_1\widetilde{p} + \left(v_x'' - k^2 v_x\right) = 0$$

$$-\mathrm{i}k_2\widetilde{p} + \left(v_y'' - k^2 v_y\right) + Ra_{\mathrm{v}}\left[\theta\cos^2\alpha - \mathrm{i}(k_1\sin\alpha + k_2\cos\alpha)\varphi\right] = 0$$

$$-\widetilde{p}' + \left(v_z'' - k^2 v_z\right) = 0 \tag{1.64}$$

$$\mathrm{i}\left(k_1 v_x + k_2 v_y\right) + v_z' = 0$$

$$\left(\varphi'' - k^2\varphi\right) = \mathrm{i}(k_1\sin\alpha + k_2\cos\alpha)\theta$$

Here $k^2 = k_1^2 + k_2^2$, and $\widetilde{p}$ is the amplitude of the renormalized pressure

$$\widetilde{p} = p + Ra_{\mathrm{v}}\left(w_0 w_x - y\theta\sin^2\alpha\right)$$

The amplitudes satisfy the homogeneous conditions on the rigid heat conducting boundary plates:

$$z = \pm\tfrac{1}{2}: \qquad v_x = v_y = v_z = 0, \qquad \theta = 0, \qquad \varphi' = 0 \tag{1.65}$$

Let us consider first the special case of one-dimensional perturbations:

$$k_2 = 0, \qquad k^2 = k_1^2, \qquad v_x = v_z = 0, \qquad v_y = v_y(z) \tag{1.66}$$

After introducing the notation $\psi = ik\varphi$, the spectral amplitude problem for one-dimensional perturbations takes the form

$$\begin{gathered}\left(v'' - k^2 v\right) + Ra_{\mathrm{v}}\left(\theta\cos^2\alpha - \psi\sin\alpha\right) = 0\\ \left(\theta'' - k^2\theta\right) - v = 0\\ \left(\varphi'' - k^2\varphi\right) - k^2\theta\sin\alpha = 0\\ z = \pm\tfrac{1}{2}: \qquad v = \theta = \varphi' = 0\end{gathered} \tag{1.67}$$

This problem determines the critical value Ra_{v} as a function of the wavenumber k with α as a parameter.

It might be demonstrated that the instability with respect to the mode selected is of the long-wave character, i.e. the neutral curve $Ra_{\mathrm{v}}(k)$ has a minimum at $k_m = 0$.† To determine the stability border, the asymptotic perturbative technique is applied with the wavenumber as a small parameter. The solution of the

† The situation is analogous to the one encountered in the theory of thermogravitational convective instability in a vertical layer heated from below (the "passing" mode) — see [2].

problem is sought in the form of expansions:

$$
\begin{aligned}
v &= v_0 + k^2 v_1 + k^4 v_2 + \cdots \\
\theta &= \theta_0 + k^2 \theta_1 + k^4 \theta_2 + \cdots \\
\psi &= \psi_0 + k^2 \psi_1 + k^4 \psi_2 + \cdots \\
Ra_v &= R_0 + k^2 R_1 + k^4 R_2 + \cdots
\end{aligned}
\tag{1.68}
$$

Substituting these expansions into (1.67), one obtains the sets of equations to determine the amplitudes in each order. In the zeroth order, it is

$$
\begin{gathered}
v_0'' + R_0 \theta_0 \cos^2 \alpha - R_0 \psi_0 \sin \alpha = 0 \\
T_0'' - v_0 = 0 \\
\psi_0'' = 0
\end{gathered}
\tag{1.69}
$$

with the appropriate boundary conditions. From here it follows that $\psi_0 =$ constant, or $\psi_0 = 1$ under an appropriate normalization. The amplitudes θ_0 and v_0 are (we consider the main level of instability)

$$
\begin{aligned}
\theta_0 &= A \cos \gamma z \cosh \gamma z + B \sin \gamma z \sinh \gamma z + \frac{\sin \alpha}{\cos^2 \alpha} \\
v_0 &= 2\gamma^2 (-A \sin \gamma z \sinh \gamma z + B \cos \gamma z \cosh \gamma z)
\end{aligned}
\tag{1.70}
$$

Here we use the notations

$$
\gamma = \left(\frac{R_0 \cos^2 \alpha}{4} \right)^{1/4}, \qquad \delta = \frac{1}{2}(\cos \gamma + \cosh \gamma)
$$

$$
A = -\frac{\sin \alpha}{\delta \cos^2 \alpha} \cos \frac{\gamma}{2} \cosh \frac{\gamma}{2}, \qquad B = -\frac{\sin \alpha}{\delta \cos^2 \alpha} \sin \frac{\gamma}{2} \sinh \frac{\gamma}{2}
$$

In the first order one arrives at the set of inhomogeneous equations

$$
\begin{gathered}
v_1'' + R_0 \left(\theta_1 \cos^2 \alpha - \psi_1 \sin \alpha \right) = v_0 - R_1 \left(\theta_0 \cos^2 \alpha - \psi_0 \sin \alpha \right) \\
\theta_1'' - v_1 = \theta_0 \\
\psi_1'' = 1 - \theta_0 \sin \alpha
\end{gathered}
\tag{1.71}
$$

To find the solvability condition, the last equation is integrated across the layer and the boundary condition $\psi_1'(\pm\frac{1}{2}) = 0$ is taken into account:

$$
\sin \alpha \int_{-1/2}^{1/2} \theta_0 \, \mathrm{d}z = 1 \tag{1.72}
$$

Substitution of θ_0 gives the formula

$$\frac{1}{\gamma}\frac{\sin\gamma + \sinh\gamma}{\cos\gamma + \cosh\gamma} = 1 - \mathrm{cotan}^2\,\alpha \tag{1.73}$$

The transcendental relation (1.73) allows the critical Rayleigh number to be determined for the long-wave mode $Ra_{\mathrm{vm}} = R_0$ as a function of the angle α between the temperature gradient and the vibration axis. The instability exists in the range $\alpha > \alpha_2$ where $\alpha_2 = 45°$. When α varies from α_2 to $90°$, the critical value Ra_{vm} decreases monotonically (Fig. 8, curve 2). In the limit $\alpha = 90°$, the stability is minimal and $Ra_{\mathrm{vm}} = 120$. When α approaches $45°$ from above,

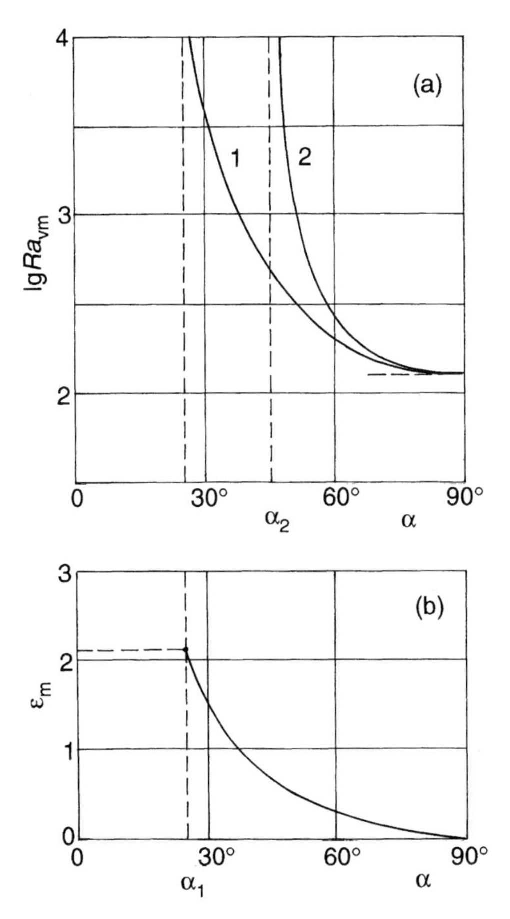

Fig. 8 Stability borders with respect to one-dimensional perturbations (a) and variational parameter ε (b) as functions of the angle α

$\alpha \to \alpha_2 + 0$, the critical Rayleigh number Ra_{vm} grows infinitely and close to $\alpha = 90°$ assumes the asymptotic behavior

$$Ra_{\mathrm{vm}} = \frac{4}{(\alpha - \pi/4)^2}$$

The mode (1.66) is not the most dangerous one. The class of perturbations might be generalized by introducing perturbations of the type

$$k_1 \neq 0, \quad k_2 \neq 0, \quad \varepsilon = \frac{k_2}{k_1}, \quad k^2 = k_1^2 + k_2^2; \qquad v_z = 0, \quad v_x(z), \quad v_y(z) \tag{1.74}$$

From the continuity equation, the relation for v_x and v_y follows: $k_1 v_x + k_2 v_y = 0$, i.e. $v_x = -\varepsilon v_y$. It is clear that a one-dimensional perturbation is obtained once again but, in contrast to (1.66), now the velocity direction is not parallel to the y axis but makes the angle arctan ε with it. The analysis shows that, within this class, the long-wave mode is also the most dangerous $(k \to 0)$ with the parameter to be varied $\varepsilon = k_2/k_1$, which determines the velocity direction. Minimization of $Ra_{\mathrm{v}}(\varepsilon)$ when $k \to 0$ provides the minimal value Ra_{vm}, yielding the instability threshold and as well the corresponding value of ε_{m}. These characteristics are presented in Fig. 8. It may be seen that perturbations of this kind are more dangerous than (1.66), where the latter is a particular case corresponding to $k_2 = 0$, i.e. $\varepsilon = 0$. The instability within the set described by (1.74) exists in the range $\alpha > \alpha_1$ where $\alpha_1 \approx 25°$. If α increases in the range $\alpha_1 < \alpha < 90°$, the critical value Ra_{vm} decreases monotonically. In the limit $\alpha \to 90°$, there are tendencies for $\varepsilon_m \to 0$ and $Ra_{\mathrm{vm}} \to 120$, i.e. one-dimensional perturbations (1.66) whose velocity is parallel to the temperature gradient again become the most dangerous ones.

Further, the equilibrium state is discussed, which is sketched in Fig. 9. The vibration axis is longitudinal: $\mathbf{n}(1, 0, 0)$. The temperature gradient is perpendicular to $\mathbf{n}$ but is not transversal: it forms the angle β with the z axis. Therefore the dimensionless components of the temperature gradient are $\nabla T_0(0, \sin\beta$, $\cos\beta)$. In the equilibrium field there is $\mathbf{w}_0(w_0, 0, 0)$, where $w_0 = y\sin\beta + z\cos\beta$. The limiting cases of the stability problem are known. At $\beta = 0$ the situation of Section 3 is recovered where the transversal temperature gradient combines with longitudinal vibrations. The stability boundary is determined by the critical value $Ra_{\mathrm{vm}} = 2129$ and is connected with the system of two-dimensional rolls aligned along the y axis. The case $\beta = 90°$ has been considered earlier in this section. It yields vectors ∇T_0 and $\mathbf{n}$ that are both longitudinal and make the angle $90°$ to one another. The crisis of the equilibrium is due to the long-wave perturbations of the 'passing' type; the critical value is $Ra_{\mathrm{vm}} = 120$.

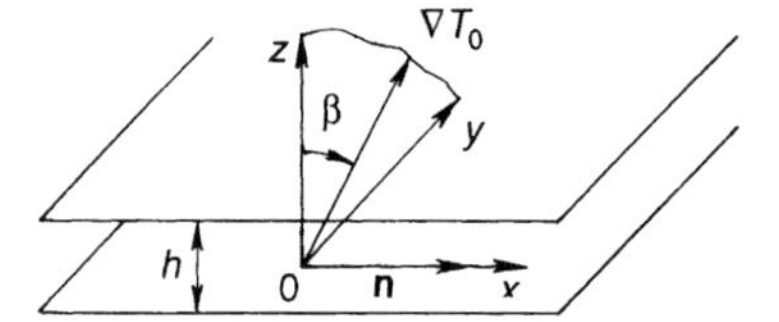

Fig. 9 Coordinate axes and the mutual arrangement of the vectors ∇T_0 and **n**

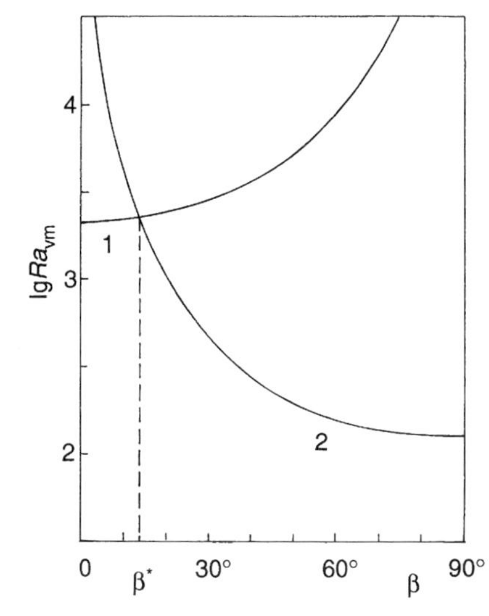

Fig. 10 Stability border for two modes: 1, two-dimensional rolls; 2, one-dimensional 'passing' mode

From the symmetry considerations it is evident that in the intermediate region $0° < \beta < 90°$, the behavior of two modes of the spectrum is to be studied. The first of them has the form of two-dimensional rolls aligned with the y axis. The stability problem may be easily reduced to that of Section 3 and the critical value is $Ra_{\text{vm}} = 2129/\cos^2\beta$. The second mode is a one-dimensional perturbation with the velocity parallel to the y axis; $k \to 0$, the limiting case of the 'passing' mode. This stability problem amounts to that described at the beginning of the section; the critical value is $Ra_{\text{vm}} = 120/\sin^2\beta$.

The competition between the two modes is illustrated in Fig. 10. It is clear that at $\beta < \beta_*$ the two-dimensional roll mode is the most dangerous. At $\beta > \beta_*$ the one-dimensional 'passing' mode becomes the most dangerous. The calculation gives $\beta_* \approx 13.5°$.

Finally, let us consider the equilibrium situation where the vibration axis is transverse to the layer and the temperature gradient is oriented arbitrarily. Upon the appropriate choice of the coordinate axes, the temperature gradient lies in the xz plane and makes the angle γ with the x axis, i.e. $\nabla T_0(\cos\gamma, 0, \sin\gamma)$ (see Fig. 11). The mechanical quasi-equilibrium is possible with the field $\mathbf{w}(w_0, 0, 0)$;

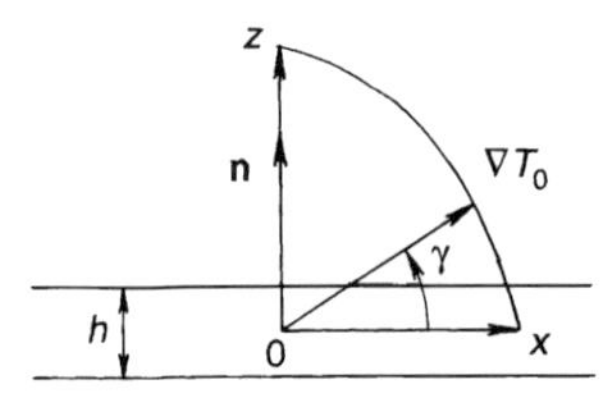

Fig. 11 The coordinate system and the orientation of the vectors ∇T_0 and $\mathbf{n}$

$w_0 = -z \cos\gamma$. The amplitude equations may be written in the form

$$-\lambda v = -\nabla p + \Delta v + Ra_v \, w_z \nabla T_0 \tag{1.75}$$

$$-\lambda \, Pr \, T + v_\gamma = \Delta T \tag{1.76}$$

$$\text{div } \mathbf{v} = 0, \qquad \text{div } \mathbf{w} = 0, \qquad \text{curl } \mathbf{w} = \nabla T \times \mathbf{n} \tag{1.77}$$

with the boundary conditions

$$z = \pm \tfrac{1}{2}: \qquad \mathbf{v} = 0, \qquad T = 0, \qquad w_z = 0 \tag{1.78}$$

Here w_z is the transversal component of the $\mathbf{w}$ perturbation and v_γ is the projection of $\mathbf{v}$ on to the equilibrium gradient direction.

From the boundary-value problem (1.75) to (1.78), the integral relations may be obtained, showing that the equilibrium is stable at arbitrary values of Ra_v and γ. The treatment is quite similar to the one presented at the end of Section 2. The periodic harmonic perturbations (1.51) are assumed, and the convective cell is chosen as the integration domain. On the 'horizontal' boundaries, the conditions (1.78) are valid and on the 'vertical' lateral boundaries the symmetry conditions are imposed, i.e. the normal components of vectors $\mathbf{v}$, $\mathbf{w}$ as well as ∇T and ∇w_z equal zero.

We multiply both sides of (1.7) by $\mathbf{v}^*$ and integrate over the volume of the cell. After simple transformations one gets

$$-\lambda \int |\mathbf{v}|^2 \mathrm{d}V = -\int |\,\text{curl } \mathbf{v}\,|^2 \mathrm{d}V + Ra_v \int w_z v_\gamma^* \, \mathrm{d}V \tag{1.79}$$

Both sides of (1.76) are multiplied by $w_z^* = (\mathbf{w}^* \mathbf{n})$ and integrated over the cell volume:

$$-\lambda \, Pr \int T w_z^* \, \mathrm{d}V + \int v_\gamma w_z^* \, \mathrm{d}V = \int \Delta T w_z^* \, \mathrm{d}V \tag{1.80}$$

The first integral on the left-hand side of equation (1.80) is transformed, taking into account the equation $\text{div } \mathbf{w}^* = 0$ and the boundary condition $w_n^* = 0$:

$$\int T w_z^* \, \mathrm{d}V = \int T \mathbf{n} \mathbf{w}^* \, \mathrm{d}V = \int (\mathbf{w} + \nabla \varphi) \mathbf{w}^* \, \mathrm{d}V$$
$$= \int |\mathbf{w}|^2 \, \mathrm{d}V + \int \nabla \varphi \cdot \mathbf{w}^* \, \mathrm{d}V = \int |\mathbf{w}|^2 \, \mathrm{d}V$$

When proceeding to the right-hand side of equation (1.80), it should be noted that

$$\Delta \mathbf{w} = -\text{curl}\,\text{curl}\,\mathbf{w} = -\text{curl}\,(\nabla T \times \mathbf{n}) = \mathbf{n}\,\Delta T - (\mathbf{n}\nabla)\nabla T$$

This means that

$$\Delta w_z = \Delta T - \frac{\partial^2 T}{\partial z^2} = \Delta_2 T$$

where Δ_2 is the two-dimensional Laplace operator in the xy plane. Integration of the right-hand side of (1.80) gives

$$\int \Delta T w_z^* \, \mathrm{d}V = \int T \Delta w_z^* \, \mathrm{d}V = \int T \Delta_2 T^* \, \mathrm{d}V = -\int |\nabla_2 T|^2 \, \mathrm{d}V$$

where ∇_2 is the two-dimensional ∇ operator in the xy plane. The resulting transformation of (1.80) is written as

$$-\lambda \, Pr \int |\mathbf{w}|^2 \, \mathrm{d}V + \int v_\gamma w_z^* \, \mathrm{d}V = -\int |\nabla_2 T|^2 \, \mathrm{d}V \tag{1.81}$$

If a set of four relations, viz. (1.79), (1.81) and their two complex-conjugate forms, is considered, then one finds that

$$\lambda + \lambda^* = \frac{2\int \left(|\text{curl}\,\mathbf{v}|^2 + |\nabla_2 T|^2 \right) \mathrm{d}V}{\int \left(|\mathbf{v}|^2 + Ra_\text{v}\, Pr |\mathbf{w}|^2 \right) \mathrm{d}V} \tag{1.82}$$

$$(\lambda - \lambda^*) \int \left(|\mathbf{v}|^2 - Ra_\text{v}\, Pr |\mathbf{w}|^2 \right) \mathrm{d}V = 0 \tag{1.83}$$

Both integrals in equation (1.82) are positive, so $\lambda_\text{r} = (\lambda + \lambda^*)/2 > 0$, which means that the perturbations decay. The integral in equation (1.83) does not have a definite sign, so it is not possible to draw any certain conclusion on whether the perturbation damping is of the monotonic or oscillatory type. Therefore, if the axis of vibrations is perpendicular to the layer, the situation is stable.†

In the cited papers by Braverman, the stability of equilibrium is investigated for some other situations in a plane layer. In addition, in [18] a general analysis is given for the problem of a quasi-equilibrium existence in a plane layer for the set of $\mathbf{w}_0$ fields which have no transversal component ($w_{0z} = 0$).

† It may also be shown that in this case the stability takes place even with respect to finite perturbations, so the absolute stability of the configuration is ensured.

4.2 Cylindrical Layer with the Radial Temperature Gradient and Longitudinal Axis of Vibration

All the afore-studied equilibrium states exemplified a constant temperature gradient. However, in the vibration field, contrary to the static gravity field, the constancy of the temperature gradient is not the necessary condition of quasi-equilibrium. Below we consider an example of a quasi-equilibrium situation with a non-uniform temperature gradient.

Let the fluid fill a cylindrical layer between two infinite coaxial cylinders with the internal radius R_1 and the external radius R_2. On the rigid surfaces different constant temperatures are maintained: $T_0(R_1) = \Theta$ and $T_0(R_2) = 0$. The whole system vibrates with a high frequency along the axial direction. The cylindrical coordinates are used and the dimensionless variables are introduced. For those, $d = R_2 - R_1$ and Θ are chosen as the length and temperature scales, respectively. It is easy to prove that in the case under discussion the mechanical quasi-equilibrium is described by the following T_0 and $\mathbf{w}_0$ distributions:

$$T_0 = \frac{1}{\ln\rho}\ln\frac{\rho}{(\rho-1)r}, \qquad w_{0r} = w_{0\varphi} = 0, \qquad w_{0z} = w_0(r) = T_0 + C \quad (1.84)$$

Here $\rho = R_2/R_1$ and the integration constant C is found from the condition of closeness of the oscillatory flow:

$$\int_{1/(\rho-1)}^{\rho/(\rho-1)} w_0 r\,\mathrm{d}r = 0, \qquad C = \frac{1}{\rho^2-1} - \frac{1}{2\ln\rho}$$

To study the stability of the quasi-equilibrium (1.84), let us introduce axisymmetrical perturbations of the type

$$v_\varphi = 0, \qquad v_r(r,z), \qquad v_z(r,z), \qquad w_\varphi = 0,$$
$$w_r(r,z), \qquad w_z(r,z), \qquad T(r,z)$$

and formulate the problem for neutral perturbations in terms of stream functions ψ and f for vectors $\mathbf{v}$ and $\mathbf{w}$, respectively:

$$v_r = -\frac{1}{r}\frac{\partial\psi}{\partial z}, \qquad v_z = \frac{1}{r}\frac{\partial\psi}{\partial r}, \qquad w_r = -\frac{1}{r}\frac{\partial f}{\partial z}, \qquad w_z = \frac{1}{r}\frac{\partial f}{\partial r} \quad (1.85)$$

Substituting (1.85) into (1.35), one gets the set of equations for neutral perturbations:

$$\frac{1}{r}D^2\psi + Ra_{\mathrm{v}}\,T_0'\left(\frac{\partial T}{\partial z} - \frac{1}{r}\frac{\partial^2 f}{\partial r\,\partial z}\right) = 0$$

$$DT + \frac{2}{r}\frac{\partial T}{\partial r} = -\frac{1}{r}T_0'\frac{\partial\psi}{\partial z} \quad (1.86)$$

$$\frac{1}{r}Df = \frac{\partial T}{\partial r}$$

Here $Ra_v = (\beta b \Omega \Theta d)^2/(2\nu\chi)$ is the vibrational Rayleigh number, prime means differentiation with respect to the radial coordinate and D is the Stokes operator

$$D = \frac{\partial^2}{\partial r^2} - \frac{1}{r}\frac{\partial}{\partial r} + \frac{\partial^2}{\partial z^2}$$

The perturbations are assumed to be periodic along the z direction and proportional to $\exp(ikz)$. Then the system of the amplitude equations takes the form

$$\begin{aligned} D^2\psi + \mathrm{i}k\,Ra_v\,T_0'(r\theta - f') &= 0 \\ DT + \frac{2}{r}\theta' &= -\frac{\mathrm{i}k}{r}T_0'\psi \\ Df &= r\theta' \end{aligned} \qquad (1.87)$$

with the boundary conditions

$$r = \frac{1}{\rho - 1}, \qquad r = \frac{\rho}{\rho - 1}: \qquad \psi = \psi' = 0, \qquad \theta = 0, \qquad f = 0 \quad (1.88)$$

Here the D-operator is modified by replacing $\partial^2/\partial z^2 \to -k^2$.

The boundary-value problem (1.87) and (1.88) is integrated numerically by the Runge–Kutta method in combination with shooting procedure. The critical value of Ra_v is evaluated as a function of the parameters k and ρ [12]. In Fig. 12

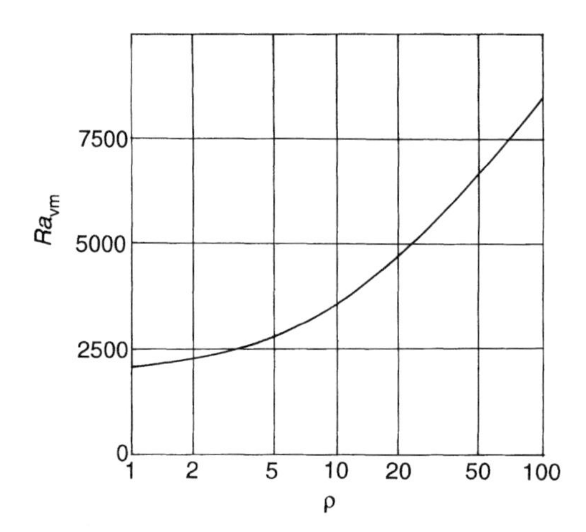

Fig. 12 Critical Rayleigh number as a function of the radii ratio

the critical value Ra_{vm} minimized with respect to k is plotted against ρ. It can be seen that, in the limiting case of a thin layer, the critical value $Ra_{\text{vm}} = 2129$ is recovered, which corresponds to that of a plane layer with longitudinal vibrations (Section 3). The critical value of Ra_{vm} increases monotonically with ρ, whereas the critical wavenumber does not vary in practice; while ρ changes from 1 to 100, it grows from 3.23 to 3.30.

We remark that the quasi-equilibrium considered here is a particular case of a wider class of states that may be characterized by a two-dimensional field of quasi-equilibrium temperature. Let the fluid fill a straight cylinder of an arbitrary cross-section. The cylinder performs vibrations along its axis. Choosing the coordinate system, we direct its z axis along the axis of the cylinder so that the xy plane coincides with that of the normal cross-section. If the boundary temperature does not depend on z and varies arbitrarily along the perimeter of the cross-section, then the equilibrium exists with the temperature field $T_0(x, y)$ being the solution of a two-dimensional Dirichlet problem while the field $\mathbf{w}_0$ has the structure $\mathbf{w}_0(0, 0, w_0)$, $w_0 = T_0(x, y) + C$, where C is found from the condition of the $\mathbf{w}_0$ field closeness.

4.3 Infinite Cylinder of a Circular Cross-section in the Presence of a Transversal Temperature Gradient and Vibrations

An example is considered of an equilibrium situation where the cavity size is limited in the directions of the temperature gradient and vibration axis.

Let the fluid fill an infinite cylinder of the circular cross-section of radius a. The Cartesian coordinates are introduced with the origin at the center of the cross-section as well as the cylindrical coordinates with the polar angle θ counted from the x axis. For dimensionless variables, a and Θ are taken as the length and temperature scales, respectively. The temperature at the cylinder lateral surface does not depend on the axial variable z and varies with the angle as

$$T_0 = x = r\cos\theta \tag{1.89}$$

In this case the temperature gradient is constant and directed along the x axis.

Let the vibration axis be perpendicular to the cylinder axis, i.e. the unit vector $\mathbf{n}$ lies in the xy plane. According to conditions (1.29) to (1.32), a quasi-equilibrium is possible in two cases. In the first case, when the vibration axis is parallel to the temperature gradient this quasi-equilibrium is stable (see Section 2). In the second case, the vibration axis is perpendicular to the temperature gradient and, as shown below, the equilibrium loses its stability when the vibrational Rayleigh number Ra_{v} exceeds some critical value.

Let the y axis be the vibration direction. Then the temperature gradient and the vibration axis are perpendicular to one another and also perpendicular to the cylinder axis. The equilibrium field $\mathbf{w}_0$ has the structure

$$w_{0r} = w_{0z} = 0, \qquad w_{0\theta} = \frac{r}{2} \tag{1.90}$$

Small plane perturbations of the equilibrium state (1.89) and (1.90) are considered. Let us introduce the stream functions ψ and F' for the average velocity $\mathbf{v}$ and the perturbation $\mathbf{w}'$ by the relations

$$v_r = \frac{1}{r}\frac{\partial \psi}{\partial \theta}, \qquad v_\theta = -\frac{\partial \psi}{\partial r}; \qquad w'_r = \frac{1}{r}\frac{\partial F'}{\partial \theta}, \qquad w'_\theta = -\frac{\partial F'}{\partial r} \tag{1.91}$$

The set of equations for the neutral perturbations ψ, T' and F' is written as follows:

$$\Delta^2 \psi = Ra_{\mathrm{v}}\left(\frac{\partial^2 F'}{\partial x \partial y} + \frac{1}{2}\frac{\partial T'}{\partial y}\right) \tag{1.92}$$

$$\Delta T = \frac{\partial \psi}{\partial y} \tag{1.93}$$

$$\Delta F = -\frac{\partial T'}{\partial x} \tag{1.94}$$

Here T' is the temperature perturbation, Δ is the two-dimensional Laplace operator and

$$\frac{\partial}{\partial x} = \cos\theta \frac{\partial}{\partial r} - \frac{\sin\theta}{r}\frac{\partial}{\partial \theta}, \qquad \frac{\partial}{\partial y} = \sin\theta \frac{\partial}{\partial r} + \frac{\cos\theta}{r}\frac{\partial}{\partial \theta}$$

The vibrational Rayleigh number is defined through the amplitude Θ of the temperature distribution at the boundary and the cylinder radius a. One needs to find a solution that is finite inside the circle and obeys the conditions

$$r = 1: \qquad \psi = \frac{\partial \psi}{\partial r} = 0, \qquad T' = 0, \qquad F' = 0 \tag{1.95}$$

The boundary-value problems (1.92) to (1.95) determines the spectrum of critical values of Ra_{v} and pertinent eigenfunctions describing the critical flow patterns. To solve the problem, the Galerkin method has been applied [12, 20, 21]. With the objective to find the lowermost spectrum level, the stream function is approximated by a set of three trial functions possessing an appropriate symmetry and satisfying the boundary conditions

$$\psi = c_1\psi_1 + c_2\psi_2 + c_3\psi_3 = \left(1 - r^2\right)^2\left(c_1 + c_2 x^2 + c_3 y^2\right) \tag{1.96}$$

The approximations for T' and F' are of the form (primes are omitted)

$$T = c_1T_1 + c_2T_2 + c_3T_3, \qquad F = c_1F_1 + c_2F_2 + c_3F_3 \tag{1.97}$$

where the trial functions F_{i} and T_{i} are found as the exact solutions of the inhomogeneous equations

$$\Delta T_{\mathrm{i}} = \frac{\partial \psi_{\mathrm{i}}}{\partial y}, \qquad \Delta F_{\mathrm{i}} = -\frac{\partial T_{\mathrm{i}}}{\partial x}, \qquad i = 1, 2, 3 \tag{1.98}$$

with the boundary conditions

$$r = 1: \qquad T_i = 0, \qquad F_i = 0 \tag{1.99}$$

These trial functions may be written in the form

$$\begin{aligned}
T_1 &= \tfrac{1}{6} r(1-r^2)(2-r^2)\sin\theta \\
T_{2,3} &= -\frac{1}{1152}\left(1-r^2\right)\left(13+13r^2-23r^4+9r^6\right) \\
&\quad \mp \frac{1}{480} r^2\left(1-r^2\right)\left(9-11r^2+4r^4\right)\cos 2\theta \\
F_1 &= -\frac{1}{96} r^2\left(1-r^2\right)\left(3-r^2\right)\sin 2\theta \\
F_2 &= \frac{1}{11\,520} r\left(1-r^2\right)\left(11+65r^2-55r^4+15r^6\right) \\
&\quad \times \cos\theta + \frac{1}{5760} r^3\left(1-r^2\right)\left(8-7r^2+2r^4\right)\cos 3\theta \\
F_3 &= \frac{1}{11\,520} r\left(1-r^2\right)\left(47-7r^2-7r^4+3r^6\right) \\
&\quad \times \cos\theta - \frac{1}{5760} r^3\left(1-r^2\right)\left(8-7r^2+2r^4\right)\cos 3\theta
\end{aligned}$$

To evaluate the coefficients c_1 to c_3, expansions (1.96) and (1.97) are substituted into (1.92) and then for the residual the orthogonality conditions of the Galerkin method are constructed. As a result, a linear homogeneous set of algebraic equations for c_i is derived. Its non-trivial solution exists if and only if the determinant equals zero. Satisfying this requirement, one arrives at the equation to determine Ra_{v}. The lowermost root gives the main level of the vibrational convective instability that has the form of a one-vortex critical flow. According to computations, $Ra_{\mathrm{v}} = 1029$. The estimated accuracy is a few percent since the approximations (1.96) and (1.97), containing just one trial function, yield the critical value $Ra_{\mathrm{v}} = 1182$.

A natural generalization for the considered problem is the case of a cylinder of an elliptic cross-section with arbitrary aspect ratio under the condition that the temperature gradient and the vibration axis are directed along different semiaxes. If ∇T_0 and $\mathbf{n}$ lie along the x and y axes, which coincide with the cross-section semiaxes a and b, respectively, then the equilibrium fields are $T_0 = Ax$, $w_{0x} = -A\kappa y$ and $w_{0y} = A(1-\kappa)x$. Here A is the value of gradient and $\kappa = (1 + b^2/a^2)^{-1}$.

5 Interaction of the Thermogravitational and Thermovibrational Mechanisms of Instability: Horizontal Layer

In the previous sections equilibrium states and their stability under pure weightlessness, i.e. when the static gravity field is absent, were considered. For that case the loss of the mechanical quasi-equilibrium stability and appearance of convection are solely due to the thermovibrational mechanism. Now we address the situation where the static gravity and high-frequency vibrations act simultaneously on a non-isothermal fluid. The onset of convection is caused by both mechanisms, gravitational and vibrational.

To describe the convective phenomena one needs to go back to the complete set of equations (1.25) where the effect of convective forces in both gravitational and vibrational fields is taken into account. The quasi-equilibrium conditions for T_0, p_0, $\mathbf{w}_0$ may be derived from (1.25) setting $\mathbf{v} = 0$ and $\partial/\partial t = 0$:

$$
\begin{gathered}
-\nabla p_0 + Ra\, T_0 \boldsymbol{\gamma} + Ra_{\mathrm{v}}(\mathbf{w}_0 \nabla)(T_0 \mathbf{n} - \mathbf{w}_0) = 0 \\
\Delta T_0 = 0 \qquad (1.100) \\
\operatorname{div} \mathbf{w}_0 = 0, \qquad \operatorname{curl} \mathbf{w}_0 = \nabla T_0 \times \mathbf{n}
\end{gathered}
$$

Here Ra and Ra_{v} are the customary and vibrational Rayleigh numbers (see (1.27)). Application of the curl operation to the first equation of (1.100) yields the hydrostatic condition in the form

$$
Ra(\nabla T_0 \times \boldsymbol{\gamma}) + Ra_{\mathrm{v}}\, \nabla(\mathbf{w}_0 \mathbf{n}) \times \nabla T_0 = 0 \qquad (1.101)
$$

Let us consider the convective stability of equilibrium in a fluid plane layer in the presence of a transversal temperature gradient. Let us go back to Fig. 1 and suppose that the layer is horizontal, the z axis is pointing vertically upwards, the equilibrium temperature gradient is $\nabla T_0 = -A\boldsymbol{\gamma}$, where $A = \Theta/h$, and the unit vector $\boldsymbol{\gamma}$ has the components $\boldsymbol{\gamma}(0, 0, 1)$. The unit vector $\mathbf{n}$ is aligned with the axis of vibration and has the components $\mathbf{n}(\cos\alpha, 0, \sin\alpha)$. One may see that there exists the equilibrium solution with $\mathbf{w}_0(w_0, 0, 0)$; $w_0 = w_0(z)$. Relations (1.100) and (1.101) are satisfied when the dimensionless profiles T_0 and w_0, as in the case of weightlessness (Section 3), are given by formulas (1.47).

The plane normal perturbations are introduced. By analogy with the weightlessness case, one might anticipate that perturbations of this kind are the most dangerous. Besides, the calculations show that the critical disturbance is monotonic. The amplitude problem differs from (1.56) only by the additional term due to thermogravitational force:

$$D^2 v - k^2 Ra\,\theta + Ra_v\left(-k^2 \cos^2\alpha \cdot \theta + \mathrm{i}k \cos\alpha \cdot w' + k^2 \sin\alpha \cdot w\right) = 0$$

$$D\theta = -v \tag{1.102}$$

$$Dw = -\left(\mathrm{i}k \cos\alpha \cdot \theta' + k^2 \sin\alpha \cdot \theta\right)$$

$$z = 0, z = 1: \qquad v = v' = 0, \qquad \theta = 0, \qquad w = 0 \tag{1.103}$$

Here v, w, θ are, respectively, the perturbation amplitudes of the vertical components of the average and pulsation velocities and temperature, $D = \mathrm{d}^2/\mathrm{d}z^2 - k^2$ and k is the wavenumber along the x axis.

The solution of the spectral amplitude problem is illustrated in Fig. 13 where the stability borders are shown in the plane (Ra, Ra_v). Minimization has been done with respect to the wavenumber, the angle of the vibration axis inclination being a parameter. The critical point on the Ra axis is the solution of the classic

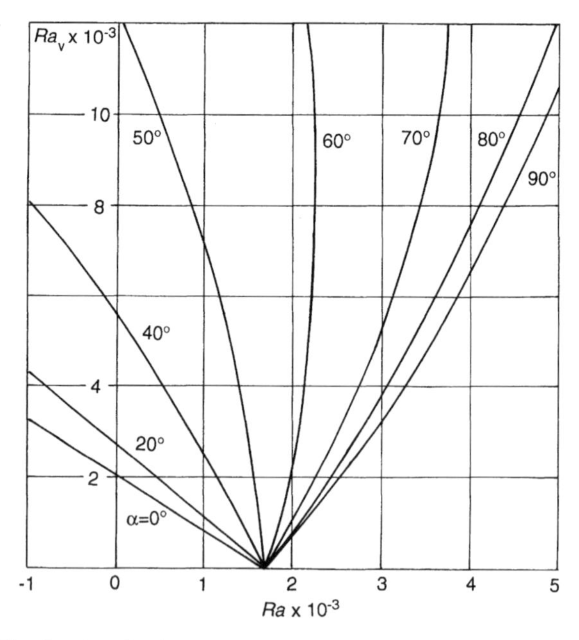

Fig. 13 Stability borders in the plane (Ra, Ra_v) for different inclination angles of the vibration axis

Rayleigh–Benard problem for a horizontal layer between rigid isothermal boundaries in the absence of vibration. This critical Rayleigh number $Ra = 1708$ is connected with the most dangerous perturbation of the wave-number $k_m = 3.116$. This point on the Ra_v axis corresponds to the case of pure weightlessness. The instability characteristics Ra_{vm} and k_m depend on α and are summarized in Table 1 in Section 3. It is clear that the least stable situation takes place under longitudinal vibration ($\alpha = 0°$). There the stability border is practically a straight line in the plane (Ra, Ra_v). The deflection occurs only in the range of very large negative Ra. In the range of inclination angles $0 \leqslant \alpha < 90°$, the stability curves intersect the Ra_v axis and go to the negative Ra. This means that the instability takes place when a system is heated from above. Since at $Ra < 0$ the Rayleigh thermogravitational mechanism is not operative, this fact proves the occurrence of a new instability mechanism—the thermovibrational one.[†]

Inclination of the vibration axis to the horizontal, i.e. the occurrence of transversal component of vibration, has a stabilizing effect. In the limiting case of transverse vibrations ($\alpha = 90°$), the specific vibrational mechanism of instability is not operative (see Section 2) and the effect of vibration becomes purely stabilizing. In the range of large Ra and Ra_v the following asymptotic takes place

$$Ra = C Ra_v^{1/2} \tag{1.104}$$

where $C = 45.5$ is a constant. This asymptotic means that there exists a limiting value of the dimensionless vibrational parameter:

$$\kappa = \frac{b\Omega\sqrt{\nu\chi}}{gh^2} \tag{1.105}$$

which does not depend on the temperature difference. When κ exceeds the critical value $\kappa_* = 0.0311$, the absolute stabilization sets in, and the equilibrium becomes stable at an arbitrary value of the temperature difference. Thus, the limiting value of the vibration intensity is $(b\Omega)_* = 0.031 gh^2/\sqrt{\nu\chi}$, i.e. the effective damping of convection by means of vibration is present when h is small and ν is large. The dependences of the critical Rayleigh number and critical wavenumber on κ are shown in Fig. 14 for the case of vertical vibration. At $\kappa \to \kappa_*$, sharp stabilization takes place and the critical wavenumber is shifted in the long-wave instability direction.

[†] Zen'kovskaya [22] has found the exact solution for the convection onset problem in a horizontal layer under vibration with a tilted axis for model boundary conditions admitting periodicity along the z direction. The convection threshold in weightlessness was not determined. Nevertheless, it was shown that convection excitation is possible when the system is heated from above, i.e. the existence of a specific vibrational mechanism of convection has been demonstrated.

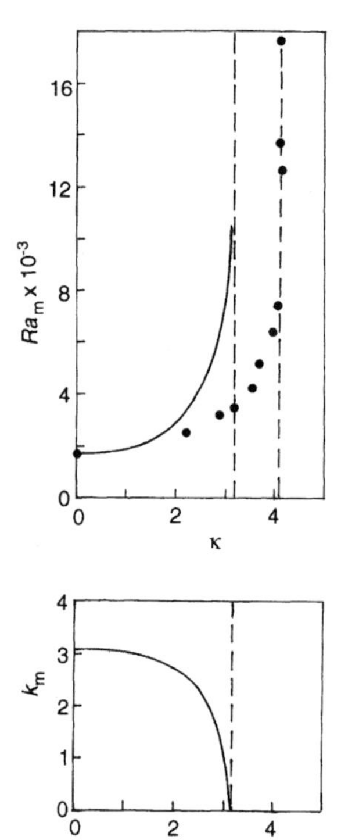

Fig. 14 Critical Rayleigh number and critical wavenumber depending on the vibrational parameter κ; the case of transversal vibrations. Points show experiment [23]

Let us recall the results of experimental study of convection excitation in a horizontal fluid layer under vibration, reported by Putin and co-workers [23–25]. As working fluids, they used distilled water, ethyl alcohol and transformer oil. The area of the horizontal plane enclosure was $35 \times 35\,\mathrm{mm}^2$ and the thickness of the layer varied from 0.8 mm to 5 mm. The horizontal boundaries were metal plates that ensured the isothermality condition. A mechanical vibrator with a constant amplitude of displacement $b = 4\,\mathrm{cm}$ was used with a variable frequency up to 25 Hz. In Fig. 15 the results are displayed concerning longitudinal (horizontal) vibration. Curve 1 gives the experimentally obtained stability border in the plane (Ra, Ra_v). The point on the Ra axis indicates the stability threshold in the absence of vibrations; this value of $Ra = 1.7 \times 10^3$ is in a good agreement with the solution of the classic Rayleigh–Benard problem $Ra = 1708$. As Ra_v increases, the convective stability border decreases since the additional thermovibrational mechanism of destabilization has an effect. The stability boundary, according to the theory, is a straight line. The experimental data in the range $Ra < 0$ correspond to the heating from above, when the only reason for instability must be the thermovibrational mechanism. Heating from above leads to a stable stratification and hence to stabilization. The point on the Ra_v axis cannot be achieved under terrestrial conditions

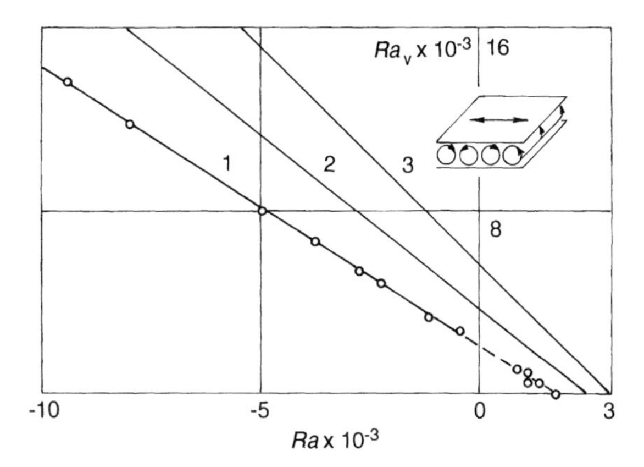

Fig. 15 Experimental results [23] for the case of longitudinal vibrations ($\alpha = 0°$): 1, theoretical stability border; 2 and 3, isolines of the Nusselt number: $Nu = 1.2, 1.4$

because it refers to a pure weightlessness. However, this critical value of the vibrational Rayleigh number may be obtained by interpolation, yielding $Ra_v = 2.1 \times 10^3$, which agrees well with the theoretical value $Ra_{vm} = 2129$.

To determine some non-linear overcritical characteristics of a vibrational convection in weightlessness, the interpolation technique may be applied. In Fig. 15 the isolines of the dimensionless heat flux—the Nusselt number—are shown. The Nusselt number is defined as the ratio of the heat flux through the layer to its equilibrium value. On the stability border apparently $Nu = 1$.

As experiments show, the crisis of equilibrium stability is associated with a formation of a set of two-dimensional rolls whose axes are perpendicular to the vibration axis. The transversal size of the rolls decreases as Ra_v grows. These results are also in agreement with the theory.

The experimental results on the equilibrium stability border under vertical vibration are presented in Fig. 14 together with the corresponding theoretical curve. A strong stabilizing effect takes place. We note a qualitative agreement of theoretical and experimental results. However, the limiting value (from the viewpoint of absolute stabilization) of the vibrational parameter $\kappa_* = 0.041$ deduced from experimental data is greater than the theoretical one $\kappa_* = 0.031$. As reported in [14], a hard excitation of convection was observed under conditions that correspond to the vertical part of the curve. Therefore, one may speak of a 'backward' bifurcation (in the direction of equilibrium). It is possible that in this case the vertical segment of the experimental curve is caused by the 'lower' critical point.

All those details correspond to a fluid layer between two highly conductive plates. If the heat conductivity coefficients of the fluid and ambient solid mass are comparable, it is necessary to take into account the temperature perturbations penetrating the mass, i.e. to solve a conjugated problem as was done in

Section 3 for the weightlessness conditions. For a symmetrical case $\widetilde{\kappa}_1 = \widetilde{\kappa}_2 = \kappa$ in the presence of longitudinal vibrations, the spectral amplitude problem involves two-dimensional neutral perturbations:

$$D^2 v - k^2(Ra + Ra_v)\theta + Ra_v \widetilde{w}' = 0$$

$$D\theta = -v \tag{1.106}$$

$$D\widetilde{w} = k^2\theta'$$

$$z = 0, z = 1: \qquad v = v' = 0, \qquad \widetilde{w} = 0, \qquad \kappa\theta' = \pm k\theta \tag{1.107}$$

Here $\widetilde{w} = ikw$. In the limiting case of heat-insulated masses ($\kappa \to \infty$), the boundary condition assumes the form $\theta'(0) = \theta'(1) = 0$. As might have been expected, long-wave perturbations play an important role in this situation. The stability boundary may be found asymptotically. Indeed, let amplitudes and critical values be presented in the following way:

$$\begin{aligned} v &= v_0 + k^2 v_1 + k^4 v_2 + \cdots \\ \theta &= \theta_0 + k^2\theta_1 + k^4\theta_2 + \cdots \\ \widetilde{w} &= \widetilde{w}_0 + k^2\widetilde{w}_1 + k^4\widetilde{w}_2 + \cdots \\ Ra &= Ra_0 + k^2\, Ra_1 + k^4\, Ra_2 + \cdots \\ Ra_v &= Ra_{v0} + k^2\, Ra_{v1} + k^4\, Ra_{v2} + \cdots \end{aligned} \tag{1.108}$$

After substituting these expansions into (1.106), in the zeroth order one gets

$$v_0^{IV} + Ra_{v0}\,\widetilde{w}_0' = 0$$

$$\theta_0'' = -v \tag{1.109}$$

$$\widetilde{w}_0'' = 0$$

They give $v_0 = 0$, $\widetilde{w}_0 = 0$, $\theta = \text{constant} = 1$ under an appropriate normalization.

With the first-order accuracy, the set takes the form

$$v_1^{IV} - 2v_0 - (Ra_0 + Ra_{v0})\theta_0 + Ra_{v0}\,\widetilde{w}_1' + Ra_{v1}\widetilde{w}_0' = 0$$

$$\theta_1'' - \theta_0 = -v_1 \tag{1.110}$$

$$\widetilde{w}_1'' - \widetilde{w}_0 = \theta_0'$$

The solution is

$$\widetilde{w}_1 = 0, \qquad v_1 = \frac{1}{24}(Ra_0 + Ra_{v0})z^2(z-1)^2$$

The solvability condition for the inhomogeneous set is obtained by integrating the second equation of the set over z from $z = 0$ to $z = 1$:

$$\int_0^1 v_1 \, dz = 1$$

Thus, the relationship between Ra_0 and Ra_{v0} affects the boundary of the long-wave instability:

$$Ra_0 + Ra_{v0} = 720 \tag{1.111}$$

This formula describes a straight line in the plane (Ra, Ra_v) passing through the point $Ra_v = 0$, $Ra = 720$ (corresponding to the long-wave instability in the layer with thermoinsulated boundaries in the absence of vibration) and through the point $Ra = 0$, $Ra_v = 720$ (corresponding to the weightlessness).

By means of higher order approximations it may be shown that all the corrections to Ra_0 and Ra_{v0} are positive. Thus the long-wave mode is the most dangerous.

In Fig. 16 the stability borders are drawn in the plane (Ra, Ra_v) for several values of the parameter κ as a result of minimization with respect to k. As in the

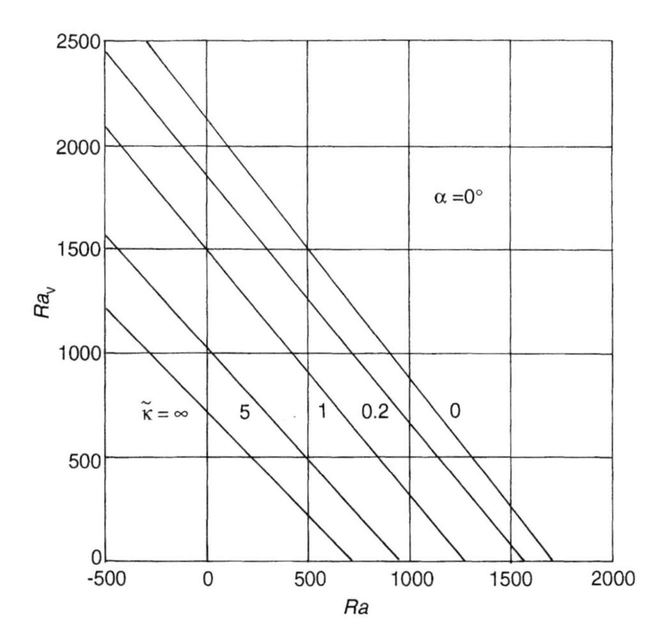

Fig. 16 Stability borders in the plane (Ra, Ra_v) for several values of the heat conductivity coefficient ratio (longitudinal vibrations, $\alpha = 0°$)

case of no vibration, the penetration of perturbations into masses lowers the stability border, and the critical perturbations shift to the long-wave region.

6 Interaction of the Thermogravitational and Thermovibrational Mechanisms of Instability: Tilted Layer

In this section we discuss the existence of the equilibrium and its stability in a tilted plane layer [26]. Geometrical arrangements and the choice of the coordinate system are clear from Fig. 17. The layer is inclined at the angle α to the vertical, and this angle is considered as a continuously varying parameter. The vibration axis, as usual, is characterized by the unit vector **n** locating in the (x, z) plane: $\mathbf{n}(n_x, 0, n_z)$. Quasi-equilibrium situations are studied with a constant temperature gradient, i.e. $\nabla T_0 = A\mathbf{m}$, where **m** is the unit vector lying inside the xz plane as well: $\mathbf{m}(m_x, 0, m_z)$. Therefore, all three vectors γ, **m**, **n** are coplanar. With respect to **m** and **n**, the configurations are selected where each of these vectors may have, independently of one another, either of four discrete orientations: vertical (v), longitudinal (l), horizontal (h) and transversal (t). Thus, in total, sixteen discrete configurations listed below and shown in Fig. 18 have been studied. The system of notations is as follows: the first symbol corresponds to the gradient orientation and the second one to the vibration axis orientation. Thus, for example, (v, l) means the configuration with a vertical temperature gradient and a longitudinal axis of vibration. The possible configurations are

$$\begin{array}{llll} (\mathrm{v},\mathrm{v}) & (\mathrm{v},\mathrm{l}) & (\mathrm{v},\mathrm{h}) & (\mathrm{v},\mathrm{t}) \\ (\mathrm{l},\mathrm{v}) & (\mathrm{l},\mathrm{l}) & (\mathrm{l},\mathrm{h}) & (\mathrm{l},\mathrm{t}) \\ (\mathrm{h},\mathrm{v}) & (\mathrm{h},\mathrm{l}) & (\mathrm{h},\mathrm{h}) & (\mathrm{h},\mathrm{t}) \\ (\mathrm{t},\mathrm{v}) & (\mathrm{t},\mathrm{l}) & (\mathrm{t},\mathrm{h}) & (\mathrm{t},\mathrm{t}) \end{array} \tag{1.112}$$

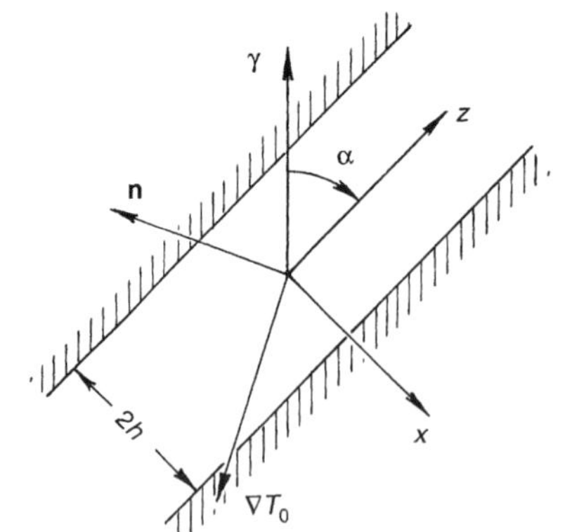

Fig. 17 Geometric arrangements and the system of coordinates

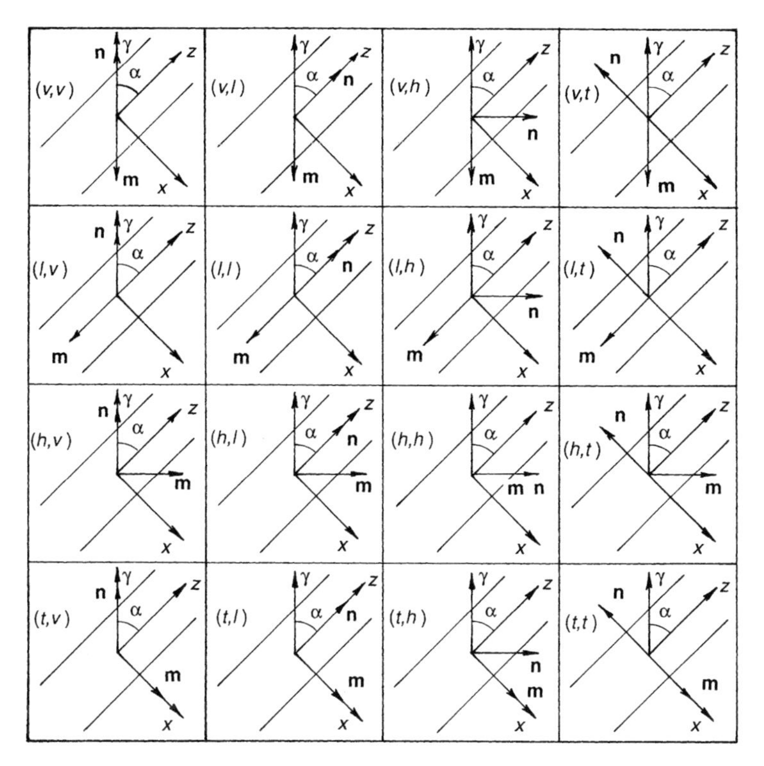

Fig. 18 List of the configurations under consideration

It should be noted that in the case of no vibration the necessary condition of equilibrium is that the temperature gradient is constant and vertical. Here, due to the average vibrational force, the hydrostatic condition may be satisfied for some other equilibrium gradient orientations. The conditions of equilibrium (1.100) are kept, of course.

Let us define the dimensionless parameters Ra and Ra_v through the half-thickness of the layer h and the temperature gradient A:

$$Ra = \frac{g\beta A h^4}{\nu\chi}, \qquad Ra_v = \frac{(\beta b \Omega A h^2)^2}{2\nu\chi} \tag{1.113}$$

and rewrite the hydrostatic relation (1.101) as follows:

$$Ra(\mathbf{m} \times \gamma) + Ra_v \nabla(\mathbf{mn}) \times \mathbf{m} = 0 \tag{1.114}$$

Suppose that in the quasi-equilibrium state the oscillatory velocity component is longitudinal: $\mathbf{w}_0(0, 0, w_0)$, where $w_0 = w_0(x)$. The profile $w_0(x)$ may be readily found from the last equation of (1.101) taking into account the closeness condition

$$w_0(x) = (m_x n_z - m_z n_x)x \tag{1.115}$$

Then the hydrostatic relation may be written as

$$Ra(m_z \sin\alpha + m_x \cos\alpha) + Ra_v(m_x n_z - m_z n_x)m_z n_z = 0 \tag{1.116}$$

To study the stability, the small plane perturbations of equilibrium are relevant and stream functions ψ and F are introduced for $\mathbf{v}$ and $\mathbf{w}'$:

$$v_x = \frac{\partial\psi}{\partial z}, \qquad v_z = -\frac{\partial\psi}{\partial x}; \qquad w'_x = \frac{\partial F}{\partial z}, \qquad w'_z = -\frac{\partial F}{\partial x} \tag{1.117}$$

Introducing the normal modes

$$(\psi, F, T') = [\varphi(x), f(x), \theta(x)] \exp(-\lambda t + \mathrm{i}kz) \tag{1.118}$$

the amplitude equations are obtained in the form

$$\begin{gathered} -\lambda D\varphi = D^2\varphi - Ra(\mathrm{i}k \sin\alpha \cdot \theta + \cos\alpha \cdot \theta') \\ -Ra_v\left[\mathrm{i}kn_z(m_x n_z - m_z n_x)\theta - m_z n_z f'' + k^2 m_x n_x f + \mathrm{i}k(m_x n_z + m_z n_x) f'\right] \\ -\lambda\, Pr\, \theta + (\mathrm{i}km_x\varphi - m_z\varphi') = D\theta \\ Df = \mathrm{i}kn_x\theta - n_z\theta' \end{gathered} \tag{1.119}$$

On the rigid high-conductive plates, the homogeneous boundary conditions are valid:

$$x = \pm 1: \qquad \varphi = \varphi' = 0, \qquad \theta = 0, \qquad f = 0 \tag{1.120}$$

Here, as usual, the prime denotes differentiation with respect to the transversal coordinate and $D = \mathrm{d}^2/\mathrm{d}x^2 - k^2$. It should be emphasized that the spectral problem (1.119) and (1.120) has a meaning not at arbitrary values of parameters but only at α, Ra, Ra_v, n_x, n_z, m_x, m_z connected by condition (1.116) of the equilibrium existence.

It is known that in the case of no vibration in some parameter range the neutral curves assume minimal values at $k = 0$, i.e. a long-wave mode is the most dangerous. One might expect that in the present situation the long-wave perturbations as well play an important role. Setting $k = 0$ and $\lambda = 0$ in the general problem leads to the following set of equations:

$$\begin{gathered} \varphi^{\mathrm{IV}} - \left(Ra \cos\alpha - Ra_v m_z n_z^2\right)\theta' = 0 \\ m_z\varphi' + \theta'' = 0 \end{gathered} \tag{1.121}$$

with the boundary conditions

$$x = \pm 1: \qquad \varphi = \varphi' = 0, \qquad \theta = 0 \tag{1.122}$$

The solution of this problem yielding the main level of the instability spectrum has the form

$$\varphi = 1 + \cos \pi x, \qquad \theta = -\frac{m_z}{\pi} \sin \pi x \tag{1.123}$$

and the existence condition for a non-trivial solution is written as

$$Ra \cos \alpha + Ra_v\, m_z n_z^2 = -\frac{\pi^4}{m_z} \tag{1.124}$$

The latter establishes a relationship between parameters at the border of the monotonic long-wave instability. To find the stability border for an arbitrary (finite) value of the wavenumber, the complete amplitude problem (1.119) and (1.120) needs to be used. The solution may be obtained by a straight numerical integration using the Runge–Kutta method with a shooting procedure.

Let us analyze the possible configurations.

6.1 Case (v, v)

First, let us consider the situation where both vectors (the temperature gradient and the vibration axis) are parallel and vertical. Under pure weightlessness such an equilibrium is absolutely stable (see Section 2). However, now the instability of the thermogravitational origin is possible.

In this case $m_x = \sin \alpha$, $m_z = -\cos \alpha$, $n_x = -\sin \alpha$, $n_z = \cos \alpha$ and $w_0 = 0$. So here we are indeed dealing with the equilibrium where the average and pulsation velocity components equal zero. From condition (1.116) it follows that the equilibrium exists for arbitrary α, Ra, Ra_v.

If there are no vibrations, the solution of a convective stability problem in a tilted layer heated from below (see [2], Section 16) is known. In the limit of a vertical layer, the convection threshold is due to the long-wave perturbations and this property holds under weak inclinations up to the critical value of the angle: $\alpha_c \approx 21°$. If $\alpha > \alpha_c$, the transition to the cellular mode takes place with a wavenumber $k_m \neq 0$. In the limit $\alpha \to 90°$, one recovers the classic Rayleigh–Benard problem for a horizontal layer.

In Fig. 19 the examples of neutral curves in the plane (k, Ra) for two values of the inclination angle and different Ra_v values are given. This family of curves demonstrates the effect of vibration on the instability characteristics. In the range of small α (Fig. 19a), the crisis is due to the long-wave mode if Ra_v is relatively small and to the cellular mode if Ra_v is large enough. The situation is different at $\alpha = 50°$ (Fig. 19b). The instability has the cellular form at relatively

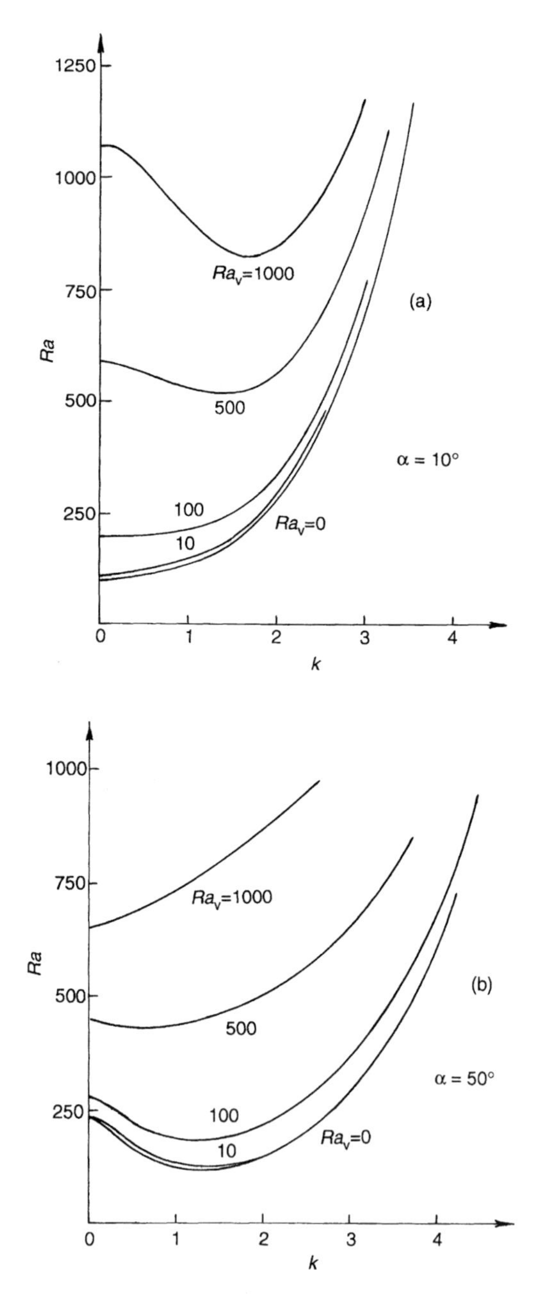

Fig. 19 (v, v) Neutral curves for different values of Ra_v: (a) $\alpha = 10°$, (b) $\alpha = 50°$

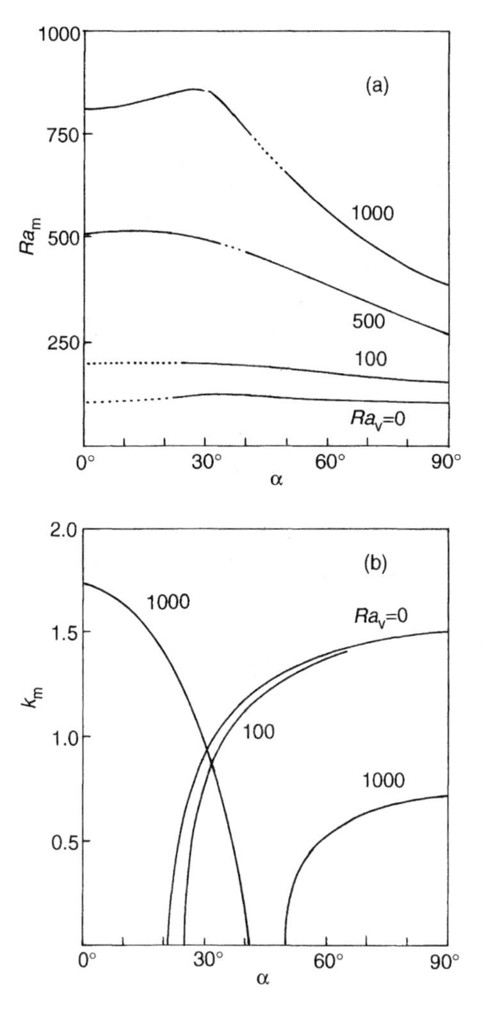

Fig. 20 (v, v) Critical characteristics as functions of the inclination angle for different Ra_v values. (a) Critical Ra_m values, where parts of the curves corresponding to the long-wave mode are dotted; (b) critical wavenumbers

small Ra_v and becomes long-wave at large Ra_v. If the orientation is close to the horizontal one, the cellular instability survives up to large enough Ra_v values.

In Fig. 20 the results are summarized in the form of dependences of the critical Rayleigh number Ra_m and critical wavenumber k_m on α for several Ra_v values. One may see that within the whole interval $0 \leqslant \alpha \leqslant 90°$ the stabilization takes place—the critical Ra_m value increases with Ra_v. The form of instability depends on α and Ra_v. At relatively small Ra_v the instability is long-wave if the orientation is close to vertical, and it becomes cellular as α increases. If Ra_v is

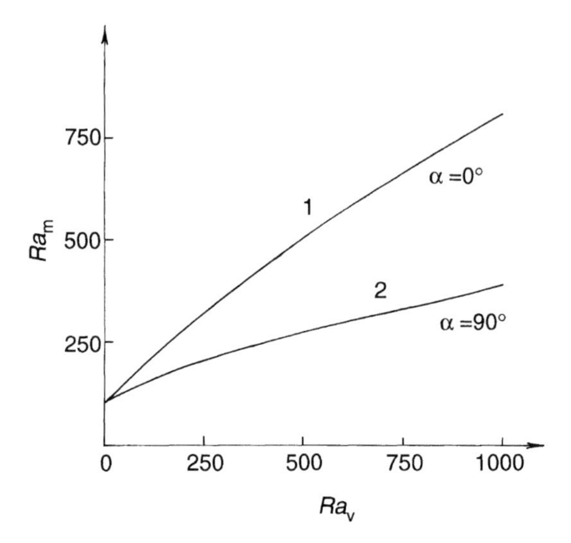

Fig. 21 (v, v) Critical Ra_m values depending on Ra_v for two limiting orientations: 1, $\alpha = 0°$; 2, $\alpha = 90°$

large enough, the intermediate interval of angles exists where the instability is caused by the long-wave mode. In this region, the computational results agree well with formula (1.124), which acquires the form

$$Ra = \frac{\pi^4}{\cos^2\alpha} + Ra_v \cos^2\alpha \tag{1.125}$$

In Fig. 21 the critical Ra_m value is displayed as a function of Ra_v for the limiting cases of the vertical ($\alpha = 0°$) and horizontal ($\alpha = 90°$) orientations. Note that the latter situation has already been discussed (see Fig. 13), unlike the situation of a vertical layer with the longitudinal gradient and vibration axis.

As might have been anticipated for the conditions under consideration, the specific thermovibrational mechanism of the convection excitation is not operative and the sole effect of vibrations is stabilization.

6.2 *Case (v, l)*

This is the situation of a vertical temperature gradient and a longitudinal vibration axis. From the hydrostatic equation (1.116) it follows that

$$Ra_v \sin\alpha \cos\alpha = 0 \tag{1.126}$$

In the absence of vibrations the equilibrium exists at any angle; the results of its stability are shown in Fig. 20, curves $Ra_v = 0$. If both parameters Ra and Ra_v

are arbitrary, there are only two limiting orientations, viz. $\alpha = 0°$ and $\alpha = 90°$, where the equilibrium is possible. The first of these cases has been already studied (see Fig. 21, curve 1), since at $\alpha = 0°$ the configuration (v, l) is coincident with (v, v). At $\alpha = 90°$ the situation of a horizontal layer in the presence of a vertical temperature gradient and longitudinal vibrations is standard and described in detail in the previous section.

6.3 Case (v, h)

This situation implies a vertical temperature gradient and a horizontal axis of vibrations. The hydrostatic relation again has the form (1.126). If there are no vibrations, the problem is already solved (see Fig. 20, $Ra_v = 0$). At arbitrary Ra, Ra_v values, only two orientations are feasible: $\alpha = 0°$ and $\alpha = 90°$. The latter one is already known—it is a horizontal layer subject to a vertical gradient and longitudinal (horizontal) vibrations. The case $\alpha = 0°$ (a vertical layer with a vertical temperature gradient and transverse vibrations) is new and needs to be analyzed. The stability border with respect to the long-wave mode is described by equation (1.124) and the critical value $Ra_m = \pi^4$ is independent of Ra_v. The neutral curves for several Ra_v values are given in Fig. 22. It can be seen that the long-wave mode is the most dangerous and the effect of vibration is stabilizing.

6.4 Case (v, t)

The quasi-equilibrium is available at arbitrary values of Ra, Ra_v and α. The critical characteristics as functions of the inclination angle are given in Fig. 23. In the range $0 \leqslant \alpha \leqslant \alpha_c$, the instability is of long-wave origin; at $\alpha > \alpha_c$ it transforms to the cellular form. The critical value $\alpha_c \approx 21°$ at $Ra_v = 0$ increases

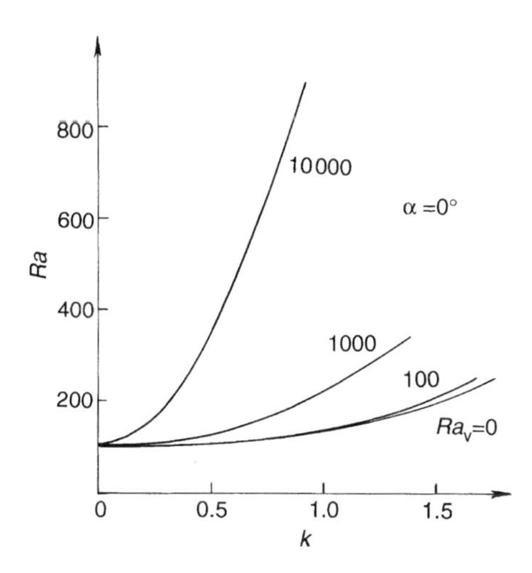

Fig. 22 (v, h). Neutral curves for different Ra_v values, $\alpha = 0°$

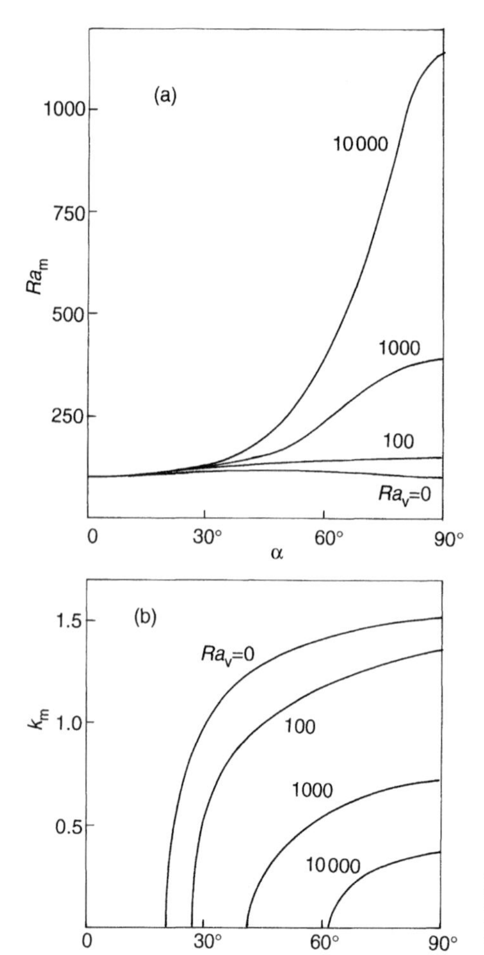

Fig. 23 (v, t) Critical characteristics as functions of the inclination angle. (a) Critical Ra_{m} values; (b) critical wavenumbers

monotonically with Ra_{v}. When $\alpha < \alpha_{\mathrm{c}}$, formula (1.124) is valid for the minimal critical Ra value. For the considered case $Ra_{\mathrm{m}} = \pi^4/\cos^2\alpha$.

6.5 Case (l, v)

Let us address the configurations listed in the second line of (1.112).† For the (l, v) variant, the hydrostatic condition may be written in the form

$$\sin\alpha(Ra - Ra_{\mathrm{v}}\cos\alpha) = 0 \tag{1.127}$$

† The long-wave instability under a longitudinal temperature gradient was studied by Birikh [27].

The first root of this equation, $\alpha = 0°$, corresponds to the already analyzed case of a vertical layer with a vertical temperature gradient and axis of vibrations; the stability border is shown in Fig. 21, curve 1. The second root requires that $Ra = Ra_v \cos\alpha$. This means that the equilibrium is possible at arbitrary values of the inclination angle but only under a certain relationship between Ra and Ra_v. At $\alpha = 0°$ obviously $Ra = Ra_v$, and curve 1 in Fig. 21 yields $Ra = Ra_v = 530$. In the limiting case $\alpha = 90°$, it is evident that $Ra = 0$, i.e. the equilibrium is available only under pure weightlessness and, according to [18], it is stable (see Section 4). The numerical results are displayed in Fig. 23, curve 1. In the interval $0 \leqslant \alpha < 90°$, the instability is caused by the cellular mode with the wavenumber k_m decreasing monotonically as α grows.

6.6 Case (l, l)

The equilibrium occurs only for the vertical orientation $\alpha = 0°$; the stability border is curve 1 in Fig. 21.

6.7 Case (l, h)

The condition of the equilibrium existence takes the form

$$\sin\alpha(Ra + Ra_v \cos\alpha) = 0 \tag{1.128}$$

Its first solution, $\alpha = 0°$, corresponds to a vertical layer with a vertical temperature gradient and transverse vibrations at arbitrary Ra and Ra_v values. The stability borders are shown in Fig. 22.

The second solution is $Ra = -Ra_v \cos\alpha$. This means that within the angular interval $0° \leqslant \alpha \leqslant 90°$, the equilibrium is possible only for $Ra < 0$. This indicates that the heating is from above and the temperature gradient points upwards, along the positive direction of the z axis. If $\alpha = 0°$ and $Ra < 0$, the equilibrium is stable (see Fig. 22). The opposite limiting case, $\alpha = 90°$, corresponds to weightlessness with $Ra = 0$. This state is stable when the gradient is longitudinal and vibrations are transverse (see Section 4).

To examine the stability, the calculations have been performed of characteristic decrements within the interval $0° \leqslant \alpha \leqslant 90°$. Some of the results are presented in Fig. 23. As an example, the fragment of the lower part of the decrement spectrum is shown for $Pr = 1$, $\alpha = 30°$ and several values of the wavenumber. The solid lines render the real parts of λ and the dashed lines show the common real part of the complex-conjugated pairs of the decrements. It can be seen that the real parts of all the levels are positive. Thus, the quasi-equilibrium states are stable. The feature of interest in this figure is merging of monotonic branches with the Ra_v growth and formation of complex-conjugated pairs. Therefore, in the range of large Ra_v the spectrum consists of damped oscillatory modes.

6.8 Case (l, t)

The hydrostatic condition gives

$$Ra \sin \alpha = 0 \tag{1.129}$$

In the case of weightlessness $Ra = 0$, the configuration corresponds to a longitudinal temperature gradient and transverse vibrations. This state is stable. Another opportunity is $\alpha = 0°$ at arbitrary Ra and Ra_v. For this case the stability borders are shown in Fig. 22.

6.9 Case (h, v)

This situation means a horizontal temperature gradient and a vertical axis of vibrations. The necessary condition of equilibrium existence is given by equation (1.116):

$$Ra + Ra_v \sin \alpha \cos \alpha = 0 \tag{1.130}$$

For definiteness, let $0° \leqslant \alpha \leqslant 90°$. Then the equilibrium exists only at negative Ra values when heated from the left (see the scheme in Fig. 18). If $\alpha = 0°$ or $\alpha = 90°$, the equilibrium is encountered only under weightlessness ($Ra = 0$). The latter configuration is absolutely stable when the gradient is longitudinal and vibrations are transverse. At $\alpha = 0°$ the standard situation occurs (see Section 3) where the gradient is transversal and vibrations are longitudinal. In the intermediate range of angles, the computations show that the critical Ra_{vm} value increases monotonically and k_m decreases with α (Fig. 24, curves 2).

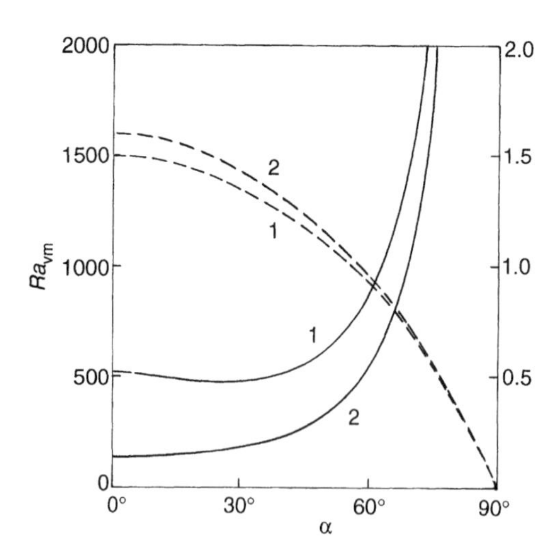

Fig. 24 Critical parameters as functions of the angle. Solid lines, $Ra_{vm}(\alpha)$; dashed lines, $k_m(\alpha)$. 1, (l, v) configuration; 2, (h, v) configuration

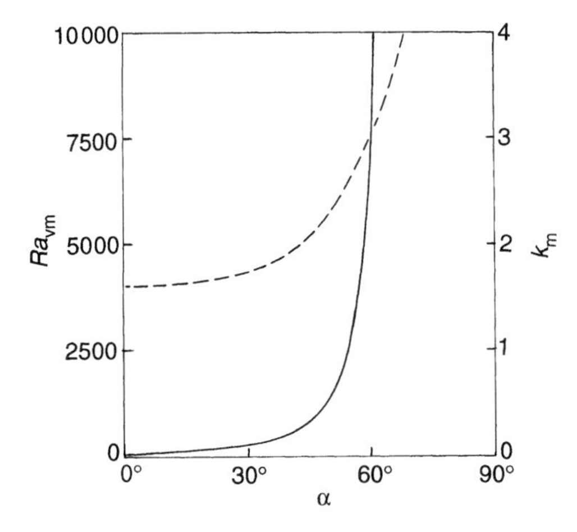

Fig. 25 (h, l) Critical parameters as functions of the angle. Solid lines, $Ra_{\mathrm{vm}}(\alpha)$; dashed lines, $k_{\mathrm{m}}(\alpha)$

6.10 Case (h, l)

This is in some sense close to the previous one. The hydrostatic condition coincides with (1.130). The limiting case $Ra = 0$, $\alpha = 0°$ is equivalent to the standard problem considered in detail in Section 3. When $Ra = 0$, $\alpha \to 90°$, contrary to the (h, v) configuration, the equilibrium is absolutely stable, with the gradient and vibrations being longitudinal. The numerical results are shown in Fig. 25. As α increases, a sharp stabilization takes place and the crisis shifts towards short-wave modes.

6.11 Case (h, h)

This situation is degenerative. It yields, from equation (1.116):

$$Ra = 0$$

i.e. the equilibrium is possible only in the case of weightlessness but the equilibrium is absolutely stable when the temperature gradient and the vibration axis are parallel (see Section 2).

6.12 Case (h, t)

This is also trivial: equation (1.116) takes the form $Ra = 0$, i.e. the condition of weightlessness. This configuration is stable in accordance with Section 4 because, vibrations are transverse, the temperature gradient may be oriented arbitrarily.

6.13 Case (t, v)

The configurations listed in the fourth line are discussed. For this case there is the relation

$$Ra \cos \alpha = 0 \tag{1.131}$$

The first cofactor gives $Ra = 0$, i.e. pure weightlessness, and there apply the standard conditions of Section 3 for a transversal temperature gradient and arbitrarily oriented vibration axis. The second cofactor requires $\cos \alpha = 0$, which corresponds to a horizontal layer with parallel **m** and **n** at arbitrary values of Ra and Ra_v. The stability border is presented in Fig. 21, curve 2.

6.14 Case (t, l)

As in the previous situation, the necessary condition of equilibrium existence is of the form (1.131). The solution $Ra = 0$ is in accordance with the standard problem of a transversal temperature gradient and longitudinal vibrations. If $\cos \alpha = 0$, i.e. $\alpha = 90°$, we deal with the joint action of thermogravitational and thermovibrational convection in a horizontal layer under longitudinal vibrations (see Fig. 15).

6.15 Case (t, h)

Once again the condition of equilibrium can be written as (1.131), similar to all the other variants of the fourth line of (1.112). Adopting the weightlessness requirement $Ra = 0$, the situation coincides with that of the case (t, v). The configuration $\alpha = 90°$ infers the superposition of both mechanisms of the convection excitation (see Fig. 15).

6.16 Case (t, t)

Finally, the configuration is considered in which both vectors, viz. the temperature gradient and vibration axis, are transversal. From (1.131) at $Ra = 0$ the absolutely stable situation follows: **m** and **n** are parallel. If $\alpha = 90°$, the configuration occurs where vertical vibrations stabilize the usual thermogravitational Rayleigh–Benard instability in a horizontal layer(see Fig. 21, curve 2).

7 Binary Mixture with the Soret Effect

In the present section we study the problem of the convection onset in a plane horizontal layer of binary mixture with the Soret thermodiffusion effect. The problem is formulated as follows. An infinite plane horizontal layer is bounded by two parallel rigid plates $z = 0$ and $z = h$; the coordinate system is shown in Fig. 1. The plates are assumed to be isothermal, i.e. maintained at constant different temperatures:

$$z = 0: \qquad T = \Theta, \qquad z = h: \qquad T = 0$$

Both variants are possible: either $\Theta > 0$, or $\Theta < 0$, allowing the layer to be heated from below or from above. The layer is filled with a binary mixture with the Soret effect. This means that mutual diffusion is caused not only by concentration inhomogeneity but by temperature as well. In the normal Soret effect, the light component diffuses in the direction of higher temperature, i.e. along the temperature gradient. In the anomalous Soret effect, the light component moves to the lower temperature region and the heavy component concentrates in hot places. Evidently, thermodiffusion gives rise to an additional stratification, thus changing the stability. In the considered case there is not any external difference of concentration, i.e. all the concentration inhomogeneities are due to the Soret effect under the temperature gradient. The boundaries of the layer are taken to be impermeable to both mixture components.

In the theory of gravitational convective instability, the Soret effect was first taken into account in [28] (see also [2]). In the 1970s it was studied in detail both theoretically and experimentally [29], and extensive bibliography is given in [30]. In this section the phenomenon is investigated in the presence of high-frequency vibrations whose axis is longitudinal and directed along the x axis: $\mathbf{n}(1, 0, 0)$ [31].

Convective phenomena in mixtures are complicated by two factors. First, the additional forces appear in the gravitational and vibration fields caused by the inhomogeneity of concentration. Second, supplementary dissipative processes are involved, such as diffusion and thermodiffusion. The difference between the reference times of heat conduction and diffusion gives rise to some specific phenomena, the so-called double-diffusion ones.

The situation is considered in the framework of the traditional Boussinesq approximation (see [2]), i.e. it is supposed that the deviations of temperature and concentration from their basic values are relatively small and the equation of state may be written as

$$\rho = \overline{\rho}(1 - \beta_1 T - \beta_2 C) \tag{1.132}$$

Here $\overline{\rho}$ is the constant density corresponding to $T = 0$ and $C = 0$ (hereafter we omit the overbar), $\beta_1 > 0$ is the heat expansion coefficient, β_2 is the concentration coefficient of density and C is the concentration of the light component. Hence, $\beta_2 > 0$.

The equation of convective motion in the proper coordinate system is obtained by replacing the static gravity acceleration by the full one (see (1.1)).

Instead of (1.2) it takes the form

$$\frac{\partial \mathbf{v}}{\partial t} + (\mathbf{v}\nabla)\mathbf{v} = -\frac{1}{\rho}\nabla p + \nu\Delta\mathbf{v} + g(\beta_1 T + \beta_2 C)\gamma + (\beta_1 T + \beta_2 C)b\Omega^2 \cos\Omega t\,\mathbf{n} \tag{1.133}$$

The equations of heat transfer and continuity retain their standard form:

$$\frac{\partial T}{\partial t} + \mathbf{v}\nabla T = \chi\Delta T \tag{1.134}$$

$$\operatorname{div}\mathbf{v} = 0 \tag{1.135}$$

Taking into account the Soret effect, the density of the light component flux is

$$\mathbf{j} = \rho D(\nabla C + \alpha\nabla T)$$

Here D is the diffusion coefficient and α is the thermodiffusion ratio. Thus $\alpha < 0$ corresponds to the normal effect whereas $\alpha > 0$ corresponds to the anomalous one. The dependence of D and α (like ν and χ) on temperature and concentration is neglected. Then the diffusion equation can be written as

$$\frac{\partial C}{\partial t} + \mathbf{v}\nabla C = D(\Delta C + \alpha\Delta T) \tag{1.136}$$

The set of equations (1.133) to (1.136) with appropriate initial and boundary conditions describes the convection of a binary mixture in the presence of vibrations in a proper coordinate system. In the limiting case of high frequencies and small amplitudes, as for a one-component medium, a boundary-value problem can be obtained for the average fields by means of the averaging technique. When requiring the smallness of the period with respect to all the reference times (see (1.23)), the diffusion time L^2/D has to be taken into account. Repeating the consideration of Section 1, the problem is stated for the average fields:

$$\begin{gathered}
\frac{\partial \mathbf{v}}{\partial t} + (\mathbf{v}\nabla)\mathbf{v} = -\frac{1}{\rho}\nabla p + \nu\Delta\mathbf{v} + g(\beta_1 T + \beta_2 C)\gamma \\
+\frac{1}{2}b^2\Omega^2(\mathbf{w}\nabla)[(\beta_1 T + \beta_2 C)\mathbf{n} - \mathbf{w}] \\
\frac{\partial T}{\partial t} + \mathbf{v}\nabla T = \chi\Delta T \\
\frac{\partial C}{\partial t} + \mathbf{v}\nabla C = D(\Delta C + \alpha\Delta T) \\
\operatorname{div}\mathbf{v} = 0, \qquad \operatorname{div}\mathbf{w} = 0, \quad \operatorname{curl}\mathbf{w} = \nabla(\beta_1 T + \beta_2 C) \times \mathbf{n}
\end{gathered} \tag{1.137}$$

Here $\mathbf{w}$ is the solenoidal part of the vector $(\beta_1 T + \beta_2 C) \times \mathbf{n}$.

The boundary conditions have the form

$$\begin{aligned} z=0: &\quad \mathbf{v}=0, \quad T=\Theta, \quad w_z=0, \quad \frac{\partial C}{\partial z}+\alpha\frac{\partial T}{\partial z}=0 \\ z=h: &\quad \mathbf{v}=0, \quad T=0, \quad w_z=0, \quad \frac{\partial C}{\partial z}+\alpha\frac{\partial T}{\partial z}=0 \end{aligned} \tag{1.138}$$

i.e. the total flux of the light component must vanish at the boundaries.

Now the dimensionless variables are introduced on the base of the same units as in Section 1; the scale for concentration is $\beta_1\Theta/\beta_2$ and the scale for $\mathbf{w}$ is $\beta_1\Theta$. The set of equations for the dimensionless fields rewrites as

$$\frac{\partial \mathbf{v}}{\partial t}+\frac{1}{Pr}(\mathbf{v}\nabla)\mathbf{v}=-\nabla p+\Delta\mathbf{v}+Ra(T+C)\boldsymbol{\gamma}+\mathrm{Ra_v}(\mathbf{w}\nabla)[(T+C)\mathbf{n}-\mathbf{w}]$$

$$\begin{aligned} Pr\frac{\partial T}{\partial t}+\mathbf{v}\nabla T&=\Delta T \\ Sc\frac{\partial C}{\partial t}+\frac{Sc}{Pr}\mathbf{v}\nabla C&=\Delta(C-\varepsilon T) \end{aligned} \tag{1.139}$$

$$\operatorname{div}\mathbf{v}=0, \qquad \operatorname{div}\mathbf{w}=0, \qquad \operatorname{curl}\mathbf{w}=\nabla(T+C)\times\mathbf{n}$$

The parameters Ra and Ra_v are now defined by (1.27) setting $\beta=\beta_1$ and Pr is the Prandtl number. Besides, we introduce two new parameters which are the Schmidt number Sc and the dimensionless thermodiffusion parameter ε:

$$Sc=\frac{\nu}{D}, \qquad \varepsilon=-\frac{\alpha\beta_2}{\beta_1} \tag{1.140}$$

In the normal Soret effect one has $\varepsilon>0$; in the anomalous one $\varepsilon<0$. We remark that there are no concentration analogs of Ra and Ra_v. This is due to the fact that no external difference of concentration is applied.

The dimensionless form of the boundary conditions is

$$\begin{aligned} z=0: &\quad \mathbf{v}=0, \quad T=1, \quad w_z=0, \quad \frac{\partial C}{\partial z}-\varepsilon\frac{\partial T}{\partial z}=0 \\ z=1: &\quad \mathbf{v}=0, \quad T=0, \quad w_z=0, \quad \frac{\partial C}{\partial z}-\varepsilon\frac{\partial T}{\partial z}=0 \end{aligned} \tag{1.141}$$

As in the case of a one-component medium, the necessary conditions of quasi-equilibrium may be found. The equations for the equilibrium fields T_0, C_0, $\mathbf{w}_0$ are

$$Ra\,\nabla(T_0+C_0)\times\boldsymbol{\gamma}+Ra_v\,\nabla(\mathbf{w}_0\mathbf{n})\times\nabla(T_0+C_0)=0$$

$$\Delta T_0=0, \qquad \Delta C_0=0 \tag{1.142}$$

$$\operatorname{div}\mathbf{w}_0=0, \qquad \operatorname{curl}\mathbf{w}_0=\nabla(T_0+C_0)\times\mathbf{n}$$

The boundary conditions for the equilibrium fields coincide with (1.141).

It is easy to see that in a horizontal layer the quasi-equilibrium exists and has the structure $T_0 = T_0(z)$, $C_0 = C_0(z)$, $w_{0x} = w_0(z)$, $w_{0y} = w_{0z} = 0$, where

$$T_0 = 1 - z, \qquad \frac{\mathrm{d}C_0}{\mathrm{d}z} = -\varepsilon, \qquad w_0 = -(1+\varepsilon)\left(z - \frac{1}{2}\right) \tag{1.143}$$

Let us now formulate the problem of linear stability for the equilibrium (1.143) with respect to small plane perturbations. Introducing a stream function similar to (1.117) and the normal modes

$$(\psi, F, T', C') = [\varphi(z), f(z), \theta(z), \xi(z)] \exp(-\lambda t + \mathrm{i}kx) \tag{1.144}$$

we arrive at the following spectral amplitude problem:

$$\begin{gathered} -\lambda D\varphi = D^2\varphi + \mathrm{i}k\,Ra\,(\theta + \xi) + \mathrm{i}k\,Ra_{\mathrm{v}}(1+\varepsilon)(\theta + \xi - f') \\ -\lambda\,Pr\,\theta - \mathrm{i}k\varphi = D\theta \\ -\lambda\,Sc\,\xi - \mathrm{i}k\frac{Sc}{Pr}\varepsilon\varphi = D(\xi - \varepsilon\theta) \\ Df = \theta' + \xi' \end{gathered} \tag{1.145}$$

$$z = 0, \quad z = 1: \qquad \varphi = \varphi' = 0, \qquad \theta = 0, \qquad f = 0, \qquad \xi' - \varepsilon\theta' = 0 \tag{1.146}$$

The decrement λ is the eigenvalue of the problem. Generally speaking, it is complex and depends on all the parameters of the problem:

$$\lambda = \lambda(Ra, Ra_{\mathrm{v}}, Pr, Sc, k) = \lambda_{\mathrm{r}} + \mathrm{i}\lambda_{\mathrm{i}}$$

Due to the impermeability boundary conditions, it might be expected that the long-wave mode with $k = 0$ plays an important role in the problem. To study the behavior of small perturbations, one may apply the regular perturbation method with k as a small parameter. The solution is found in the form of expansions for all the amplitudes and the decrement:

$$\begin{aligned} \varphi &= \varphi_0 + k\varphi_1 + k^2\varphi_2 + \cdots \\ \theta &= \theta_0 + k\theta_1 + k^2\theta_2 + \cdots \\ f &= f_0 + kf_1 + k^2 f_2 + \cdots \\ \xi &= \xi_0 + k\xi_1 + k^2\xi_2 + \cdots \\ \lambda &= \lambda_0 + k\lambda_1 + k^2\lambda_2 + \cdots \end{aligned} \tag{1.147}$$

Substituting these expansions into the set (1.145) and selecting terms of the same order with regard to k, a sequence of boundary-value problems is obtained. Since the boundary conditions coincide with (1.146), we do not write them down.

In the zeroth order we have:

$$
\begin{aligned}
-\lambda_0 \varphi_0 &= \varphi_0^{\mathrm{IV}} \\
-\lambda_0 \, Pr \, \theta_0 &= \theta_0'' \\
-\lambda \, Sc \, \xi_0 &= \xi_0'' - \varepsilon \theta_0'' \\
f_0'' &= \theta_0' + \xi_0'
\end{aligned}
\tag{1.148}
$$

As it might have been anticipated, all the levels of the spectrum correspond to damping perturbations except for one level which is neutral and of the concentration type:

$$
\lambda_0 = 0, \qquad \varphi_0 = 0, \qquad \theta_0 = 0, \qquad f_0 = 0, \qquad \xi_0 = \text{constant} \tag{1.149}
$$

where the constant is determined by normalization.

Let us try to find corrections to this approximation. In the first order, the set consists of the inhomogeneous equations

$$
\begin{aligned}
\varphi_1^{\mathrm{IV}} &= -\mathrm{i}\xi_0 [Ra + (1+\varepsilon) Ra_{\mathrm{v}}] \\
\theta_1'' &= 0 \\
f_1'' &= \theta_1' + \xi_1' \\
\xi_1'' - \varepsilon \theta_1'' &= -Sc \, \xi_0 \lambda_1
\end{aligned}
\tag{1.150}
$$

The solvability condition for this set results from the last equation integrated over z from 0 to 1. It yields $\lambda_1 = 0$ and the following amplitudes:

$$
\theta_1 = 0, \qquad \xi_1' = 0, \qquad f_1 = 0, \qquad \varphi_1 = -\frac{\mathrm{i}\xi_0}{24} [Ra + (1+\varepsilon) Ra_{\mathrm{v}}] z^2 (1-z)^2 \tag{1.151}
$$

In all higher orders, inhomogeneous sets emerge, and the solvability conditions provide the decrement corrections. Therefore, in the second order

$$
\begin{aligned}
\varphi_2^{\mathrm{IV}} &= -\mathrm{i}\xi_1 [Ra + (1+\varepsilon) Ra_{\mathrm{v}}] \\
\theta_2'' &= -\mathrm{i}\varphi_1 \\
f_2'' - \theta_2' - \xi_2' &= 0 \\
\xi_2'' - \varepsilon \theta_2'' &= \xi_0 (1 - \lambda_2 \, Sc) - \mathrm{i}\varepsilon \frac{Sc}{Pr} \varphi_1
\end{aligned}
\tag{1.152}
$$

Integrating the last equation with respect to z from 0 to 1, one may deduce the solvability condition and then

$$\lambda_2 = \frac{1}{Sc}\left\{1 - \frac{Sc}{Pr}\frac{\varepsilon}{720}\left[Ra + (1+\varepsilon)\, Ra_v\right]\right\} \tag{1.153}$$

The decrement is real, so the long-wave perturbations are monotonic. The condition $\lambda_2 = 0$ enables one to find the stability border:

$$Ra + (1+\varepsilon)\, Ra_v = \frac{720\, Le}{\varepsilon} \tag{1.154}$$

where $Le = Pr/Sc = D/\chi$ is the Lewis number. It should be emphasized that, as follows from the spectral amplitude problem (1.145) and (1.146), in a general case of finite wavenumbers the border of monotonic stability does not depend separately on the Prandtl or Schmidt numbers but only on their combination, the Lewis number.

Equation (1.154) describes the straight line in the plane (Ra, Ra_v). In the limiting case of no vibration $Ra_v = 0$ and the long-wave instability, the border is

$$Ra = \frac{720\, Le}{\varepsilon} \tag{1.155}$$

i.e. the instability sets in when the layer is heated from below for the normal Soret effect and when it is heated from above for the anomalous one. In Fig. 26a the relevant instability regions are shaded. In the opposite limiting case of weightlessness $Ra = 0$, equation (1.154) gives

$$Ra_v = \frac{720\, Le}{\varepsilon(1+\varepsilon)} \tag{1.156}$$

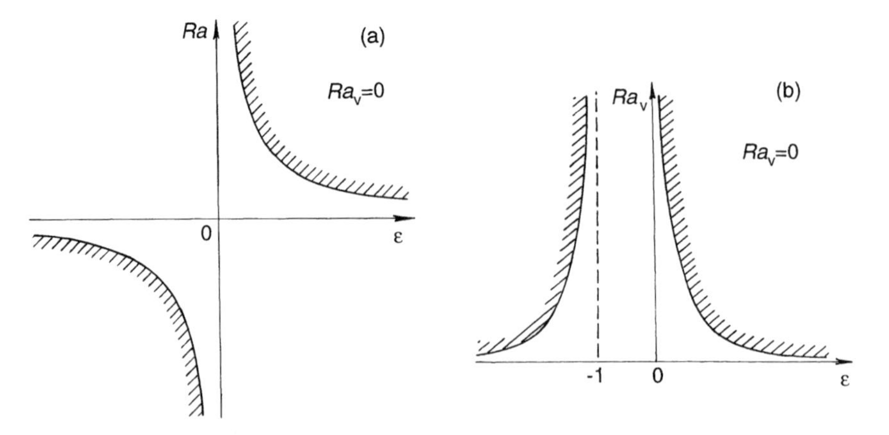

Fig. 26 Long-wave instability boundaries: (a) $Ra_v = 0$; (b) $Ra = 0$

As illustrated in Fig. 26b, the instability exists for $\varepsilon > 0$ and $\varepsilon < -1$ (a strong anomalous Soret effect). The value $\varepsilon = -1$ is in a certain sense special. Indeed, the vertical dimensional density gradient in the equilibrium is $(1 + \varepsilon)\rho\beta_1\Theta/h$, where ρ is the standard density value. At $\varepsilon = -1$ this gradient vanishes since vertical stratification is absent. If the instability exists in this situation (see Fig. 26a), then it is apparently due to the double diffusion.

To solve the complete spectral problem (1.145) and (1.146), straight numerical integrating has been applied in combination with the shooting procedure [31]. Some results of calculations are presented below. Three types of instability are distinguished: with respect to monotonic cellular perturbations (solid lines in Figs. 27 to 31), with respect to oscillatory cellular perturbations (dashed lines) and with respect to monotonic long-wave perturbations (dotted lines). The typical cases of liquid and gaseous mixtures are considered.

Let us begin with a relatively simple case of no vibrations $Ra_v = 0$. Here only the classical thermogravitational mechanism of the convection excitation works. In Fig. 27 the stability borders, i.e. the minimal critical Rayleigh number dependences on the dimensionless Soret parameter, are presented together with the critical wavenumber k_m and the frequency of the critical modes λ_{im}. Lines 1

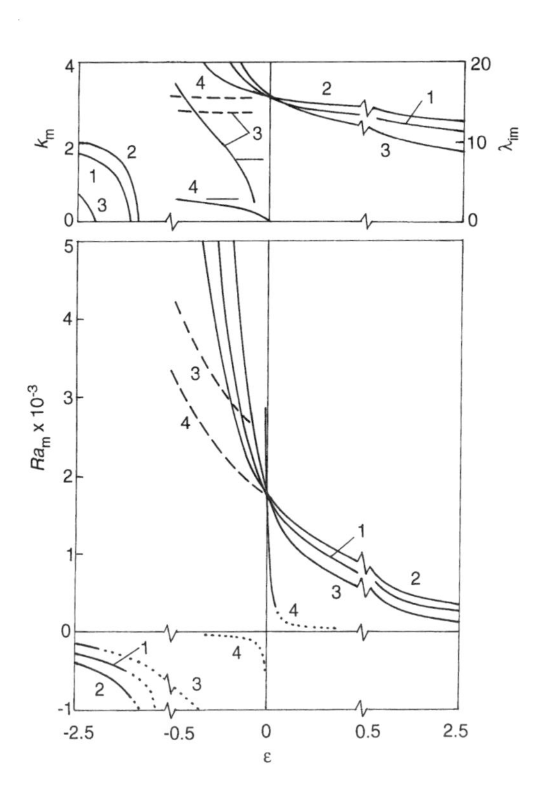

Fig. 27 Instability characteristics for the case $Ra_v = 0$. Solid lines, cellular monotonic mode; dashed lines, regular oscillatory mode; dotted lines, long-wave monotonic mode. Curves enumeration: 1, $Le = 1$; 2, $Pr = 0.75$, $Sc = 0.5$, 3, $Pr = 0.75$, $Sc = 1.5$; 4, $Pr = 6.7$, $Sc = 677$

correspond to $Le = 1$, which may be, for example, a model degenerated gaseous mixture with $Pr = Sc = 1$. In this case the oscillatory instability is evidently absent. Lines 2 correspond to $Le = 1.5$ ($Le > 1$), which is, for instance, a gaseous mixture with $Pr = 0.75$ and $Sc = 0.5$. As one can see, the oscillatory instability is possible but it is the monotonic one that is the most dangerous. For $Le = 0.5$, which is, say, a gaseous mixture with $Pr = 0.75$ and $Sc = 1.5$, one observes the exchange of instability types in the range $\varepsilon < 0$ for the anomalous Soret effect (lines 3). Lines 4 refer to a typical liquid mixture like a water–salt solution with $Pr = 6.7$ and $Sc = 677$ ($Le \approx 0.01$, $Le \ll 1$). Under these circumstances, the exchange also takes place from the monotonic mode to the oscillatory one when $\varepsilon < 0$, and in fact in the whole range of existence of the oscillatory mode, it is the most dangerous. The main features of Fig. 27 are:

(a) absence of an oscillatory instability at $\varepsilon > 0$ and the effect of destabilization especially pronounced for a liquid mixture;
(b) the effect of stabilization at $\varepsilon < 0$ and the onset of an oscillatory instability in some situations;
(c) existence of a monotonic mode for $\varepsilon < 0$ and $Ra < 0$ when the layer is heated from above.

All the stability curves at $Ra > 0$ for the layer heated from below intersect the vertical axis at the point $Ra_{\mathrm{m}} = 1708$ corresponding to the Rayleigh–Benard convection threshold in a plane layer of a one-component fluid. This is understandable because the concentration gradient arises only due to thermodiffusion, i.e. when $\varepsilon \to 0$, and the transition to a one-component medium takes place. It is worth while to note also that in the region where the long-wave mode is the most dangerous, the numerical results agree well with formula (1.154).

In the limiting case of weightlessness, $Ra = 0$ and only the vibrational mechanism is at work. The critical characteristics are displayed in Fig. 28. At $\varepsilon = 0$ the standard problem of a vibrational convective instability for a one-component fluid layer is recovered, the critical value being $Ra_{\mathrm{vm}} = 2129$ (see Section 3). If $\varepsilon > 0$, strong destabilization takes place whereas in the range $-1 < \varepsilon < 0$ strong stabilization exists. Note that along the Ra_{vm} axis a logarithmic scale is used. The instability exists in the range of strong anomalous Soret effect $\varepsilon < -1$ with domination of the long-wave mode. When $-1 < \varepsilon < 0$, the long-wave instability is impossible and the cellular mode of either the monotonic or oscillatory type takes place.

Let us consider in more detail some examples of the general case $Ra \neq 0$ and $Ra_{\mathrm{v}} \neq 0$ where both mechanisms of the convection excitation coexist. Figure 29 demonstrates the influence of vibrations on the instability characteristics for $Le = 1$ when only the monotonic mode is encountered. The regions of equilibrium stability are located in between the curves of the same numbers (1–1, 2–2, 3–3, etc.). The effect of vibrational destabilization is well pronounced. In

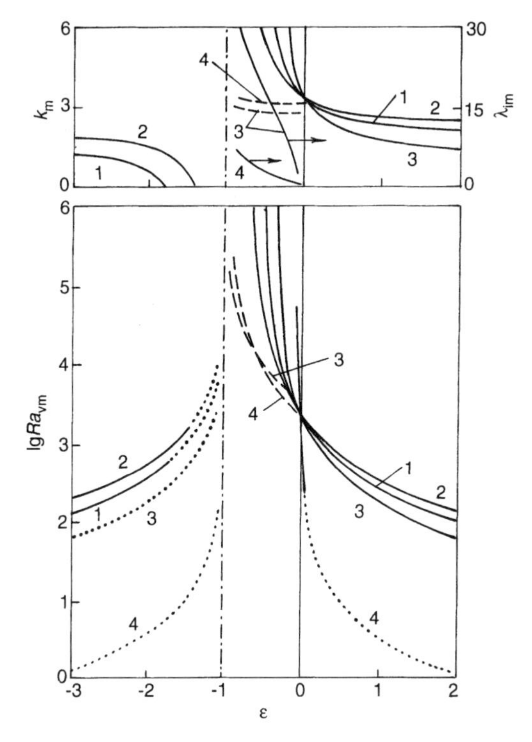

Fig. 28 Instability characteristics for the case $Ra = 0$. Notations and the curve indexing are the same as in Fig. 27

particular, it manifests itself in a contraction of the stability region in the plane (ε, Ra_m) as Ra_v increases. The curve corresponding to $Ra_v = 2129$ passes through the origin of the coordinates since it is exactly this value that meets the threshold of vibrational convection in weightlessness.

Figure 30 corresponds to a gaseous mixture with $Pr = 0.75$, $Sc = 1.5$ $(Le = 0.5,\ Le < 1)$ when the oscillatory instability is possible in the range $\varepsilon < 0$. We remark that such an instability holds even in the case of heating from above if Ra_v is large enough ($Ra_v = 5000$, curve 3). For a typical liquid mixture with $Pr = 6.7$, $Sc = 677$ (Fig. 31), only in the range of small positive and negative ε does the instability caused by the monotonic cellular mode occur. If $\varepsilon > 0$ and is not so small or if $\varepsilon < 0$ and the layer is heated from above, the long-wave instability becomes essential. In the range of negative ε, the oscillatory instability is also developed.

Finally, the stability map is presented in the plane (Ra, Ra_v) for a specific liquid binary mixture consisting of water and ethanol to the extent of 90 and 10 wt%, respectively. According to the published data, this mixture exhibits the

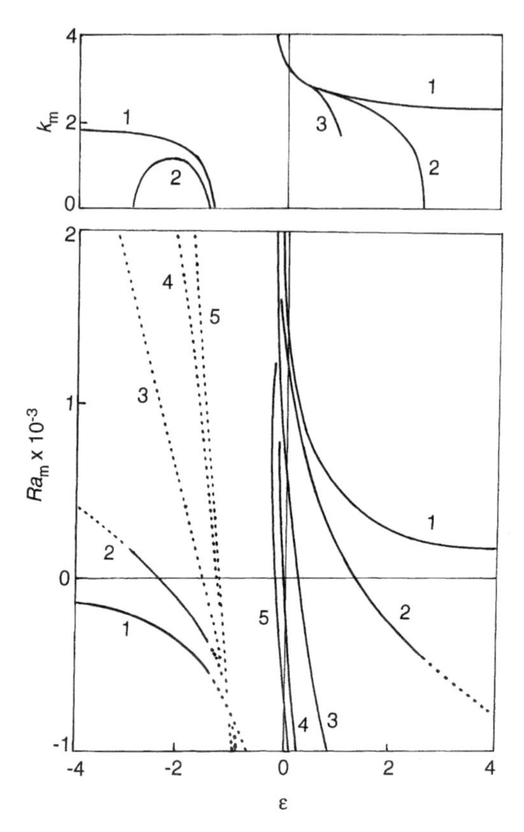

Fig. 29 Critical Ra_{m} values depending on ε for different Ra_{v}, $Le = 1$: 1, $Ra_{\text{v}} = 0$; 2, $Ra_{\text{v}} = 200$, 3, $Ra_{\text{v}} = 1000$; 4, $Ra_{\text{v}} = 2129$; 5, $Ra_{\text{v}} = 3000$

anomalous Soret effect and the actual parameter values are: $\varepsilon = -4.5 \times 10^{-2}$, $Pr = 11.3$, $Sc = 1100$. The stability borders and regions of different regimes are shown in Fig. 32, where the minimization has been made with respect to k . The line corresponding to a one-component fluid with $\varepsilon = 0$ is added for comparison. The dashed line indicates the border of the oscillatory cellular instability. The dotted line gives the stability border when the layer is heated from above and pertains to the long-wave mode of the double-diffusion origin. The region I is associated with stable quasi-equilibrium, the region II with an oscillatory convection and the region III with a steady long-wave convection. From this map a peculiarity that is both interesting and important may be seen: if the liquid mixture possesses just a weak anomalous Soret effect (modifying only slightly the position of the vibrational–convection instability border) the instability character changes qualitatively—it becomes oscillatory. This fact has

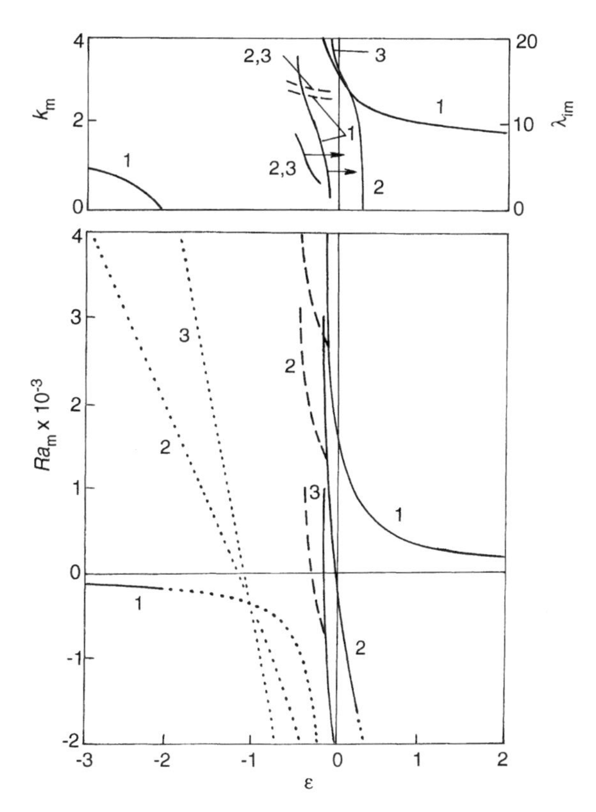

Fig. 30 Critical Ra_m values depending on ε for different Ra_v. Gaseous mixture: $Pr = 0.75$, $Sc = 1.5$. 1, $Ra_v = 0$; 2, $Ra_v = 2129$; 3, $Ra_v = 5000$

already been pointed out in Figs. 27 and 28, curves 4. The shapes of the neutral curves on the stability map are shown schematically in the respective regions.

In conclusion, the papers [32, 33] that also deal with the problem of stability under vibration must be mentioned. In the first one a horizontal layer was considered but model, physically unreal, boundary conditions were adopted. In [33] a horizontal layer was studied in weightlessness where all three vectors (the vibration axis, temperature and concentration gradients) were coplanar and located in the plane perpendicular to the layer. In both papers the Soret effect has not been taken into account.

In the present section the vibrational instability of a binary mixture in a plane horizontal layer subject to longitudinal vibrations has been discussed. This situation is the most unstable; all the other variants of the vibration axis orientation yield a vertical component of vector **n** which contributes to stabilization. The limiting case is the transverse vibration where the specific mechanism of vibrational excitation does not work and the effect of vibration is purely

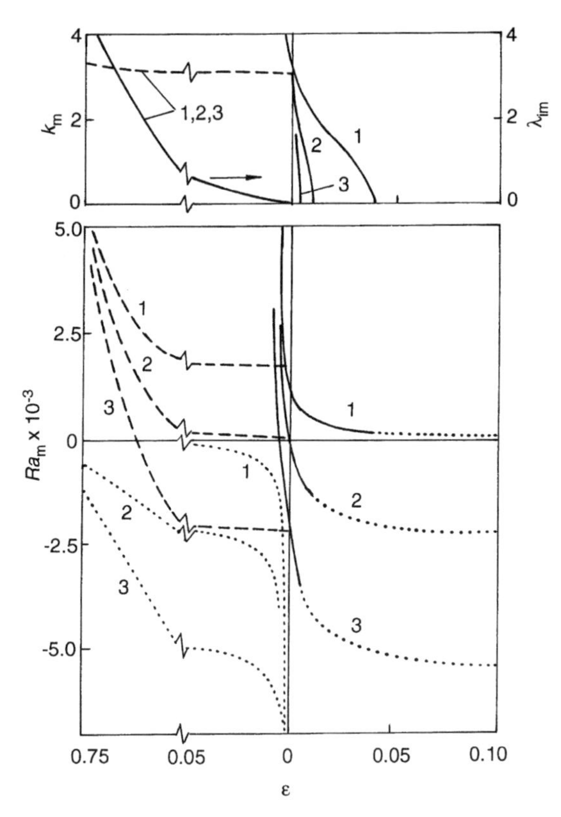

Fig. 31 Critical Ra_m values depending on ε for different Ra_v. Liquid mixture: $Pr = 6.7$, $Sc = 677$. 1, $Ra_v = 0$; 2, $Ra_v = 2129$; 3, $Ra_v = 5000$

stabilizing. The latter case has been considered in paper [34]. The main results are presented here. As the computations show, when the transverse vibration is applied at arbitrary values of the mixture parameters, the lower stability border may be found from the solution of the problem without vibration. As the vibration parameter increases, the stability border grows monotonically up to the value determined from the problem with $Ra_v \neq 0$ but $k = 0$. As for longitudinal vibrations, this long-wave limiting case is simple to analyze: the asymptotic Ra_m value does not depend on the parameter of vibration and may be written down as

$$Ra = \frac{720\, Le}{\varepsilon} \tag{1.157}$$

All of the remaining neutral curves in the plane (ε, Ra) intervene between two limiting curves; the lower corresponds to $Ra_v = 0$ and the upper one to the

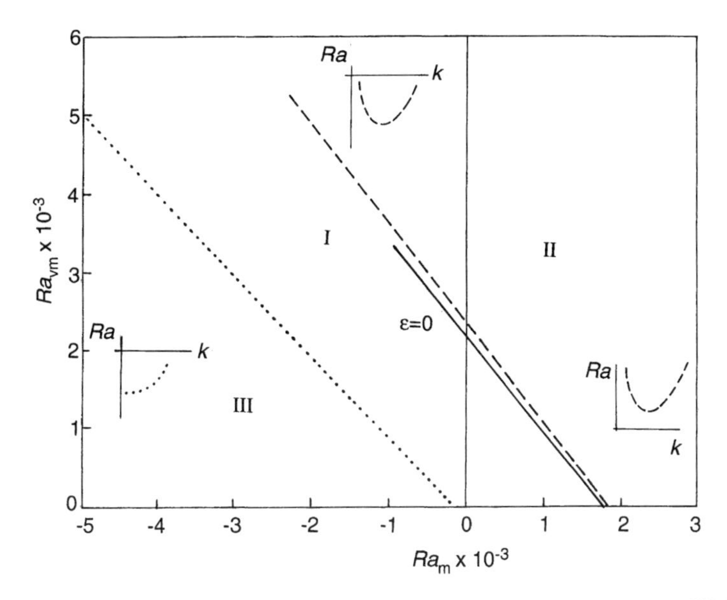

Fig. 32 Stability map for the liquid mixture of water–ethanol: I, stable equilibrium; II, oscillatory cellular instability; III, monotonic long-wave instability

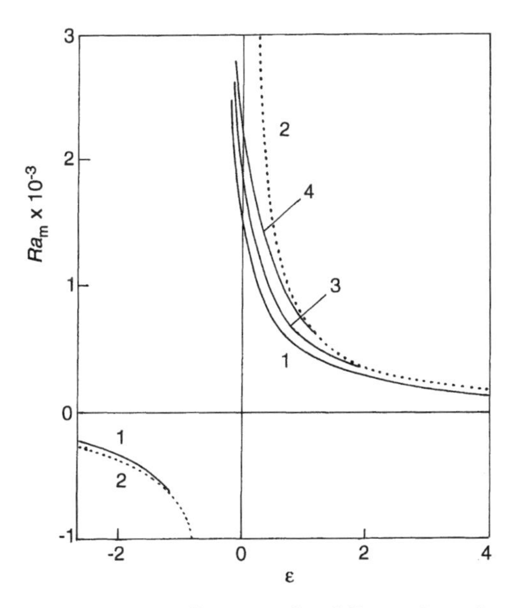

Fig. 33 Critical Ra_m values depending on ε for different Ra_v; the case of transversal vibrations. Notation and curve numbering are the same as in Fig. 29

long-wave approximation (1.157). Figure 33 demonstrates the results for the afore-considered binary mixture layer exposed to transversal vibrations.

References

1. Landau, L.D. and E.M. Lifshitz. *Hydrodynamics*, Nauka, Moscow, 1986, 736 pp.
2. Gershuni, G.Z. and E.M. Zhukhovitsky. *Convective Stability of Incompressible Fluid*, Nauka, Moscow, 1972, 392 pp.
3. Gershuni, G.Z. and E.M. Zhukhovitsky. On free thermal convection in vibrational field in weightlessness, *Commun. AN SSSR (Doklady)*, 1979, **249**(3), 580–4.
4. Landau, L.D. and E.M. Lifshitz. *Mechanics*, Nauka, Moscow, 1988, 215 pp.
5. Schlichting, H. Berechnung ebener periodischer Grenzschichtströmungen, *Phys. Z.*, 1932, **33**, 327–35.
6. Lin, C.C. Motion in the boundary layer with a rapidly oscillating external flow. *Proc. 9th Int. Congr. Appl. Mech., Brussels*, 1957, **4**, 155–67.
7. Schlichting, H. *Theory of Boundary Layer*, Nauka, Moscow, 1969, 742 pp.
8. Zen'kovskaya, S.M. and I.B. Simonenko. On the influence of high frequency vibration on the convection initiation. *Trans. Acad. Sci. SSSR (Izvestiya), Mech. Zhidk. Gasa*, 1966, **5**, 51–5.
9. Zen'kovskaya, S.M. Study of convection in a fluid layer subject to vibrational forces, *Trans. Acad. Sci. SSSR (Izvestiya), Mech. Zhidk. Gasa*, 1968, **1**, 55–8.
10. Simonenko, I.B. Foundation of averaging method for the problem of convection in the field of rapidly oscillating forces and for other parabolic equations, *Math. Sb.*, 1972, **87**(129), No. 2, 236–53.
11. Simonenko, I.B. Method of averaging in theory of nonlinear equations of parabolic type with application to problems of hydrodynamic stability, Rostov-on-Don, Rostov University, 1989, 111 pp.
12. Gershuni, G.Z. and E.M. Zhukhovitsky. On convective instability of fluid in a vibrational field in weightlessness. *Trans. Acad. Sci. SSSR (Izvestiya), Mech. Zhidk. Gasa*, 1981, **4**, 12–9.
13. Squire, H.B. On the stability for three-dimensional disturbances of viscous fluid flow between parallel walls, *Proc. Roy. Soc.*, 1933, **142**(847), 621–8.
14. Krylov, D.G. Convective instability of plane fluid layer in vibrational field at arbitrary heat conductivity of boundaries, in *Convective Flows*, Perm, 1991, 46–9.
15. Braverman, L. and A. Oron. On the oscillatory instability of a fluid layer in a high frequency vibrational field in weightlessness, *Europ. J. Mech., B*, 1994, **13**(1), 115–28.
16. Braverman, L.M. On vibrational convective instability of plane fluid layer in weightlessness, *Trans. Acad. Sci. SSSR (Izvestiya), Mech. Zhidk. Gasa*, 1984, **6**, 178–80.
17. Braverman, L.M. On some types of vibrational convective instability in a plane fluid layer in weightlessness, *Trans. Acad. Sci. SSSR (Izvestiya), Mech. Zhidk. Gasa*, 1987, **5**, 4–7.
18. Braverman, L.M. On mechanical quasiequilibrium of a plane layer of nonisothermic fluid in high frequency vibrational field, in *Viscous Fluid Dynamics*, Sverdlovsk, 1987, pp, 20–8.
19. Braverman, L.M. On vibrational convective instability of a plane fluid layer in weightlessness, in *Viscous Fluid Dynamics*, Sverdlovsk, 1987, pp. 29–35.
20. Gershuni, G.Z., E.M. Zhukhovitsky and A.N. Sharifulin. Vibrational convection in cylindrical cavity, *Numer. Meth. Cont. Med. Mech., Novosibirsk*, 1983, **14**(4), 21–33.

21. Braverman, L.M., G.Z. Gershuni, E.M. Zhukhovitsky, A.K. Kolesnikov; and V.M. Shikhov. New results of vibrational convective instability study. III, in *All-Union Sem. on Hydromech. and Heat/Mass Transf. in Weightlessness*, Abstracts, Tchernogolovka, 1984, pp. 11–13.
22. Zen'kovskaya, S.M. On the effect of vibration on convective instability. *Numer. Meth. of Viscous Flow Dynamics, Novosibirsk*, 1979, pp. 116–22.
23. Zavarykin, M.P., S.V. Zorin and G.F. Putin. On thermoconvective instability in vibrational field, *Commun. AN SSSR (Doklady)*, 1988, **299**(2), 309–12.
24. Zavarykin, M.P., S.V. Zorin and G.F. Putin. Experimental study of vibrational thermal convection. III, in *All-Union Sem. on Hydromech. and Heat/Mass Transf. in Weightlessness*, Abstracts, Tchernogolovka, 1984, pp. 34–6.
25. Zavarykin, M.P. and S.V. Zorin. Terrestrial modelling of vibrational convection in weightlessness, on *Numer. and Experim. Modelling of Hydrodynamic Phenomena in Weightlessness*, Sverdlovsk, 1988, pp. 85–92.
26. Demin, V.A., G.Z. Gershuni and I.V. Verkholantsev. Mechanical quasiequilibrium and thermovibrational convective instability in an inclined fluid layer. *Int. J. Heat Mass Transfer*, 1996, **39**, 1979–91.
27. Birikh, R.V. On vibrational convection in plane layer with longitudinal temperature gradient. *Trans. Acad. Sci. SSSR (Izvestiya), Mech. Zhidk. Gasa*, 1990, **4**, 12–15.
28. Gershuni, G.Z. and E.M. Zhukhovitsky. On convective instability of two-component mixture in gravity field. *Appl. Math. Mech.*, 1963, **27**(2), 301–8.
29. Legros, J.-C., J.K. Platten and P.G. Poty. Stability of two-component fluid layer heated from below, *Phys. Fluids*, 1972, **15**(8), 1383–9.
30. Platten, J.K. and J.-C. Legros. *Convection in Liquids*, Springer-Verlag, 1984.
31. Gershuni, G.Z., A.K. Kolesnikov, J.-C. Legros and B.I. Myznikova. On vibrational convective instability of binary mixture horizontal layer with Soret effect. *J. Fluid Mech.*, 1997, **330**, pp. 251–69.
32. Zen'kovskaya, S.M. On effect of vibration on convection appearance in binary mixture. *Moscow Inst. Sci. Inf.*, 1981, 1570.
33. Braverman, L.M. On vibrational convective instability of binary mixture plane layer in weightlessness. in *Convective Flows*, Perm, 1987, pp. 48–55.
34. Gershuni, G.Z., A.K. Kolesnikov, J.-C. Legros and B.I. Myznikova. On the convective instability of a horizontal binary mixture layer with Soret effect under transversal high frequency vibration (in press).

2 Plane-Parallel Flows and Their Stability

In this chapter the effect of high-frequency vibrations on a plane-parallel flow caused by an ordinary thermogravitational mechanism either in a vertical layer under a transversal temperature difference or in a horizontal one under a longitudinal temperature gradient is considered. The vibrational convective flow in weightlessness which occurs under some special conditions of heating and particular orientation of the vibration axis is also studied.

1 Convective Flow in a Vertical Layer: The Case of Transverse Vibrations

A plane-parallel convective flow is considered in a plane vertical layer whose boundaries are maintained at constant different temperatures in the presence of high-frequency vibrations. The coordinate system is shown in Fig. 34. The thickness of the layer is $2h$ and the boundary temperatures are $\pm\Theta$; the size of the layer along the z and y directions is quite large. The channel is supposed to be bounded from below and from above by impermeable walls. Hence, inside the layer the convective circulation takes place: the fluid goes up near the hot wall and moves down near the cold one. The vibration axis is, as usual, characterized by the unit vector $\mathbf{n}(n_x, n_y, n_z)$.

Let us return to equations (1.17) to (1.20) describing the average convective flow under the static gravity and vibrational fields and write them in the dimensionless form using as units of length, time, velocity, temperature, $\mathbf{w}$ and pressure, respectively, h, h^2/ν, $g\beta\Theta h^2/\nu$, Θ and $\rho g\beta\Theta h$:

$$\frac{\partial \mathbf{v}}{\partial t} + Gr\,(\mathbf{v}\nabla)\mathbf{v} = -\nabla p + \Delta\mathbf{v} + T\gamma + \frac{Gr_v}{Gr}(\mathbf{w}\nabla)(T\mathbf{n} - \mathbf{w})$$

$$\frac{\partial T}{\partial t} + Gr\,\mathbf{v}\nabla T = \frac{1}{Pr}\Delta T \tag{2.1}$$

$$\operatorname{div}\mathbf{v} = 0, \qquad \operatorname{div}\mathbf{w} = 0, \qquad \operatorname{curl}\mathbf{w} = \nabla T \times \mathbf{n}$$

Here Pr is the Prandtl number, Gr is the Grashof number specifying the intensity of thermogravitational convection whereas Gr_v is its vibrational analog related to the vibrational Rayleigh number:

$$Gr = \frac{g\beta\Theta h^3}{\nu^2}, \qquad Gr_v = \frac{(\beta b\Omega\Theta h)^2}{2\nu^2} = \frac{Ra_v}{Pr}$$

In the case of transverse vibration $n_y = n_z = 0$, $n_x = 1$. It is easy to verify that the mechanical equilibrium is impossible and the convective flow takes place for any arbitrarily small temperature difference. If the vertical size of the layer is large enough, the flow in its central part may be assumed to be plane-parallel. Then it is possible to find the solution of the following structure:

$$v_x = v_y = 0, \qquad v_z = v_0(x), \qquad T = T_0(x), \qquad p = p_0(z), \qquad \mathbf{w}_0 = 0 \tag{2.2}$$

The distributions $T_0(x)$, $v_0(x)$, $p_0(z)$ satisfy the equations

$$\begin{aligned} \frac{\mathrm{d}p_0}{\mathrm{d}z} &= v_0'' + T_0 = C \\ T_0'' &= 0 \end{aligned} \tag{2.3}$$

Here the prime denotes differentiation with respect to the transversal coordinate x and C is the constant coming from the separation of variables.

At solid isothermal boundaries of the layer the non-slip conditions are valid:

$$x = \mp 1: \qquad v_0 = 0, \qquad T_0 = \pm 1 \tag{2.4}$$

In addition, the condition of closeness must hold, i.e. the total flux through the layer cross-section is zero:

$$\int_{-1}^{1} v_0 \, dx = 0 \tag{2.5}$$

The solution of the problem (2.3) to (2.5) is as follows:

$$v_0 = \tfrac{1}{6}(x^3 - x), \qquad T_0 = -x, \qquad p_0 = \text{constant} \tag{2.6}$$

The temperature distribution in the basic plane-parallel regime is linear and the velocity profile is a third-order polynomial and corresponds to a flow consisting of two counter-streams (Fig. 34). Thus, in a vertical layer transverse vibrations do not influence the basic flow developed under the static gravity field due to the transversal temperature difference. However, we shall show that transverse vibrations affect considerably the evolution of perturbations in the basic regime, and consequently the stability of the latter.

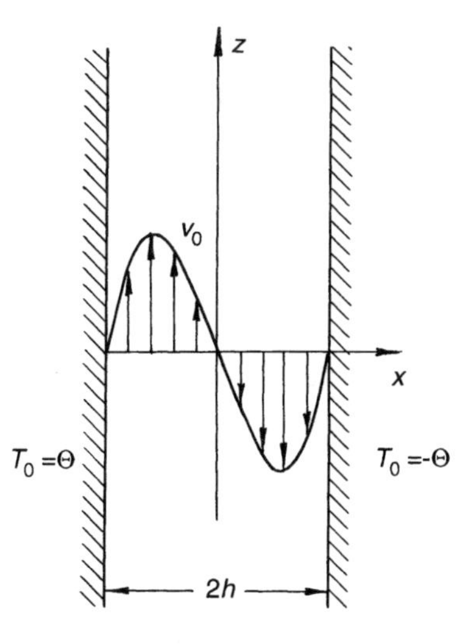

Fig. 34 Coordinate axes and the velocity profile of the basic flow

Setting out to study the plane-parallel flow stability, one could anticipate the principal qualitative result. Since the vibration axis is parallel to the unperturbed temperature gradient, the net effect would definitely be stabilizing.

Consider the perturbed fields as $\mathbf{v}_0 + \mathbf{v}$, $T_0 + T$, $p_0 + p$ and $\mathbf{w}_0 + \mathbf{w}$. The equations for perturbations $\mathbf{v}$, T, p and $\mathbf{w}$ are derived as usual by substituting the perturbed fields into the basic set (2.1). Linearization around the basic state yields

$$\begin{gathered}
\frac{\partial \mathbf{v}}{\partial t} + Gr[(\mathbf{v}_0\nabla)\mathbf{v} + (\mathbf{v}\nabla)\mathbf{v}_0] = -\nabla p + \Delta\mathbf{v} + T\gamma + \frac{Gr_{\mathrm{v}}}{Gr}(\mathbf{w}\nabla T_0)\mathbf{n} \\
\frac{\partial T}{\partial t} + Gr(\mathbf{v}_0\nabla T + \mathbf{v}\nabla T_0) = \frac{1}{Pr}\Delta T \\
\operatorname{div}\mathbf{v} = 0, \qquad \operatorname{div}\mathbf{w} = 0, \qquad \operatorname{curl}\mathbf{w} = \nabla T \times \mathbf{n}
\end{gathered} \tag{2.7}$$

Normal perturbations, periodic in the plane of the layer, are introduced:

$$(\mathbf{v}, T, p, \mathbf{w}) \sim \exp\left[-\lambda t + \mathrm{i}\left(k_y y + k_z z\right)\right], \tag{2.8}$$

where λ is the characteristic decrement and k_x and k_y are the wavenumbers along the corresponding axes. The following set of equations is obtained for the perturbation amplitudes:

$$
\begin{gathered}
-\lambda v_x + \mathrm{i}k_z Gr\, v_0 v_x = -p' + \left(v_x'' - k^2 v_x\right) + \frac{Gr_{\mathrm{v}}}{Gr} T_0' w_x \\
-\lambda v_y + \mathrm{i}k_z\, Gr\, v_0 v_y = -\mathrm{i}k_y p + \left(v_y'' - k^2 v_y\right) \\
-\lambda v_z + \mathrm{i}k_z\, Gr\, v_0 v_z + Gr\, v_0' v_x = -\mathrm{i}k_z p + \left(v_z'' - k^2 v_z\right) + \theta \\
-\lambda\theta + \mathrm{i}k_z\, Gr\, v_0\theta + Gr\, T_0' v_x = \frac{1}{Pr}\left(\theta'' - k^2\theta\right) \\
v_x' + \mathrm{i}\left(k_y v_y + k_z v_z\right) = 0 \\
w_x' + \mathrm{i}\left(k_y w_y + k_z w_z\right) = 0 \\
w_y' - \mathrm{i}k_y w_x = -\mathrm{i}k_y\theta \\
w_z' - \mathrm{i}k_z w_x = -\mathrm{i}k_z\theta, \\
k_y w_z - k_z w_y = 0
\end{gathered} \tag{2.9}
$$

Here $k^2 = k_y^2 + k_z^2$. These amplitudes satisfy the homogeneous boundary conditions:

$$
x = \pm 1: \qquad v_x = v_y = v_z = 0, \qquad \theta = 0, \qquad w_x = 0 \tag{2.10}
$$

Below we demonstrate that the most dangerous perturbations are the two-dimensional ones with $v_y = 0$, $w_y = 0$, $k_y = 0$. For the fields $\mathbf{v}$ and $\mathbf{w}$ of such perturbations, the stream functions ψ and F can be written as follows:

$$
v_x = -\frac{\partial\psi}{\partial z}, \qquad v_z = \frac{\partial\psi}{\partial x}; \qquad w_x = -\frac{\partial F}{\partial z}, \qquad w_z = \frac{\partial F}{\partial x} \tag{2.11}
$$

with the amplitudes φ and f, respectively. After the pressure is eliminated, the spectral amplitude problem is derived in terms of φ, f and θ:

$$
\begin{gathered}
\Delta^2\varphi + \mathrm{i}k_z\, Gr\left(v_0''\varphi - v_0\Delta\varphi\right) - \frac{k_z^2\, Gr_{\mathrm{v}}}{Gr}\, T_0'\, f = -\lambda\Delta\varphi \\
\frac{1}{Pr}\Delta\theta + \mathrm{i}k_z\, Gr\left(T_0'\varphi - v_0\theta\right) = -\lambda\theta \\
\Delta f = -\mathrm{i}k_z\theta;
\end{gathered} \tag{2.12}
$$

$$
x = \pm 1: \qquad \varphi = \varphi' = 0, \qquad \theta = 0, \qquad f = 0 \tag{2.13}
$$

where $\Delta = (d^2/dx^2) - k_z^2$.

The boundary-value problem (2.12) and (2.13) contains the characteristic decrement $\lambda = \lambda_r + i\lambda_i$ as an eigenvalue where the stability border is defined by the condition $\lambda_r = 0$. In this case the instability may be caused by monotonic modes ($\lambda_i = 0$) or, in general, by oscillatory ones ($\lambda_i \neq 0$). The parameters of the problem are the wavenumber k_z, the Grashof numbers Gr and Gr_v and the Prandtl number Pr. To study numerically the problem (2.12) and (2.13), the Runge–Kutta–Merson method was used [1] in combination with the step-by-step orthogonalization procedure.

First, we show some results of the solution [2] of the stability problem for the flow (2.6) where vibrations are absent ($Gr_v = 0$). Asymptotic as well as numerical analyses reveal two instability mechanisms. In the limiting case $Pr \to 0$, the reference heat conduction time is small compared to the viscous one, and the process of perturbations development may be approximately treated as isothermal. Then the spectral amplitude problem reduces to the classic Orr–Sommerfeld one with the antisymmetrical velocity profile (2.6). The instability is due to the pure hydrodynamic (non-viscous) mechanism associated with the point of inflection at the velocity distribution. At $Pr = 0$ the critical instability parameters are $Gr_m = 495.63$, $k_m = 1.344$ (minimization is done with respect to k). The instability is of a monotonic kind ($\lambda_i = 0$) and has the form of a periodic in z direction steady vortices located on the mutual boundary of the two counter-flows. As the Prandtl number increases, the characteristics of this mode vary slowly and non-monotonically. The instability arises at $Gr_m \approx 5 \times 10^2$ within the whole interval of Pr variation (Fig. 35, curve 1). The hydrodynamic mechanism is responsible for the instability at small and moderate values of the Prandtl number up to $Pr \approx 10$.

At some threshold value of the Prandtl number $Pr^* = 11.562$, the new instability mode evolves, originating from amplification of temperature waves

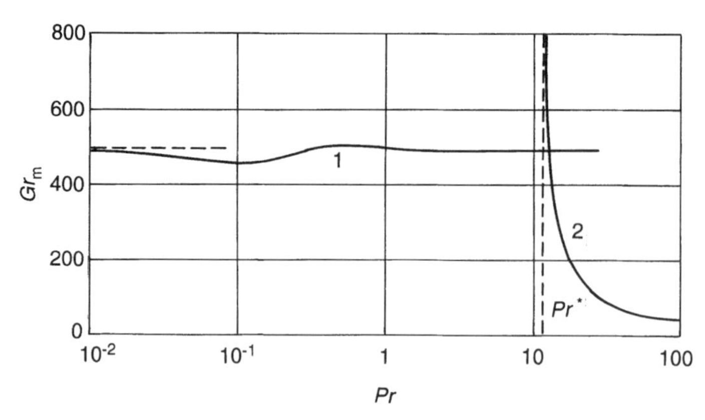

Fig. 35 The minimal critical Grashof number as a function of the Prandtl number: 1, hydrodynamic mode; 2, wave mode

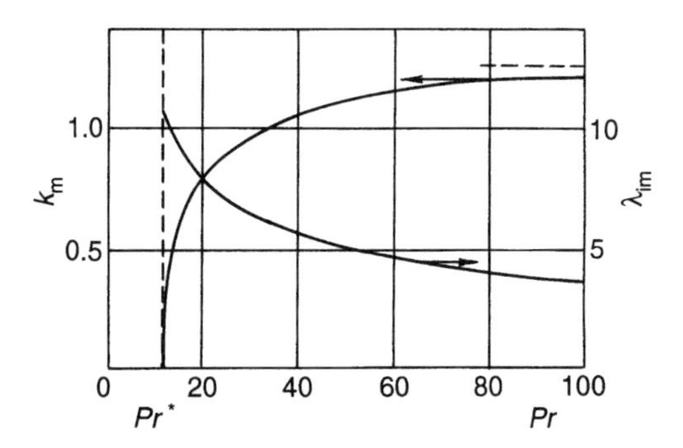

Fig. 36 The wavenumber and the neutral frequency of critical perturbations for the wave mode

in counter-flows. These waves differ from each other by the phase velocity sign and by their localization close to either hot or cold walls, respectively. From the viewpoint of stability, both of them are equivalent due to the symmetry of the problem. As Pr increases in the range $Pr > Pr^*$, the minimal critical Grashof number decreases monotonically (Fig. 35, curve 2) and in the limit $Pr \to \infty$ the asymptotic $Gr_m = 590/\sqrt{Pr}$ takes place. This allows the critical temperature difference to be estimated: $\Theta_m = 590\sqrt{\nu^3\chi}/(g\beta h^3)$. When $Pr > 12.45$, the wave instability becomes more dangerous than the hydrodynamic one. The wave instability characteristics, namely the wavenumber k_m and the neutral frequency λ_{im} corresponding to the minimum point of the neutral curve, are presented in Fig. 36 as functions of Pr.

Figures 37 to 43 display the results of the transverse vibration effect on the stability borders and instability characteristics for hydrodynamic and wave modes. As in the case of no vibrations, at small and moderate values of the Prandtl number the hydrodynamic mechanism is responsible for instability. In Fig. 37 the family $Gr(k)$ of neutral curves is shown for different Gr_v. One can see that as the vibration intensity grows, the instability threshold increases. At the same time, the wavelength of the critical perturbations is shifted to the long-wave range. The shape of curves $Gr_m(Gr_v)$ and $k_m(Gr_v)$ given in Fig. 38, weakly depends on Pr. This is due to the pure hydrodynamic nature of the crisis. Therefore, the curves for $Pr = 1$ and for $Pr = 10$ are practically coincident. At large Gr_v, the asymptotic sets in:

$$Gr_v \to \infty: \qquad Gr_m = g\,Gr_v^{1/4}, \qquad k_m = \kappa\,Gr_v^{-1/4} \tag{2.14}$$

with $g = 108.0$ and $\kappa = 6.84$ at $Pr = 1$.[†]

[†] In the limiting case $Gr_v \to \infty$, the basic spectral amplitude problem (2.12) and (2.13) simplifies (see [1]).

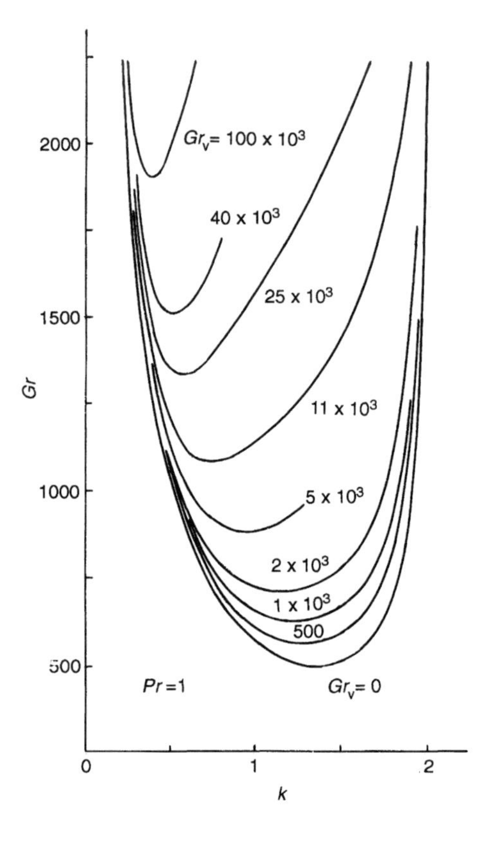

Fig. 37 The family of neutral curves for different values of Gr_v; hydrodynamic mode; $Pr = 1$

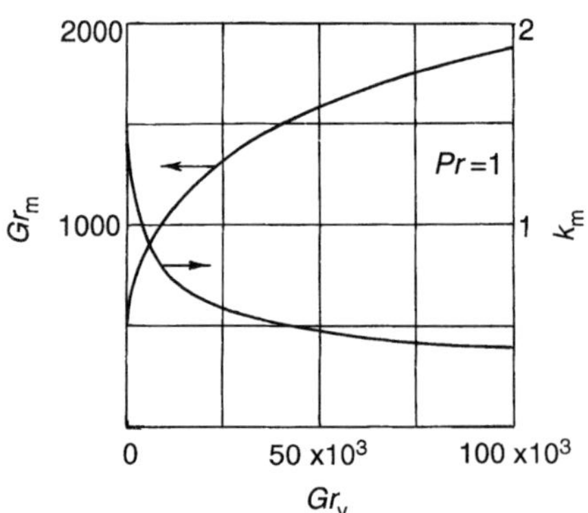

Fig. 38 Critical parameters of hydrodynamic mode as functions of Gr_v; $Pr = 1$

As the Prandtl number increases, the wave mode of instability appears. At the point Pr^* it is induced at infinitely large Gr_m and infinitely small k_m, meaning that in this range the term in (2.12) describing the vibration effect vanishes. Thus the threshold value Pr^* does not change under vibrations.

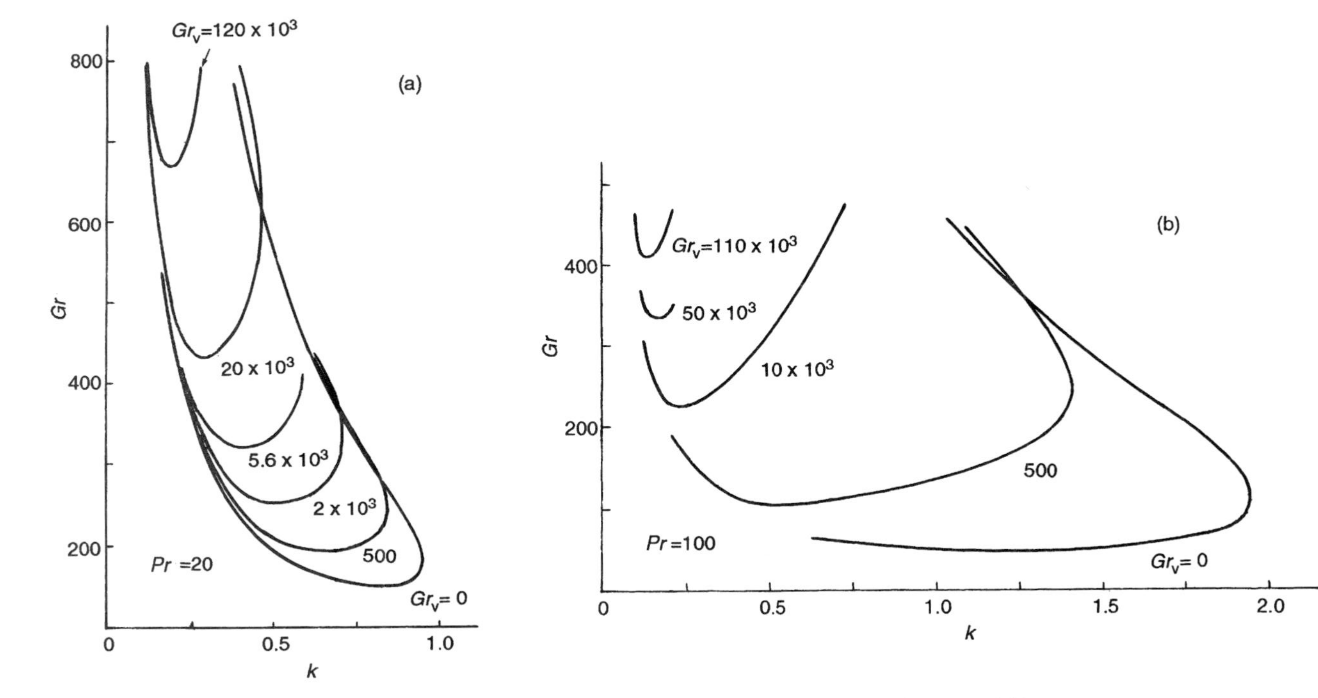

Fig. 39 Family of neutral curves for different Gr_v values; wave mode: (a) $Pr = 20$, (b) $Pr = 100$

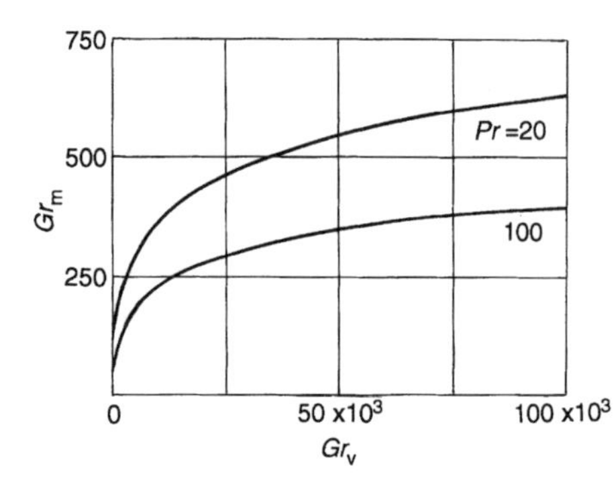

Fig. 40 The critical Grashof number as a function of Gr_v; wave mode

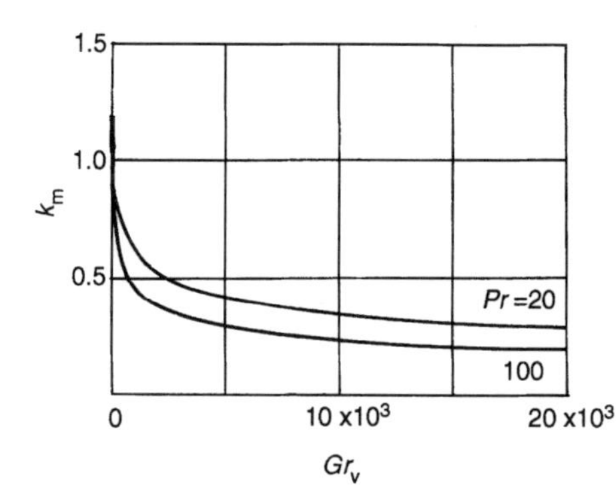

Fig. 41 The critical wavenumber as a function of Gr_v; wave mode

In Fig. 39 the family of neutral curves is shown for the wave instability situation. When the vibration intensity increases, the stabilization is observed, and the critical wavenumber grows as for the hydrodynamic mode. The dependences of Gr_m and k_m on Gr_v are given in Figs. 40 and 41. In the limit $Gr_v \to \infty$, the asymptotic (2.14) holds. However, in the case of wave instability, the coefficients g and κ depend on Pr. Therefore, for $Pr = 15$, 20, 50, 100 there are, respectively, $g = 41.2$, 35.6, 26.4, 22.2 and $\kappa = 3.75$, 3.49, 2.76, 2.37. It follows from the asymptotic (2.14) that the critical temperature difference is $\Theta_m \sim b\Omega\nu^3/(\beta g^2 h^5)$, i.e. it is directly proportional to the vibration intensity $b\Omega$. Figures 42 and 43 illustrate the main characteristics of the wave instability Gr_m and k_m as the functions of the Prandtl number for three values of the vibrational parameter Gr_v.

All the results presented in Figs. 37 to 43 correspond to the case of plane perturbations described by the spectral amplitude problem (2.12) and (2.13). Let us go back to the complete amplitude problem (2.9) and (2.10) for three-dimensional perturbations and demonstrate that the plane ones are the most dangerous.

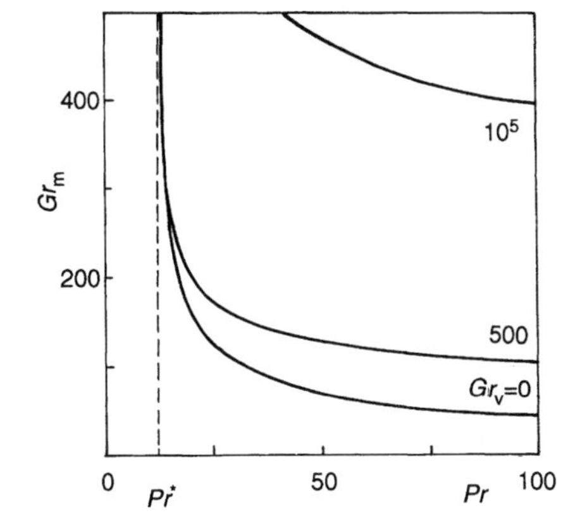

Fig. 42 The critical Grashof number as a function of Pr; wave mode

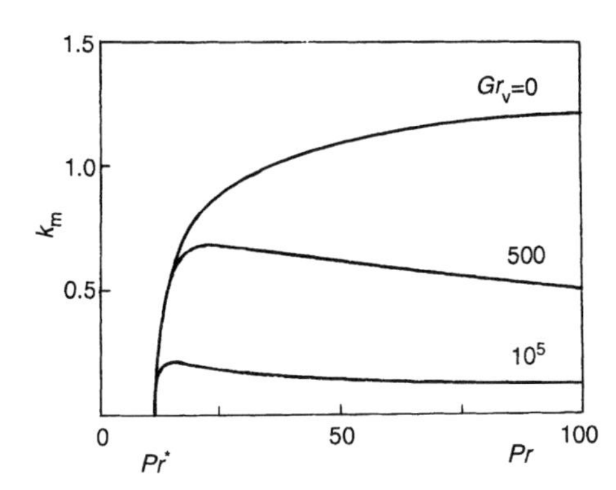

Fig. 43 The critical wavenumber as a function of Pr; wave mode

Setting $v_y = 0$, $w_y = 0$ and $k_y = 0$, $k_z = k$, just as in the two-dimensional case, the general set (2.9) and (2.10) yields

$$
\begin{aligned}
-\overline{\lambda}\overline{v}_x + \mathrm{i}\overline{k}\,\overline{Gr}\,v_0\overline{v}_x &= -\overline{p}' + \overline{\Delta}\overline{v}_x + \frac{\overline{Gr}_\mathrm{v}}{\overline{Gr}}T_0'\,\overline{w}_x \\
-\overline{\lambda}\overline{v}_z + \mathrm{i}\overline{k}\,\overline{Gr}\,v_0\overline{v}_z + \overline{Gr}\,v_0'\,\overline{v}_x &= -\mathrm{i}\overline{k}\overline{p} + \overline{\Delta}\overline{v}_z + \overline{\theta} \\
-\overline{\lambda}\overline{\theta} + \mathrm{i}\overline{k}\,\overline{Gr}\,v_0\overline{\theta} + \overline{Gr}\,T_0'\,\overline{v}_x &= \frac{1}{\overline{Pr}}\overline{\Delta}\,\overline{\theta}
\end{aligned}
\tag{2.15}
$$

$$
\begin{aligned}
\overline{v}_x' + \mathrm{i}\overline{k}\overline{v}_z &= 0 \\
\overline{w}_x' + \mathrm{i}\overline{k}\,\overline{w}_z &= 0 \\
\overline{w}_z' - \mathrm{i}\overline{k}\overline{w}_x &= -\mathrm{i}\overline{k}\,\overline{\theta}
\end{aligned}
\tag{2.16}
$$

$$x = \pm 1: \qquad \overline{v}_x = \overline{v}_z = 0, \qquad \overline{\theta} = 0, \qquad \overline{w}_x = 0$$

where $(\bar{\Delta} = (d^2/dx^2) - \bar{k}^2)$. When this result is compared with (2.9) and (2.10), it is apparent that the three-dimensional problem may be reduced to a two-dimensional one by the following transformation:

$$\lambda = \bar{\lambda}, \qquad v_x = \bar{v}_x, \qquad k_y v_y + k_z v_z = \bar{k}\,\bar{v}_z$$
$$w_x = \alpha \bar{w}_x, \qquad k_y w_y + k_z w_z = \alpha \bar{k}\,\bar{w}_z \tag{2.17}$$
$$p = \bar{p}, \qquad k_z \theta = \bar{k}\,\bar{\theta}$$
$$k_z\,Gr = \bar{k}\,\overline{Gr}, \qquad k \equiv \sqrt{k_y^2 + k_z^2} = \bar{k}, \qquad \alpha \frac{Gr_v}{Gr} = \frac{\overline{Gr}_v}{\overline{Gr}}, \qquad Pr = \overline{Pr}$$

where the constant α is to be evaluated.

Let us introduce the parameter a characterizing the wavevector direction for the three-dimensional perturbations:

$$a = \frac{k_z}{\sqrt{k_y^2 + k_z^2}} = \frac{k_z}{k} \tag{2.18}$$

In the limiting case of plane perturbations $k_y = 0$ and $a = 1$, whereas the opposite limiting case $k_z = 0$ corresponds to three-dimensional spiral perturbations with $a = 0$. From relations (2.17) at $\alpha = 1/a$, follow the formulas that connect the parameters of spiral and plane perturbations:

$$Gr = \frac{\overline{Gr}}{a}, \qquad Gr_v = \overline{Gr}_v, \qquad k = \bar{k} \tag{2.19}$$

Thus, once the critical values $\overline{Gr}$, $\overline{Gr}_v$ and k determining the stability border with respect to plane perturbations are known, from (2.19) for Gr, Gr_v and k one may find the stability border for three-dimensional perturbations. So far as $0 \leqslant a \leqslant 1$ for given k and Gr_v is concerned the relation $Gr > \overline{Gr}$ between the critical values shows that the plane perturbations are the most dangerous. Therefore, one arrives at the generalization of the Squire theorem for the problem under study [2].

2 Convective Flow in a Vertical Layer: The Case of Longitudinal Vibrations

Let longitudinal vibrations be in action with the vector $\mathbf{n}$ in the plane of the layer and directed at an angle α to the vertical z axis, i.e. having the components

$\mathbf{n}(0, \sin\alpha, \cos\alpha)$. The qualitative distinction from the situation with transverse vibrations is that the vibration axis is now perpendicular to the temperature gradient, and thus the thermovibrational mechanism is operative. In the limiting case of weightlessness, the standard problem is recovered of the quasi-equilibrium stability in the presence of a transversal temperature gradient and longitudinal vibrations (see Section 2 in Chapter 1). In the general case, the mechanisms giving rise to instabilities in a plane-parallel convective flow must come into interaction with the thermovibrational one.

2.1 The Basic Flow and the Amplitude Problem

As in the preceding section, let us begin from the basic set (2.1) and demonstrate that for the case under consideration there exists the solution describing the flow with the plane-parallel structure:

$$\begin{gathered} v_x = v_y = 0, \qquad v_z = v_0(x), \qquad T = T_0(x), \qquad p = p_0(z) \\ w_x = 0, \qquad w_y = w_{0y}(x), \qquad w_z = w_{0z}(x) \end{gathered} \tag{2.20}$$

Indeed, in such a regime, for the average and pulsation velocities, temperature and pressure the following equations are valid:

$$\begin{gathered} \frac{\mathrm{d}p_0}{\mathrm{d}z} = v_0'' + T_0 = C, \qquad T_0'' = 0 \\ w_{0y}' = T_0' \sin\alpha, \qquad w_{0z}' = T_0' \cos\alpha \end{gathered} \tag{2.21}$$

Taking into account the boundary conditions and the condition of closeness, one obtains the distributions:

$$\begin{gathered} v_0 = \tfrac{1}{6}\left(x^3 - x\right), \qquad T_0 = -x, \qquad p_0 = \text{constant} \\ w_{0y} = -\sin\alpha \cdot x, \qquad w_{0z} = -\cos\alpha \cdot x \end{gathered} \tag{2.22}$$

Therefore, as in the case of transverse vibrations, the effect on the velocity and temperature profiles is absent. However, in this case the pulsation velocity component is non-zero and is parallel to vector $\mathbf{n}$.

Let us introduce the normal three-dimensional perturbations of the type (2.8) and write down the set of amplitude equations that is analogous to (2.9) but differing in **n** orientation:

$$-\lambda v_x + \mathrm{i}k_z\, Gr\, v_0 v_x = -p' + \left(v_x'' - k^2 v_x\right) - \frac{\mathrm{i}\, Ra_\mathrm{v}}{Gr\, Pr}\left(k_y w_{0y} + k_z w_{0z}\right) w_x$$

$$-\lambda v_y + \mathrm{i}k_z\, Gr\, v_0 v_y = -\mathrm{i}k_y p + \left(v_y'' - k^2 v_y\right) + \frac{\mathrm{i}\, Ra_\mathrm{v}}{Gr\, Pr}\left(k_y w_{0y} + k_z w_{0z}\right)\left(\theta \sin\alpha - w_y\right)$$

$$-\lambda v_z + \mathrm{i}k_z\, Gr\, v_0 v_z + Gr\, v_0' v_x = -\mathrm{i}k_z p + \left(v_z'' - k^2 v_z\right) + \theta + \frac{\mathrm{i}\, Ra_\mathrm{v}}{Gr\, Pr}\left(k_y w_{0y} + k_z w_{0z}\right)\left(\theta \cos\alpha - w_z\right),$$

$$-\lambda\theta + \mathrm{i}k_z\, Gr\, v_0\theta + Gr\, T_0' v_x = \frac{1}{Pr}\left(\theta'' - k^2\theta\right) \tag{2.23}$$

$$v_x' + \mathrm{i}\left(k_y v_y + k_z v_z\right) = 0$$

$$w_x' + \mathrm{i}\left(k_y w_y + k_z w_z\right) = 0$$

$$w_y' - \mathrm{i}k_y w_x = \sin\alpha \cdot \theta'$$

$$w_z' - \mathrm{i}k_z w_x = \cos\alpha \cdot \theta'$$

Here $k^2 = k_y^2 + k_z^2$ and the vibrational Rayleigh number $Ra_\mathrm{v} = (\beta b\Omega\Theta h)^2/(2\nu\chi) = Gr\, Pr$ is entered instead of Gr_v. When the static vibrational instability mechanism is operative, it is more convenient to use the vibrational Rayleigh number. The unknown functions in (2.23) must satisfy the homogeneous boundary conditions

$$x = \pm 1: \qquad v_x = v_y = v_z = 0, \qquad \theta = 0, \qquad w_x = 0 \tag{2.24}$$

2.2 Plane Perturbations

Below we show that for a layer under vertical vibrations the problem of plane perturbations is very important. It turns out that all the information about the stability borders with respect to three-dimensional perturbations in the layer with arbitrarily oriented vibration axis can be deduced from the solution of the boundary-value problem obtained from (2.23) and (2.24) at $\alpha = 0$, $w_{0y} = 0$, $w_{0z} = -x$, $v_y = 0$, $w_y = 0$, $k_y = 0$ and $k_z = k$. Let us present the basic amplitude problem, marking all the entering variables and parameters with overbars:

$$
\begin{gathered}
-\overline{\lambda}\overline{v}_x + \mathrm{i}\overline{k}\,\overline{Gr}\; v_0\overline{v}_x = -\overline{p}' + \left(\overline{v}_x'' - \overline{k}^2\,\overline{v}_x\right) - \frac{\mathrm{i}\,\overline{Ra}_{\mathrm{v}}}{\overline{Gr\,Pr}}\overline{k}\,w_{0z}\overline{w}_x \\
-\overline{\lambda}\overline{v}_z + \mathrm{i}\overline{k}\,\overline{Gr}\; v_0\overline{v}_z + \overline{Gr}\; v_0'\overline{v}_x = -\mathrm{i}\overline{k}\overline{p} + \left(\overline{v}_z'' - \overline{k}^2\overline{v}_z\right) + \overline{\theta} + \frac{\mathrm{i}\,\overline{Ra}_{\mathrm{v}}}{\overline{Gr\,Pr}}\overline{k}w_{0z}\left(\overline{\theta} - \overline{w}_z\right) \\
-\overline{\lambda}\overline{\theta} + \mathrm{i}\overline{k}\,\overline{Gr}\; v_0\overline{\theta} + \overline{Gr}\,T_0'\,\overline{v}_x = \frac{1}{\overline{Pr}}\left(\overline{\theta}'' - \overline{k}^2\overline{\theta}\right) \\
\overline{v}_x' + \mathrm{i}\overline{k}\overline{v}_z = 0 \\
\overline{w}_x' + \mathrm{i}\overline{k}\overline{w}_z = 0 \\
\overline{w}_z' - \mathrm{i}\overline{k}\overline{w}_x = \overline{\theta}' \\
x = \pm 1: \qquad \overline{v}_x = \overline{v}_z = 0, \qquad \overline{\theta} = 0, \qquad \overline{w}_x = 0
\end{gathered}
\tag{2.25}
$$

This problem may be rewritten in terms of stream function amplitudes for the average and pulsation velocities (see (2.11)):

$$
\begin{gathered}
\Delta^2\varphi + \mathrm{i}\overline{k}\,\overline{Gr}\left(v_0''\varphi - v_0\Delta\varphi\right) + \mathrm{i}\overline{k}\frac{\overline{Ra}_{\mathrm{v}}}{\overline{Gr\,Pr}}w_{0z}'\left(\overline{\theta} - f'\right) = -\overline{\lambda}\Delta\varphi \\
\frac{1}{\overline{Pr}}\Delta\overline{\theta} + \mathrm{i}\overline{k}\,\overline{Gr}\left(T_0'\varphi - v_0\overline{\theta}\right) = -\overline{\lambda}\,\overline{\theta} \\
\Delta f = \overline{\theta}' \\
x = \pm 1: \qquad \varphi = \varphi' = 0, \qquad \overline{\theta} = 0, \qquad f = 0
\end{gathered}
\tag{2.26}
$$

The problem (2.26) has been solved by Sharifulin [3, 4]; the amplitude equations were integrated numerically with the aid of a step-by-step orthogonalization procedure. Some additional data on the stability characteristics may be taken from [5]. Actually this paper deals with a different physical problem: the stability of a plane-parallel convective flow in a vertical layer of a liquid dielectric with the boundaries maintained at different temperatures. A transverse electric field is imposed and a linear dependence of the dielectric constant ε on temperature is taken into account. It may be proved that if the space inhomogeneity of ε is weak, the corresponding amplitude problem is formally equivalent to (2.26). The stability borders in the plane $\left(\overline{Ra}_{\mathrm{v}}, \overline{Gr}\right)$ are given in Fig. 44 for different values of the Prandtl number; the minimization is done with respect to k. The stability regions about the origin of the coordinate system. The solid lines describe the monotonic perturbations and the dashed lines the oscillatory ones. In the case of no vibration ($\overline{Ra}_{\mathrm{v}} = 0$) at small and

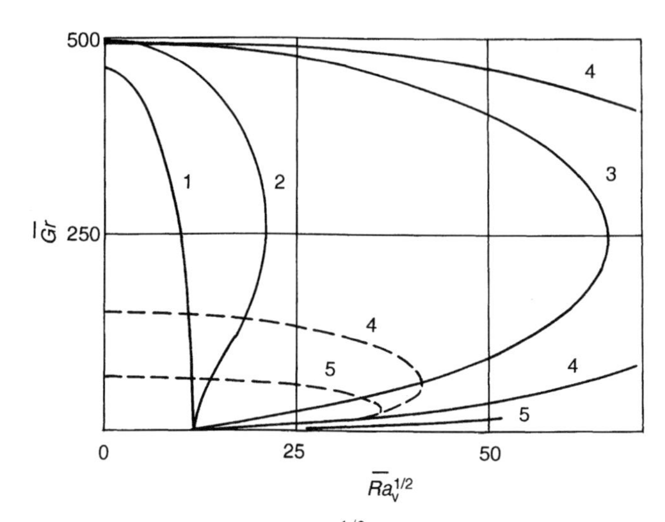

Fig. 44 Stability borders in the plane $(\overline{Ra}_v^{1/2}, \overline{Gr})$. Solid lines, monotonic instability; dashed lines, wave mode. Curves 1–5 correspond to the Prandtl numbers, respectively, of $Pr = 0.1$, 1, 10, 20, 50

moderate values of Pr, the instability is caused by the hydrodynamic mode, whereas at $Pr > 12.45$ it is of the wave origin. Another limiting case $\overline{Gr} = 0$ corresponds to weightlessness. Quasi-equilibrium represents the main state, its instability sets in at $\overline{Ra}_v = 133.1$ and is caused by the perturbation with the wavenumber $\bar{k}_m = 1.61$. We would like to add a reminder that the layer half-thickness is used here as the length scale. The stability curves in Fig. 44 describe the interaction of different instability mechanisms—hydrodynamic and wave on the one hand and vibrational convective on the other. It can be seen that the presence of vibrations leads to destabilization of convective flows at any value of the Prandtl number. As for the influence of the convective flow on the vibrational instability, it depends on the Prandtl number. At small values ($Pr < 0.27$), the convective flow acts as a destabilizing factor. At $Pr > 0.27$ and small $\overline{Gr}$ values, i.e. when the intensity of the flow is weak, the threshold of the vibration instability increases with $\overline{Gr}$. At large $\overline{Gr}$ values the destabilization occurs naturally, due to predominance of the mechanisms of flow instability.

The interference of different instability mechanisms is reflected in the structure of the neutral curves. In Fig. 45 the neutral curves are given for $Pr = 20$. At this value, the wave mode of instability exists. The upper regions are related to the usual wave instability, i.e. to amplification of the temperature waves due to the vibration effect. The lower areas adjacent to the abscissa axis refer to the vibrational convective mode. At $\overline{Ra}_v = 1040$, on the upper boundary of the vibrational convective instability region the zone of wave instability is

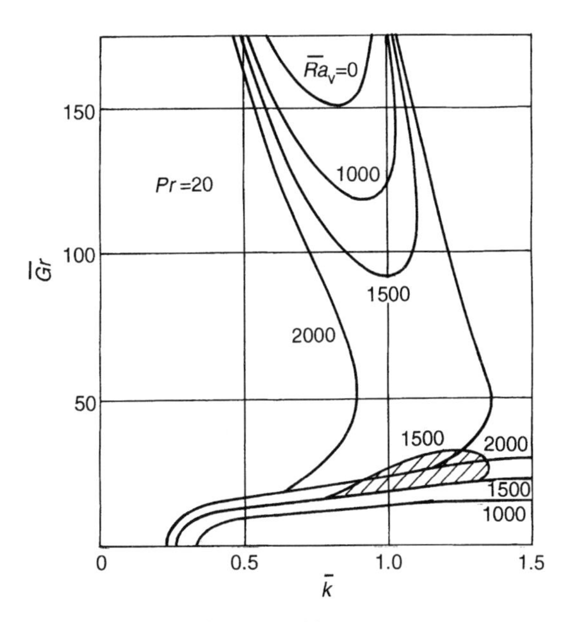

Fig. 45 Family of neutral curves for $Pr = 20$

conceived to expand with $\overline{Ra}_v$. The fusion of the upper and lower regions of the wave instability occurs at $\overline{Ra}_v = 1740$.

2.3 Three-dimensional (Spiral) Perturbations

Let us address the general case of three-dimensional perturbations in a layer with an arbitrary orientation of the vibration axis. It has been shown [6] that the spectral problem (2.23) and (2.24) may be reduced to the basic two-dimensional problem (2.25). Direct comparison of the amplitude equations of those sets testifies that the reduction takes place under the following relations among the variables and parameters:

$$\lambda = \overline{\lambda}, \qquad v_x = \overline{v}_x, \qquad k_y v_y + k_z v_z = \overline{k}\,\overline{v}_z$$

$$w_x = c\overline{w}_x, \qquad k_y w_y + k_z w_z = c\overline{k}\,\overline{w}_z$$

$$p = \overline{p}, \qquad k_z \theta = \overline{k}\,\overline{\theta}, \qquad (k_y \sin\alpha + k_z \cos\alpha)\theta = c\overline{k}\overline{\theta} \tag{2.27}$$

$$k_z\, Gr = \overline{k}\,\overline{Gr}, \qquad Pr = \overline{Pr}, \qquad \frac{Ra_v}{Gr}(k_y \sin\alpha + k_z \cos\alpha) w_x = \frac{\overline{Ra}_v}{\overline{Gr}}\overline{k}\,\overline{w}_x$$

where the constant c is to be found.

From (2.27), the formulas may be derived linking the characteristics of the three- and two-dimensional perturbations:

$$Gr = \frac{\overline{Gr}}{a}, \qquad Ra_{\mathrm{v}} = \frac{\overline{Ra}_{\mathrm{v}}}{\left(\sqrt{1-a^2}\sin\alpha + a\cos\alpha\right)^2}, \qquad k \equiv \sqrt{k_y^2 + k_z^2} = \overline{k} \tag{2.28}$$

Here

$$a = \frac{k_z}{\sqrt{k_y^2 + k_z^2}} = \frac{k_z}{k} \tag{2.29}$$

is the parameter defining the orientation of the wavevector and the constant c equals

$$c = \frac{1}{k_z}\left(k_y \sin\alpha + k_z \cos\alpha\right) \tag{2.30}$$

The obtained formulas enable the critical values of Gr and Ra_{v} to be determined for three-dimensional perturbations with wavenumbers k_y and k_z in the layer with the vibration axis inclined at the angle α to the z coordinate if the critical values $\overline{Gr}$ and $\overline{Ra}_{\mathrm{v}}$ are known for two-dimensional perturbations with the wavenumber $\overline{k}$ for the layer with the vertical vibration axis.

We begin the consideration with the case of vertical vibrations, i.e. $\alpha = 0$. Formulas (2.28) yield

$$Gr = \frac{\overline{Gr}}{a}, \qquad Ra_{\mathrm{v}} = \frac{\overline{Ra}_{\mathrm{v}}}{a^2} \tag{2.31}$$

These relations map the critical point $(\overline{Gr}, \overline{Ra}_{\mathrm{v}})$ into the similar point (Gr, Ra_{v}) for three-dimensional perturbations with parameter a. Since, according to (2.29), the values of the parameter a lie within the interval (0, 1), it follows from (2.31) that $Gr > \overline{Gr}$ and $Ra_{\mathrm{v}} > \overline{Ra}_{\mathrm{v}}$. This means that the plane perturbations are the most unstable ones.

Consider now the situation where the vibration axis is tilted, i.e. $\alpha \neq 0$. The case of no vibration $Ra_{\mathrm{v}} = 0$ is trivial. Formula (2.31) delivers the known result: in the absence of vibrations, plane perturbations are the most dangerous for the stability of convective flow in a vertical layer. The opposite limiting case $\overline{Gr} = 0$, or $Gr = 0$, corresponds to weightlessness. The critical value of the vibrational Rayleigh number is described by the second ratio of (2.28). The instability threshold is determined as the minimal Ra_{v} value with respect to a. From the extremum condition $\partial Ra_{\mathrm{v}}/\partial a = 0$, it follows that the minimal value equals $\overline{Ra}_{\mathrm{v}}$ and is reached at $a_{\mathrm{m}} = \cos\alpha$. This result is clear. As has been demonstrated in Section 2 in Chapter 1, the mechanical equilibrium under

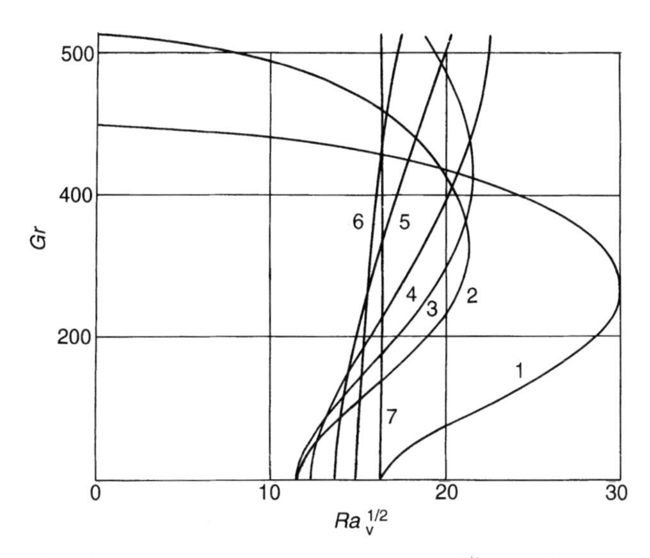

Fig. 46 Family of stability curves in the plane $(Ra_v^{1/2}, Gr)$ for $\alpha = 45°$, $Pr = 1$. Curves 1–7 correspond to the parameter a values, respectively, of 1, 0.8, 0.6, 0.4, 0.2, 0.1, 0

longitudinal vibrations loses its stability at $Ra_v = \overline{Ra}_v = 133.1$, which is connected with the most dangerous perturbation in the shape of rolls with their axes perpendicular to the vibration axis and the equilibrium temperature gradient. The value $a_m = \cos\alpha$ just matches such a case.

To illustrate the general situation, let the inclination angle be $\alpha = 45°$. In Fig. 46 the family of the stability curves is displayed in the plane $(Ra_v^{1/2}, Gr)$ for the Prandtl number value $Pr = 1$ and different values of the parameter a. The latter varies between 0 and 1, i.e. from the spiral perturbations with vertical axes to the plane ones being the rolls with axes parallel to the y coordinate. From the parameters mentioned only monotonic instability is possible. The stability region abuts the origin of the coordinates. On the upper boundary the stability breaks due to the action of the hydrodynamic mechanism and is implied by plane perturbations. On the right boundary the instability has the vibrational convective origin and is connected with three-dimensional perturbations. This border is found as the envelope of the family of curves for different a values. Along the envelope, a varies continuously. As Gr decreases from the intersection point of the envelope and the curve 1, parameter a grows from 0.08 to $\cos\alpha = 0.707$.

In Fig. 47a, b, c the stability borders are presented in the plane $(Ra_v^{1/2}, Gr)$ for three values of α and several values of Pr. Figure 47a pertains to the case $\alpha = 30°$. Curve 1 ($Pr = 0.1$) is the envelope for the family of curves depending

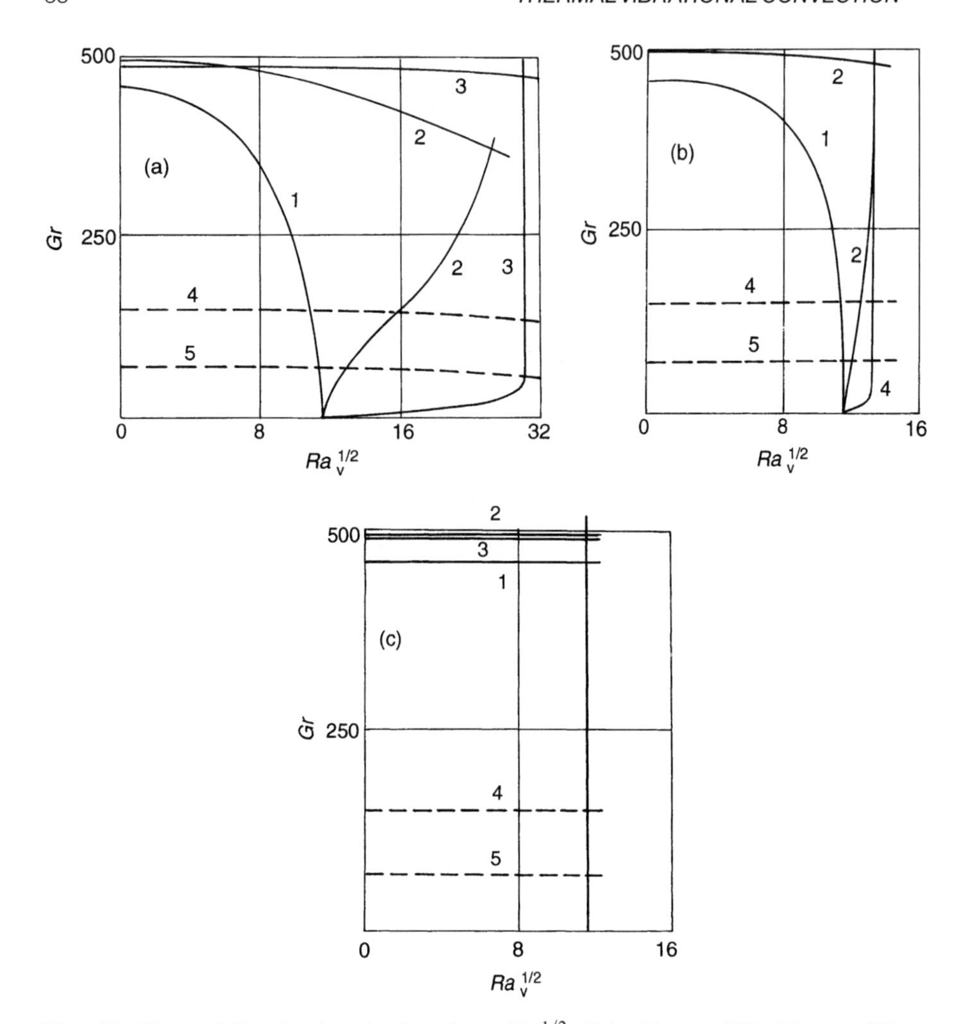

Fig. 47 The stability borders in the plane $(Ra_v^{1/2}, Gr)$. (a) $\alpha = 30°$, (b) $\alpha = 60°$, (c) $\alpha = 90°$. Curves 1–5 correspond to the Prandtl numbers, respectively, of $Pr = 0.1, 1, 10, 20, 50$. Solid lines, monotonic mode, dashed lines, wave mode

on parameter a. Along this curve, as Gr goes down, parameter a gradually decreases from $a = 1$ (plane perturbations, the point on the Gr axis) to $a = \cos\alpha = 0.87$. Note that the point on the Ra_v axis is compatible with the vibrational convective loss of stability at $Ra_v = 133.1$. Curves 2 and 3 are formed by two intersecting branches. The upper lines 2 and 3 correspond to the plane hydrodynamic mode ($a = 1$). The right branches reflect the effect of the three-dimensional vibrational mode. The right line 3 ($Pr = 10$) is vertical within the range of $Gr > 60$, and in this range the most unstable is the spiral mode with $a = 0$. At large Pr values (curves 4 and 5 render the results for $Pr = 20$ and $Pr = 50$, respectively) the upper border of the stability region is determined by the plane wave mode while the right boundary is almost vertical; cf. Fig. 46 where the envelope of the family of borders for vibrational modes is plotted.

Qualitatively the same alternative occurrence of plane and three-dimensional modes takes place at $\alpha = 60°$ (Fig. 47b).

Finally, for the limiting case $\alpha = 90°$, a horizontal vibration in the plane of a layer with the axis parallel to the y coordinate should be considered. The stability borders are presented in Fig. 47c. The upper border of the stability region is caused by monotonic or wave plane modes depending on the Prandtl number. This border is horizontal, i.e. the critical value of the Grashof number is independent on the vibrational parameter Ra_v. This follows from the fact that $w_{0z} = 0$ at $\alpha = 90°$ and the vibrational force drops out of the equation of motion for plane perturbations (see (2.23)). The right border of the stability region is due to the spiral three-dimensional perturbations with $a = 0$ and is vertical, $Ra_v = 133.1$ at all Prandtl numbers. For the case under study, the spectral amplitude problem does not include the velocity of the basic state and coincides with the problem of quasi-equilibrium stability in weightlessness. If $Ra_v < 133.1$, the instability turns up when reaching the critical value $Gr_m(Pr)$ and is driven by plane perturbations, either hydrodynamic or wave in relation to Pr. If $Gr < Gr_m(Pr)$, the instability sets in when Ra_v achieves the critical value $Ra_v = 133.1$. The crisis is caused by the spiral three-dimensional perturbations

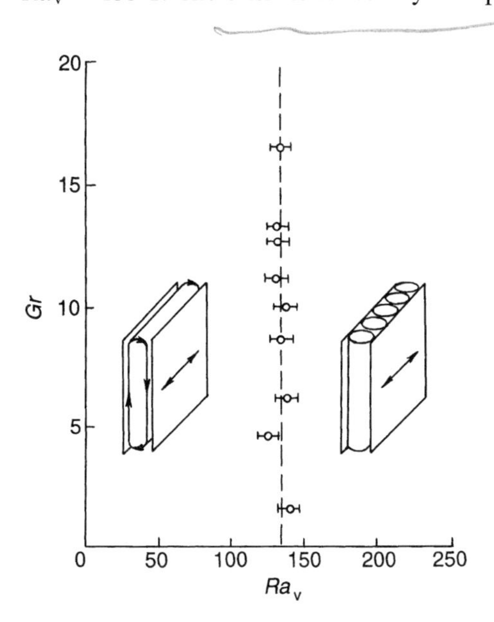

Fig. 48 The stability borders with respect to the vibrational convective mode (experiment [7])

in the form of rolls with axes parallel to the z coordinate. The same situation was observed in experiments [7]. Ethyl alcohol with $Pr = 16.1$ was used as the working liquid. As it turns out, the upper border of the stability region is related to the plane wave mode and the critical Grashof number is $Gr_m = 210$. Experiments were performed at lower Gr values. At fixed Gr, as Ra_v grows, the instability sets on taking the form of vertical rolls against the background of a plane-parallel flow (Fig. 48). The measured critical value of the Rayleigh number $Ra_v = 1.3 \times 10^2$ is very close to the theoretical one of 133.1. Thus the authors of experiments [7] succeeded, even under terrestrial conditions, to single out the pure vibrational convective instability mechanism.

3 Advective Flows in a Vibrational Field

The term 'advection' here designates fluid or gas flows in a plane horizontal layer under gravity, the flows being caused by a longitudinal temperature gradient. The peculiarity of these flows is that the velocity is perpendicular to the buoyancy force. Such flows are interesting from the general point of view because they reveal instability owing to different physical origins. As well they are attractive from the standpoint of applications (various aspects of geophysics, crystal growth, etc.). In this section the effect of high-frequency vibrations on advective flows is studied.

3.1 Plane-Parallel Flows [8]

Let us consider a plane fluid layer which is infinite in horizontal directions. The origin of coordinates is placed in the layer mid-section and the x axis points vertically upward. The layer is bounded by parallel rigid plates at $x = \pm h$. A constant longitudinal temperature gradient A is maintained along the z coordinate: $\langle \partial T/\partial z \rangle = A$, where the brackets denote averaging across the layer.

First, we consider an interesting case of longitudinal vibration with the axis parallel to the temperature gradient: $\mathbf{n}(0, 0, 1)$. Then equations (1.17) to (1.20) admit the solution describing a plane-parallel advective flow of the following structure:

$$v_x = v_y = 0, \qquad v_z = v_0(x), \qquad T_0 = Az + \tau_0(x)$$
$$w_x = w_y = 0, \qquad w_z = w_0(x), \qquad p = p_0(x, z) \tag{2.32}$$

Functions v_0, τ_0 and w_0 defining the plane-parallel flow profiles satisfy the equations

$$\nu v_0'' + \varepsilon A w_0 - g\beta A x = C_1$$
$$\chi \tau_0'' = A v_0, \qquad \tau_0 - w_0 = C_2 \tag{2.33}$$

where C_1 and C_2 are the constants resulting from the separation of variables. Conditions of closeness for the average and pulsation velocity components yield $C_1 = C_2 = 0$ and require the profiles v_0, τ_0 to be odd functions of the transversal coordinate x.

At the boundaries the non-slip conditions are valid so $v_0(\pm h) = 0$; the condition for the pulsation component $w_{0x} = 0$ holds automatically. As for τ_0, two variants of the boundary conditions are used, namely (a) the plates are of high heat conductivity with a constant temperature gradient A, i.e. $\tau_0(\pm h) = 0$, and (b) the plates are thermally insulated, i.e. $\tau_0'(\pm h) = 0$.

The solution for case (a) is

$$v_0 = \frac{g\beta A h^3}{2\nu}\frac{1}{\delta_1 r^2}\left(\frac{\cosh r\xi \sin r\xi}{\cosh r \sin r} - \frac{\sinh r\xi \cos r\xi}{\sinh r \cos r}\right) \tag{2.34}$$

$$\tau_0 = w_0 = \frac{g\beta A^2 h^5}{4\nu\chi}\frac{1}{r^4}\left[\xi - \frac{1}{\delta_1}\left(\frac{\cosh r\xi \sin r\xi}{\sinh r \cos r} + \frac{\sinh r\xi \cos r\xi}{\cosh r \sin r}\right)\right] \tag{2.35}$$

Here $\xi = x/h$ is the dimensionless transversal coordinate and $r = (\varepsilon A^2 h^4/4\nu\chi)^{1/4}$ is the dimensionless vibrational parameter related to the vibrational Rayleigh number as $\mathrm{Ra_v} = 4r^4$. In addition, the notation $\delta_1 = \tan r \,\mathrm{cotanh}\, r + \mathrm{cotan}\, r \tanh r$ is introduced for brevity.

The solution for case (b) is

$$v_0 = \frac{g\beta A h^3}{4\nu}\frac{\delta_2}{r^3}\left(\frac{\cosh r\xi \sin r\xi}{\cosh r \sin r} - \frac{\sinh r\xi \cos r\xi}{\sinh r \cos r}\right) \tag{2.36}$$

$$\tau_0 = w_0 = \frac{g\beta A^2 h^5}{4\nu\chi}\frac{1}{r^4}\left[\xi - \frac{\delta_2}{2r}\left(\frac{\cosh r\xi \sin r\xi}{\sinh r \cos r} + \frac{\sinh r\xi \cos r\xi}{\cosh r \sin r}\right)\right] \tag{2.37}$$

where $\delta_2 = \sinh 2r \sin 2r/(\sinh 2r + \sin 2r)$.

In the absence of vibrations (the limit $r \to 0$) expressions (2.34) and (2.36) yield the following velocity profile of the advective flow:

$$v_0 = \frac{g\beta A h^3}{6\nu}(\xi^3 - \xi) \tag{2.38}$$

which describes two counter-flows: the upper one ($\xi > 0$) opposed to the temperature gradient and the lower one ($\xi < 0$) along it. Formulas (2.35) and (2.37) provide the known temperature profiles, respectively, for highly conductive boundaries and for thermally insulated ones:

$$\tau_0 = \frac{g\beta A^2 h^5}{360\nu\chi}(3\xi^5 - 10\xi^3 + 7\xi) \tag{2.39a}$$

$$\tau_0 = \frac{g\beta A^2 h^5}{360\nu\chi}(3\xi^5 - 10\xi^3 + 15\xi) \tag{2.39b}$$

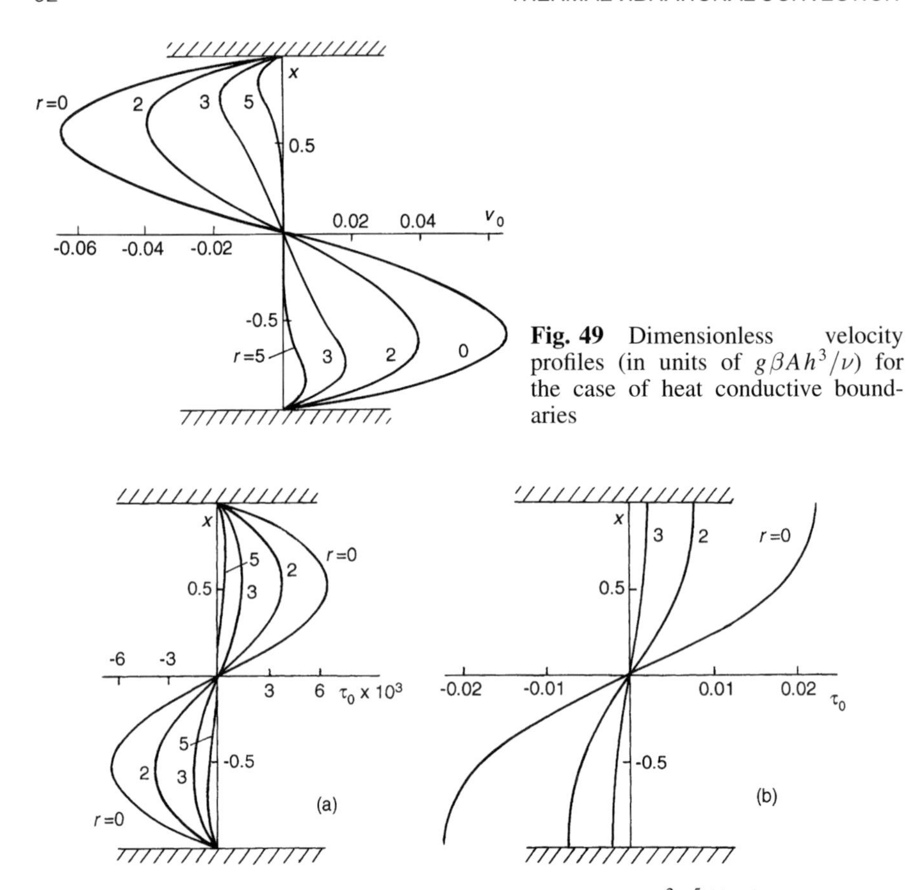

Fig. 49 Dimensionless velocity profiles (in units of $g\beta A h^3/\nu$) for the case of heat conductive boundaries

Fig. 50 Dimensionless temperature profiles (in units of $g\beta A^2 h^5/(\nu\chi)$) for the layer with (a) boundaries of high heat conductivity and (b) thermally insulated boundaries

The following figures demonstrate the deformation of the velocity and temperature profiles under high-frequency longitudinal vibrations. Figures 49 and 50a cover the case of highly conductive boundaries. One sees that as the vibrational parameter r increases, the flow slows down and the transverse temperature inhomogeneity decreases. For vibrations that are intense enough (large r values), the flow acquires the structure of non-interacting boundary layers whose thickness decreases as h/r. The temperature profiles show that the central part of the channel is potentially stably stratified whereas close to boundaries there exist zones of unstable stratification.

For the layer with thermally insulated boundaries (case (b)), relations (2.34) and (2.36) indicate that the velocity profiles are similar to those in Fig. 49, the only distinction being that the scale factor depends on r. The flow braking due to vibrations manifests itself much more than in case (a). The temperature profiles (Fig. 50b) differ substantially from those for highly heat conductive boundaries. Now all the layer is potentially stably stratified and a relatively large transversal temperature difference appears:

$$\Theta = \tau(h) - \tau(-h) = \frac{g\beta A^2 h^5}{\nu\chi}\frac{1}{2r^4}\left(1 - \frac{4}{r}\frac{\sinh^2 r + \sin^2 r}{\sinh 2r + \sin 2r}\right) \tag{2.40}$$

If there are no vibrations, then $\Theta = 2g\beta A^2 h^5/(45\nu\chi)$. As r increases, the temperature difference decreases as $1/r^4$.

The intensity of motion may be characterized by the volumetric flow rate in one of the counter-flows:

$$Q = \int_{-h}^{0} v_0 \, \mathrm{d}x \tag{2.41}$$

For each of the two variants considered, equation (2.41) gives, respectively,

$$\frac{Q}{Q_0} = \frac{6}{r^3}\frac{\sinh r - \sin r}{\cosh r + \cos r} \tag{2.42a}$$

$$\frac{Q}{Q_0} = \frac{12}{r^4}\frac{(\sinh r - \sin r)(\cosh r - \cos r)}{\sinh 2r + \sin 2r} \tag{2.42b}$$

where $Q_0 = g\beta A h^4/(24\nu)$ is the volumetric flow rate in a flow without vibrations. As the vibrational parameter r increases, the volumetric flow rate decreases monotonically (Fig. 51). At large r, the asymptote holds for each case: (a) $Q/Q_0 \sim 6/r^3$; (b) $Q/Q_0 \sim 6/r^4$. Thus the flow intensity in the layer with thermally insulated boundaries is lower than that in the first situation and decays more rapidly as r increases.

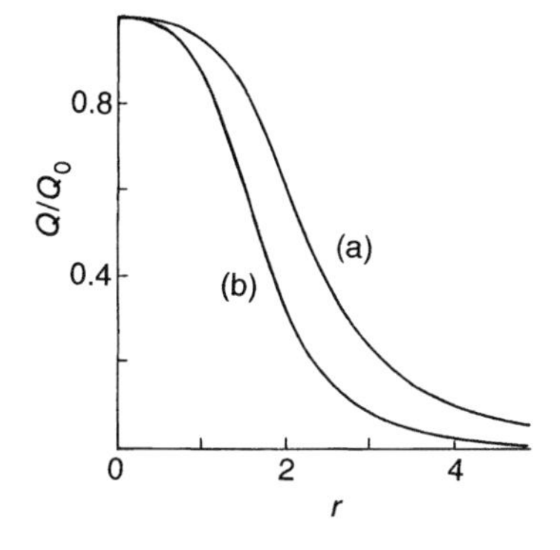

Fig. 51 The relative volumetric flow rate versus the vibrational parameter through the layer with (a) boundaries of high heat conductivity and (b) thermally insulated boundaries

Let us discuss the effect of transverse vibration with the vector $\mathbf{n}(1, 0, 0)$. There is the solution that describes the plane-parallel flow of the type (2.32). Instead of the set (2.33), one can obtain

$$\nu v_0'' - g\beta A x = 0, \qquad \chi \tau_0'' = A v_0, \qquad w_0' = -A. \tag{2.43}$$

The first two equations, together with the appropriate boundary conditions for v_0 and τ_0, give the velocity and temperature profiles which coincide with (2.38) and (2.39) in the absence of vibration. Thus, the transverse vibration does not affect a gravitational advective flow. However, it gives rise to the longitudinal pulsation velocity with the amplitude in the form $w_0 = -Ax$.

Finally, let the vibration axis be horizontal and perpendicular to the temperature gradient: $\mathbf{n}(0, 1, 0)$. Now the $\mathbf{w}_0$ field has the structure $w_x = w_z = 0$, $w_y = w_0(x, z)$. As in the case of transverse vibrations, the velocity and temperature profiles in a plane-parallel flow do not depend on r and are specified by (2.38) and (2.39). The presence of vibrations leads to the y component of the pulsation velocity in the form $w_0 = Az + \tau_0(x)$.

3.2 Stability

Even without vibrations, advective flows possess a number of instability mechanisms and shapes of critical perturbations (see [9, 10]). In the case of highly heat conductive boundaries in the range of small Prandtl numbers, the instability is attributed to the development of plane perturbations in the form of a periodic function in the z direction set of stagnant vortices located on the mutual boundary of the counter-flows. This instability is of pure hydrodynamic (inviscid) origin and as the Prandtl number grows, it is sharply suppressed due to the stable stratification occurring in the region where the perturbations develop. In the narrow interval of small Prandtl numbers, spiral oscillatory perturbations in the form of rolls with axes parallel to the basic velocity are responsible for the crisis of flow. These perturbations, by their nature, are involved in internal waves within the stable stratification region.

At moderate and high values of the Prandtl number, the instability stems from the existence of areas with unstable stratification close to horizontal plates. The instability is of the Rayleigh type and is connected both with plane drifting perturbations and with spiral stagnant ones which are the most dangerous. If the boundaries of the layer are thermally insulated, there are no zones with unstable stratification. At small Prandtl numbers, the hydrodynamic and internal wave modes are responsible for the instability.

Under longitudinal vibrations, the velocity and the temperature profiles quantitatively change (Figs. 49 and 50), but their structure remains qualitatively the same. Thus one might expect that mechanisms of instability are in evidence as well.

The stability of an advective flow under longitudinal vibrations was studied by Katanova [11]. The stability borders were determined with respect to plane perturbations in the layer with highly heat conductive plates. The amplitude problem for normal perturbations has been written in terms of (φ, f, θ), where φ and f are the amplitudes of stream functions for perturbations of the average and pulsation velocities and θ is the amplitude of the temperature perturbation:

$$\Delta^2\varphi + \mathrm{i}k\,Gr\left(v_0''\varphi - v_0\Delta\varphi\right) - \mathrm{i}k\theta + \frac{Gr_{\mathrm{v}}}{Gr}\left[\mathrm{i}k\tau_0'(\theta - f') + f''\right] = -\lambda\Delta\varphi$$

$$\frac{1}{Pr}\Delta\theta + Gr\left[\mathrm{i}k\left(\tau_0'\varphi - v_0\theta\right) - \varphi'\right] = -\lambda\theta \qquad (2.44)$$

$$\Delta f = \theta'$$

$$x = \pm 1: \qquad \varphi = \varphi' = 0, \qquad f = 0, \qquad \theta = 0 \qquad (2.45)$$

Here all the variables are dimensionless including the profiles v_0 and τ_0, as the units of length, time, velocity and temperature, h, h^2/ν, $g\beta Ah^3/\nu$, Ah, respectively, are used. The Grashof number and its vibrational analog are defined through the longitudinal temperature gradient: $Gr = g\beta Ah^4/\nu^2$, $Gr_{\mathrm{v}} = (b\beta\Omega Ah^2)^2/(2\nu^2)$. In (2.44) and (2.45), λ is the decrement, k is the wavenumber, where a prime denotes differentiation with respect to the dimensionless transversal coordinate, and Δ is the operator of the form $\Delta = d^2/dx^2 - k^2$.

The spectral amplitude problem (2.44) and (2.45) was solved in [11] by the differential factorization method. The main results are displayed in Figs. 52 and 53. At small values of the Prandtl number, the instability is of a hydrodynamic origin. In Fig. 52 the neutral curves are shown for $Pr = 0.1$. It may be seen that,

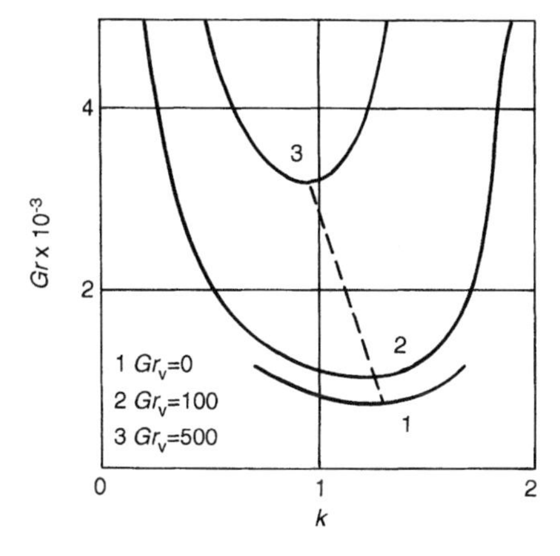

Fig. 52 Neutral curves for the hydrodynamic mode

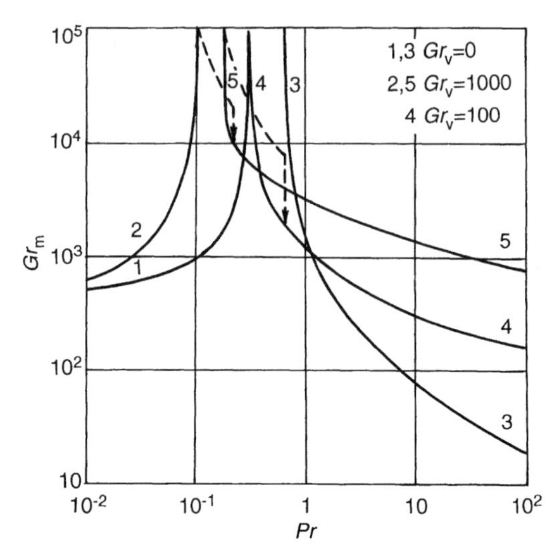

Fig. 53 Stability curves in the plane (Pr, Gr_m): 1, 2, the hydrodynamic mode; 3–5, the Rayleigh-type mode

as the intensity of vibrations increases, the stability grows and the minima of neutral curves shift to the long-wave (small k) range. The instability is monotonic: $\lambda_i = 0$. At large Prandtl numbers, the instability is of a stratification origin and is associated with formation of convective cells in the unstable parts of the layer close to its rigid boundaries (in the absence of vibrations the hydrodynamic mode is stabilized at $Pr = 0.32$). The plane Rayleigh mode is of an oscillatory form and develops as a set of cells drifting through the basic flow rightwards or leftwards. These two alternatives differ only in the sign of the phase velocity, i.e. by localization either in the heated flow (the phase velocity is negative) or in the cooled one (the phase velocity is positive). From the viewpoint of stability, these modifications are equivalent; they correspond to the same critical Grashof number. At large Pr values, as the vibration intensity grows, the stability enhances but curtailment of the critical wavelength is seen.

Figure 53 illustrates several total stability curves in the plane (Pr, Gr_m); minimization is done with respect to a wavenumber. At small and large values of the Prandtl number, the stabilization effect with regard to the hydrodynamic and Rayleigh modes is clearly visible. In the intermediate range $0.1 < Pr < 10$, the pattern is more complicated. Neutral curves inherent in the Rayleigh mode have two minima and, as Pr varies, the transition from one minimum to another takes place. Breaks and dashed parts on curves 4 and 5 reflect this fact.

3.3 Tilted Axis of Vibration

The problem considered in this section was generalized by Birikh [12] in two aspects. First, the layer was assumed to be oriented arbitrarily with respect to gravity. Second, the vibration axis was directed arbitrarily with respect to the layer. However, the temperature gradient was invariably longitudinal. The case of an inclined vibration axis is discussed below.

Let the horizontal fluid layer be subject to a longitudinal temperature gradient and vibrations. Moreover, the vibration axis is inclined at an arbitrary angle to the temperature gradient. All three vectors, viz. γ, $\mathbf{n}$ and ∇T, are assumed to be coplanar. Through its components, the unit vector $\mathbf{n}$ can be written as $\mathbf{n}(n_x, 0, n_z)$. It is easy to verify that in this situation there exists a solution of the type (2.32) describing a plane-parallel flow. Taking into account the oddness of the profiles, the set of equations for the functions v_0, τ_0 and w_0 takes the form

$$
\begin{aligned}
&\nu v_0'' + \varepsilon A n_z^2 \tau_0 - \left(g\beta A + \varepsilon A^2 n_x n_z\right)x = 0 \\
&A v_0 = \chi \tau_0'', \qquad w_0 = -A n_x x + n_z \tau_0
\end{aligned}
\tag{2.46}
$$

with appropriate boundary conditions. Evidently, in the case of longitudinal vibrations ($n_x = 0, n_z = 1$) the problem is reduced to that of Section 3.1. If $n_x = 1, n_z = 0$ (transverse vibrations), the profiles v_0 and τ_0 remain the same as in the absence of vibrations, but now one has a non-zero pulsation velocity $w_0 = -Ax$ as well.

The boundaries of the layer are assumed to be highly conductive: $\tau_0(\pm h) = 0$. Solution (2.46) yields velocity and temperature distributions formally coincident with (2.34) and (2.35). To pass from these formulas to the solution of (2.46), it is necessary to define the parameter r as $r = \left[\varepsilon A^2 n_z^2 h^4/(4\nu\chi)\right]^{1/4}$ and perform the replacement $g\beta A \to p = g\beta A + \varepsilon A^2 n_x n_z$. In other words, the existence of a transversal component of the vector $\mathbf{n}$ leads to the exchange of the velocity and temperature profiles.

It is interesting that, unlike purely longitudinal or purely transverse vibration, in the case of a tilted vibration axis for finite n_x and n_z there is a flow component of the thermovibrational origin existing in weightlessness where $g = 0$. To arrive at the solution, one has to substitute $g\beta A \to p = \varepsilon A^2 n_x n_z$ in (2.34) and (2.35). In the general case, the flow appears to be a superposition of thermogravitational (advective) and thermovibrational components. Their relative contribution is determined by the relationship between the terms $g\beta A$ and $\varepsilon A^2 n_x n_z$. Setting for certainty $n_z > 0$, one finds that at $n_x > 0$ the vibrational component intensifies advection while at $n_x < 0$ its main effect is damping.

If the boundaries of the layer are thermally insulated, the picture is analogous.

4 Vibrational Convective Flows in Weightlessness

4.1 Boundary Layer of Constant Thickness

In the present section some vibrational convective flows developing under longitudinal vibrations and a transversal temperature difference in weightlessness are examined [13, 14].

Let us consider the problem of a vibrational convective boundary layer of constant thickness. Such a layer forms close to an infinite plane plate embedded in an infinite fluid. All of the system undergoes high-frequency oscillations along the z axis directed along the plate. Far from the plate, in the fluid a constant temperature gradient is imposed in the direction of the vibration axis:

$$x \to \infty: \qquad T_\infty = Az \tag{2.47}$$

where x is the transversal coordinate; the origin of the coordinate system is on the plate. At the surface of the plate, the same temperature gradient is maintained. Besides, a constant difference of temperature is present between the plate and the fluid at infinity:

$$x = 0: \qquad T_w = Az + \Theta \tag{2.48}$$

The solution describing a flow of the plane-parallel type is available:

$$\begin{aligned} v_x = v_y = 0, \qquad v_z = v_0(x); \qquad T = Az + \tau_0(x) \\ w_x = w_y = 0, \qquad w_z = w_0(x); \qquad p = \text{constant} \end{aligned} \tag{2.49}$$

where functions v_0, τ_0 and w_0 satisfy the equations

$$\nu v_0'' + \varepsilon A \tau_0 = 0, \qquad \chi \tau_0'' - A v_0 = 0, \qquad w_0 = \tau_0 \tag{2.50}$$

with the boundary conditions

$$\begin{aligned} x = 0: \qquad v_0 = 0, \qquad \tau_0 = \Theta \\ x \to \infty: \qquad v_0 \to 0, \qquad \tau_0 \to 0 \end{aligned} \tag{2.51}$$

The solution is of the form:

$$\begin{aligned} v_0 = \Theta \sqrt{\frac{\varepsilon \chi}{\nu}} \frac{A}{|A|} \exp(-\gamma x) \sin \gamma x \\ \tau_0 = w_0 = \Theta \exp(-\gamma x) \cos \gamma x, \qquad \gamma = \left(\frac{\varepsilon A^2}{4 \nu \chi} \right)^{1/4} \end{aligned} \tag{2.52}$$

Therefore, a boundary layer of a plane-parallel flow forms in the vicinity of the plate. Its thickness is constant and of the order of $1/\gamma$, i.e. it decreases as the gradient A (or the vibrational parameter ε) grows. At $\Theta > 0$ in the wall-adjacent (main) part the flow moves along the longitudinal temperature gradient. The maximal velocity is

$$v_{\mathrm{m}} = 0.3224\Theta\sqrt{\frac{\varepsilon\chi}{\nu}} \tag{2.53}$$

and it is achieved at the distance $x_{\mathrm{m}} = \pi/4\gamma$ from the plate.

The solution may be easily generalized to account for the presence of static gravity. If the plate is vertical, the z axis and the temperature gradient point upwards, and the projections of the gravity acceleration vector are $\mathbf{g}(0, 0, -g)$. In this situation (2.52) still holds but parameter γ now equals

$$\gamma = \left(\frac{\varepsilon A^2 + g\beta A}{4\nu\chi}\right)^{1/4} \tag{2.54}$$

If there are no vibrations ($\varepsilon = 0$), the solution is derived describing the convective boundary layer of Gill [15] which arises near the plate in the stably stratified fluid. When vibrations exist and the system is heated from above ($A > 0$), the boundary layer exists at any value of A. In the opposite case of heating from below ($A < 0$) the solution in the form of a boundary layer of constant thickness persists under condition $\varepsilon|A| > g\beta$.

4.2 Antisymmetrical and Symmetrical Flows in the Layer

Below we consider a layer of finite thickness bounded by rigid parallel plates $x = \pm h$. Two exact solutions may be found describing plane-parallel thermo-vibrational convection flows. The first one is the flow under a constant temperature gradient maintained along the plates but in such a manner that the temperature difference 2Θ remains constant:

$$x = \pm h: \qquad T_0 = Az \mp \Theta \tag{2.55}$$

In this case the flow has the form (2.49). Imposing the condition of closeness, one gets the antisymmetrical profiles:

$$v_0 = \frac{\Theta}{\Delta_1}\sqrt{\frac{\varepsilon\chi}{\nu}}\left(\frac{\cosh r\xi \sin r\xi}{\cosh r \sin r} - \frac{\sinh r\xi \cos r\xi}{\sinh r \cos r}\right) \tag{2.56}$$

$$\tau_0 = w_0 = -\frac{\Theta}{\Delta_1}\left(\frac{\cosh r\xi \sin r\xi}{\sinh r \cos r} + \frac{\sinh r\xi \cos r\xi}{\cosh r \sin r}\right) \tag{2.57}$$

Here $\xi = x/h$, $r = [\varepsilon A^2 h^4/(4\nu\chi)]^{1/4}$, $\Delta_1 = \tanh r \operatorname{cotan} r + \operatorname{cotanh} r \tan r$.

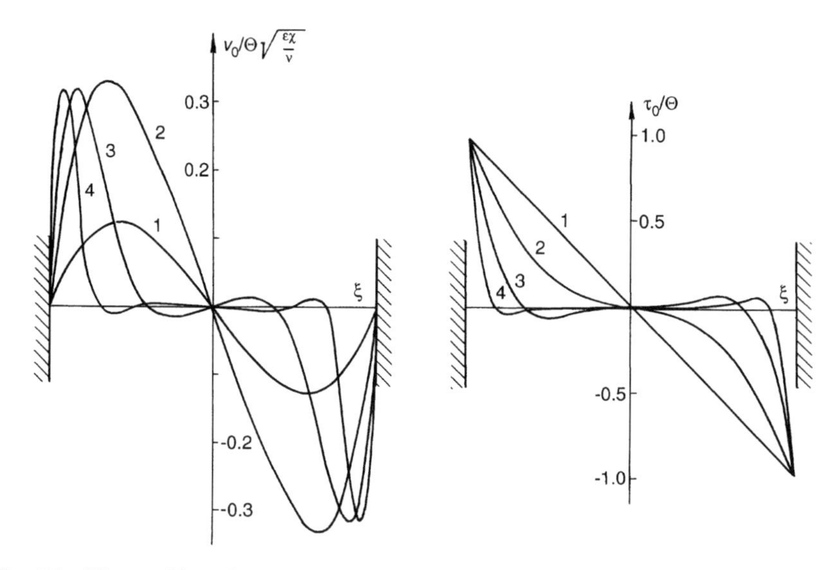

Fig. 54 The profiles of (a) the velocity and (b) temperature for the antisymmetrical flow. Curves 1–4 correspond to the parameter r values, respectively, of 1, 2.254, 5, 10

When the longitudinal gradient is absent ($r \to 0$), the solution (2.56) and (2.57) governs the equilibrium ($v_0 = 0$) of a plane fluid layer with a linear transversal temperature distribution under longitudinal vibrations. The development of temperature and velocity profiles with the parameter r is shown in Fig. 54. At large r the flow acquires the character of uncoupled boundary layers whose structure is coincident with that of section 4.1 above.

As an integral measure of intensity of a vibrational convective flow, one may choose the volumetric flow rate Q_1. It renders the flux through a half-layer section per unit length along the y axis:

$$Q_1 = \int_{-h}^{0} v_0 \, \mathrm{d}x = \Theta h \sqrt{\frac{\varepsilon\chi}{\nu}} f_1(r) \tag{2.58}$$

$$f_1(r) = \frac{1}{2r\Delta_1}\left(\tan r + \operatorname{cotan} r - \tanh r + \operatorname{cotanh} r - \frac{1}{\sinh r \cos r} - \frac{1}{\cosh r \sin r}\right)$$

At small r values, the volumetric flow rate is proportional to the longitudinal gradient A because $f_1 \approx r^2/12$. At large r values, $f_1 \approx 1/2r$, i.e. $Q_1 \sim A^{-1/2}$. Function $f_1(r)$ has its maximum at $r = 2.254$ and equals $f_{1\mathrm{m}} = 0.2086$. The corresponding velocity and temperature profiles are shown in Fig. 54, curve 2.

The second solution yields a symmetrical 'thermosyphone' flow occurring in an infinite layer with boundary conditions:

$$x = \pm h: \qquad T_0 = Az + \Theta \tag{2.59}$$

Thus, a fixed temperature gradient is imposed at the plates, the transversal temperature difference is nil, but the boundary temperature differs by Θ from that of the ambient medium. In contrast to (2.56), the flow is not closed and the velocity and temperature profiles are even functions of x:

$$v_0 = \frac{\Theta}{\Delta_2}\sqrt{\frac{\varepsilon\chi}{\nu}}\left(\frac{\cosh r\xi \cos r\xi}{\cosh r \cos r} - \frac{\sinh r\xi \sin r\xi}{\sinh r \sin r}\right) \tag{2.60}$$

$$\tau_0 = w_0 = -\frac{\Theta}{\Delta_2}\left(\frac{\cosh r\xi \cos r\xi}{\sinh r \sin r} + \frac{\sinh r\xi \sin r\xi}{\cosh r \cos r}\right) \tag{2.61}$$

Here $\Delta_2 = \tanh r \tan r + \operatorname{cotanh} r \operatorname{cotan} r$. The velocity profiles are presented in Fig. 55. At large r the uncoupled boundary layers of the Gill type are produced in the neighborhood of the plates. The integral intensity can be characterized by the total volumetric flow rate through the channel's section:

$$\begin{gathered} Q_2 = \Theta h\sqrt{\frac{\varepsilon\chi}{\nu}}\, f_2(r) \\ f_2(r) = \frac{1}{r\Delta_2}(\tan\, r + \operatorname{cotan}\, r + \tanh\, r - \operatorname{cotanh}\, r) \end{gathered} \tag{2.62}$$

At small and large r values the asymptotic behavior is given, respectively, by $f_2 \approx 4r^2/3$ and $f_2 \approx 1/r$. Hence, as in the case of antisymmetrical flow, the

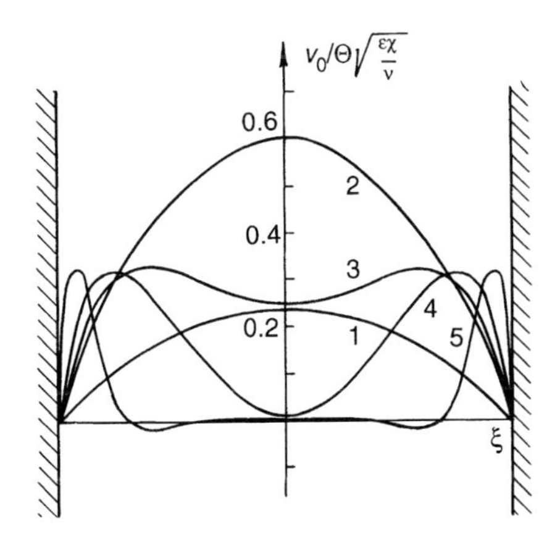

Fig. 55 The velocity profiles for a symmetrical flow. Curves 1–5 correspond to the parameter r values, respectively, of 0.5, 1.127, 2, 3, 10

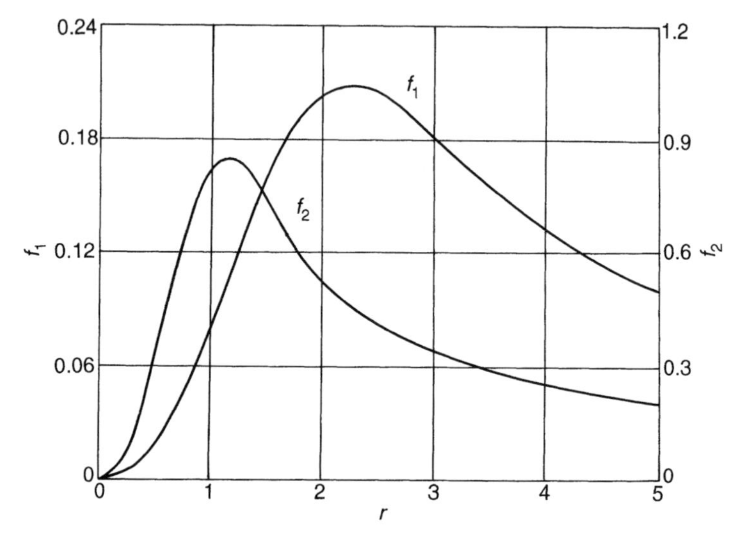

Fig. 56 Functions $f_1(r)$ and $f_2(r)$ determining the intensities of the odd and even flows, respectively

intensity increases with A at small r values and decreases with $A^{-1/2}$ at large r. The maximal value $f_{2\text{m}} = 0.8345$ is attained at $r_\text{m} = 1.127$ (Fig. 55, curve 2). The functions $f_1(r)$ and $f_2(r)$ determining the intensity of even and odd flows are plotted in Fig. 56.

Let us estimate the axial velocity of the vibrational convective flow of the 'thermosyphone' type when the volumetric flow rate is maximal, i.e. at $r = 1.127$. We set the frequency at 100 Hz, the amplitude of vibration at 1 mm, the temperature difference at 20° and the velocities for layers of water and air, respectively, at 0.04 cm/s and 2 cm/s. If the half-thickness of the channel equals 1 cm, the volumetric flow rate maximizes when the temperature gradients in the water and air layers are 1 degree/cm and 3 degree/cm, respectively. From the viewpoint of the experimental realization of the flow the obtained estimations are quite reasonable.

As in the boundary layer problem, the solutions may be generalized to account for the situation where the gravity is present. If the layer is vertical and the longitudinal temperature gradient corresponds to the stable stratification, the solutions retain their form but the parameter r is now equal to

$$r = \left(\frac{\varepsilon A^2 + g\beta A}{4\nu\chi} \right)^{1/4} h \tag{2.63}$$

4.3 Stability of an Antisymmetrical Flow

The stability of a closed vibrational convective flow with odd profiles (2.56) and (2.57) was investigated in [16]. It is impossible to find the transformation enabling the characteristics of a three-dimensional problem to be expressed through those of a two-dimensional one. However, it may be shown that the flow is stable in the limit of spiral three-dimensional perturbations. This gives a certain reason to expect that the plane mode is the most dangerous. We consider this case below.

Let us write down the spectral amplitude problem in terms of φ, f, θ which are, respectively, the amplitudes of the normal perturbations of stream functions of the average and pulsation velocity components and temperature:

$$\Delta^2\varphi + \mathrm{i}k\widetilde{Gr}_\mathrm{v}\left(v_0''\varphi - v_0\Delta\varphi\right) + 2r^2 f'' + \mathrm{i}k\widetilde{Gr}_\mathrm{v}\, Pr\,\tau_0'(\theta - f') = -\lambda\Delta\varphi$$

$$\frac{1}{Pr}\Delta\theta + \mathrm{i}k\widetilde{Gr}_\mathrm{v}\left(\tau_0'\varphi - v_0\theta\right) - \frac{2r^2}{Pr}\varphi' = -\lambda\theta \tag{2.64}$$

$$\Delta f = \theta'$$

$$x = \pm 1: \qquad \varphi = \varphi' = 0, \qquad f = 0, \qquad \theta = 0 \tag{2.65}$$

Here all the variables are dimensionless; the units of length, time, velocity and temperature are taken as h, h^2/ν, $\Theta\sqrt{\varepsilon\chi/\nu}$ and Θ, respectively. The problem includes the Prandtl number, the dimensionless parameter $r = \left[\varepsilon A^2 h^4/(4\nu\chi)\right]^{1/4}$ characterizing the longitudinal temperature gradient and the criterion $\widetilde{Gr}_\mathrm{v} = \Theta h\sqrt{\varepsilon\chi/\nu^3}$ defined through the transversal temperature difference. The parameter r is related to the vibrational Rayleigh number by the dependence $r = (Ra_\mathrm{v}/4)^{1/4}$. In its turn, $\widetilde{Gr}_\mathrm{v}$ differs from the vibrational Grashof number used in the previous sections according to $\widetilde{Gr}_\mathrm{v}^2 = Gr_\mathrm{v}/Pr$.

The problem was solved by the Runge–Kutta–Merson method with the aid of a step-by-step orthogonalization procedure. The main computational results are discussed below.

In the limiting case of no longitudinal temperature gradient ($r \to 0$), the problem is reduced to that of the mechanical quasi-equilibrium stability in a fluid layer with transversal temperature difference and longitudinal vibration axis (see Section 3 in Chapter 1). The instability is not associated with the dynamics and is of a static origin. The convection sets in when the vibrational Rayleigh number exceeds the critical value equal to 2129 (Ra_v is defined through the full temperature difference and the thickness of the layer). Hence the critical value of $\widetilde{Gr}_\mathrm{v}$ is

$$\widetilde{Gr}_\mathrm{v} = \frac{11.54}{Pr} \tag{2.66}$$

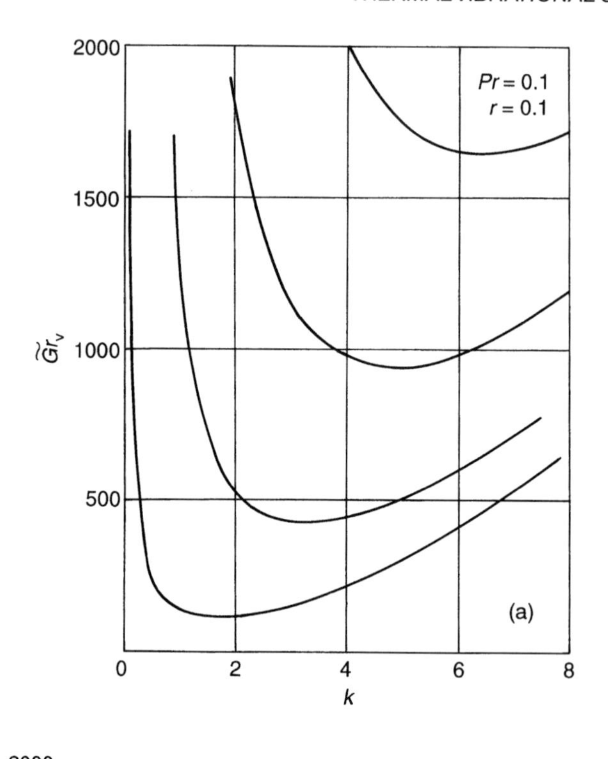

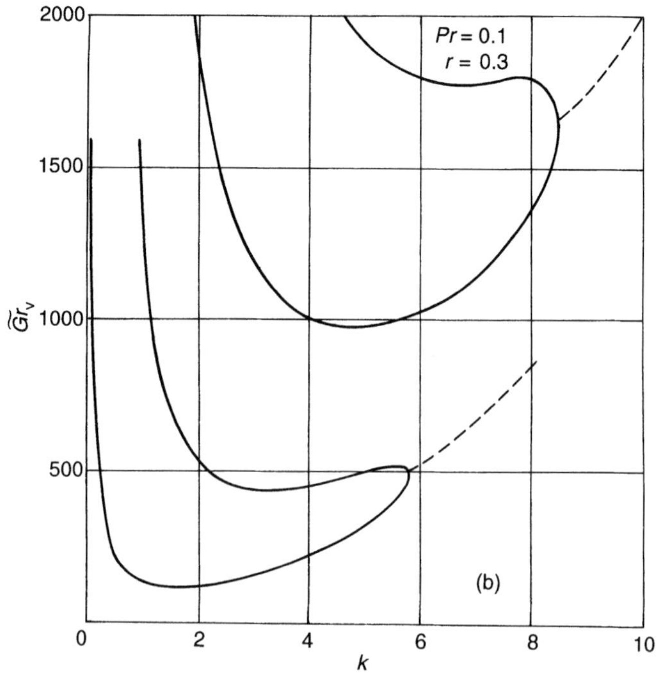

Fig. 57 (*caption opposite*)

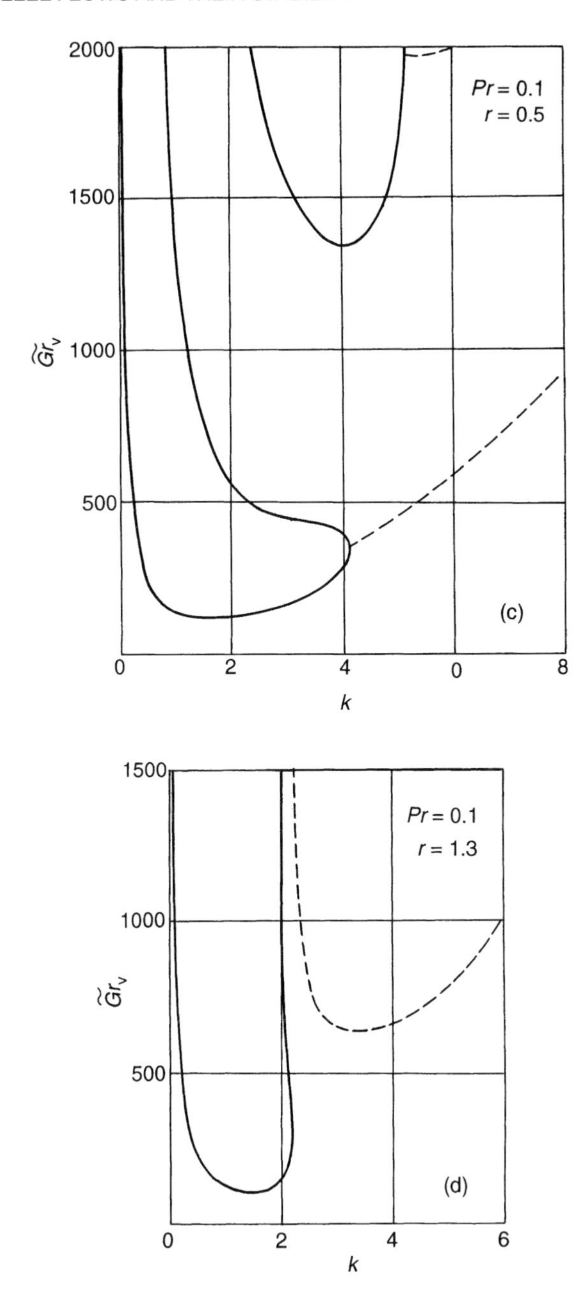

Fig. 57 The neutral curves for the lowest levels of the instability spectrum at $Pr = 0.1$. Solid lines, monotonic mode; dashed lines, oscillatory mode. (a) $r = 0.1$, (b) $r = 0.3$, (c) $r = 0.5$, (d) $r = 1.3$

At small r the instability retains its static (stratification) nature; the weak flow does not change the instability mechanism and just produces some minor shifts of the critical parameters. As r increases, the instability changes its character, since new mechanisms begin to develop: the hydrodynamic one manifesting itself by formation of vortices on the counter-flow boundary and the wave one striving to amplify the temperature waves. A complicated reorganization takes place in the spectra of the characteristic perturbations and the structure of the neutral curves. The Prandtl number is of considerable importance in the process.

Figure 57a to d exemplifies the evolution of the neutral curves with r for lower levels of the instability spectrum in the plane $(k, \widetilde{Gr}_v)$ for a relatively small fixed value of the Prandtl number ($Pr = 0.1$). When $r = 0.1$, the neutral curves do not move far from those in equilibrium $r = 0$; all the curves correspond to the monotonic instability. At $r = 0.3$, the characteristic feature of the instability spectrum turns up: as k augments, the fusion of two monotonic branches occurs, yielding the oscillatory instability border shown by dashed curves. Each point on such an oscillatory branch represents two critical wave perturbations, differing solely by the sign of their phase velocities. Upon further increase of the longitudinal gradient (the parameter r, that is the same) the oscillatory instability range extends towards the range of long wavelengths (see the figure for $r = 0.5$). Given that, the third and fourth levels of the spectrum are forced out to the range of high $\widetilde{Gr}_v$ values. Yet further, separation occurs into monotonic and oscillatory branches. At $r = 1.3$, the oscillatory curve goes above the monotonic one, and thus the monotonic mode is the most dangerous. However, later on the situation changes: for $r > 2$ the monotonic mode is sharply stabilized and the wave instability becomes the most dangerous.

The dependence of the critical parameter $\widetilde{Gr}_{vm}(r)$ minimized with respect to the wavenumber k is summarized in Fig. 58 for small values of the Prandtl number. In the range of small r, there exists a family of curves depending on Pr due to the fact that formula (2.66) is valid. If r assumes intermediate values ($r \sim 1$–2), the curves of the family come closer to one another and the significance of the parameter $\widetilde{Gr}_v$ grows, reflecting the hydrodynamic nature of the crisis. At $r \approx 2.3$–2.4, the hydrodynamic mode is stabilized, and at higher values of r the instability is of the wave origin. The phase velocity of the critical wave perturbations is of the order of the extremal velocity of the basic flow that is typical of the wave instability (see [9]).

In the range of large Prandtl numbers, the asymptotic $\widetilde{Gr}_{vm} \sim 1/Pr$ holds, i.e. the crisis takes place upon exceeding the critical value of the parameter $\widetilde{Ra}_{vm} = \widetilde{Gr}_{vm} Pr$. The existence of such an a symptote is supported by the computational results [16]; in addition, it could be deduced from the basic amplitude problem (2.65). Indeed, let us consider the limiting case $Pr \to \infty$. Let $\widetilde{Gr}_v \, Pr =$ constant and we introduce the renormalized decrement $\overline{\lambda} = \lambda Pr$.

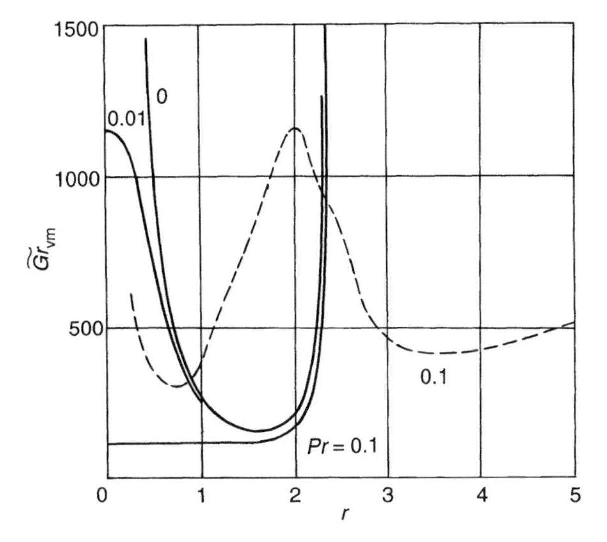

Fig. 58 The stability borders at small Prandtl numbers. Solid lines, steady mode; dashed lines, wave mode

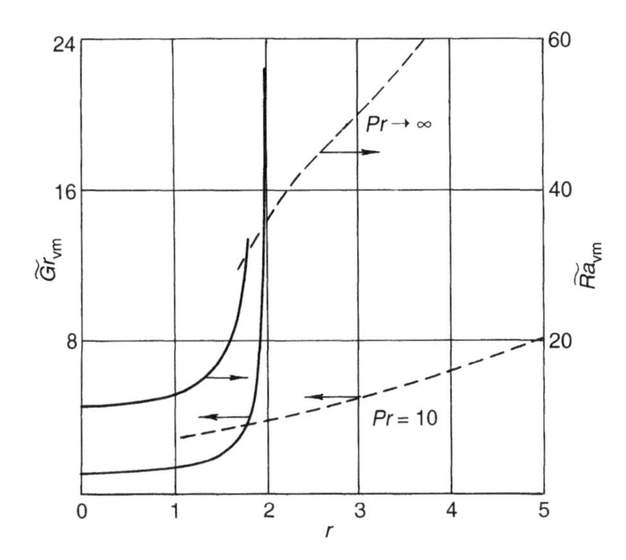

Fig. 59 The stability borders for steady and wave modes at large Prandtl numbers

Retaining the leading terms in the first equation of the set (2.65), one arrives at the asymptotic spectral problem

$$\begin{gathered}
\Delta^2\varphi + 2r^2 f'' + \mathrm{i}k\widetilde{Ra}_{\mathrm{v}}\,\tau_0'(\theta - f') = 0 \\
\Delta\theta + \mathrm{i}k\widetilde{Ra}_{\mathrm{v}}\left(\tau_0'\varphi - v_0\theta\right) - 2r^2\varphi' = -\lambda\theta \qquad (2.67) \\
\Delta f = \theta' \\
x = \pm 1: \qquad \varphi = \varphi' = 0, \qquad f = 0, \qquad \theta = 0
\end{gathered}$$

From (2.67) it is seen that $\widetilde{Ra}_{\mathrm{v}}$ is the governing parameter. In Fig. 59 the stability border is drawn as a function of $\widetilde{Ra}_{\mathrm{vm}}(r)$. This curve is obtained both from approaching the limit $Pr \to \infty$ in the numerical solution of (2.65) and from direct solution of the asymptotic problem (2.67). For comparison, the stability border at $Pr = 10$ is shown as well. From Fig. 59 it follows that for $r \approx 2$ the hydrodynamic mode is stabilized and the instability is transmitted to the wave mode.

Thus, in an antisymmetrical vibrational convective flow three types of instability might develop—static (stratification), hydrodynamic and wave—depending on the values of the governing parameters

References

1. Gershuni, G.Z., E.M. Zhukhovitsky and V.M. Shikhov. The stability of convective flow in a vertical layer subject to transversal vibration, in *Convective Flows*, Perm, 1987, pp. 18–24.
2. Squire, H.B. On the stability for three-dimensional disturbances of viscous fluid flow between parallel walls. *Proc. Roy. Soc.*, 1933, **A142** (847), 621–8.
3. Sharifulin, A.N. The stability of convective motion in a vertical layer subject to longitudinal vibration, *Trans. Acad. Sci. SSSR (Izvestiya), Mech. Zhidk. Gasa*, 1983, **2**, 186–8.
4. Sharifulin, A.N. The wavy instability of free convection flow in vibrational field, in *Nonstationary Processes in Fluids and Solids*, Sverdlovsk, 1983, pp. 58–62.
5. Takashima, M. and H. Hamabata. The stability of natural convection in a vertical layer of dielectric fluid in the presence of a horizontal ac electric field, *J. Phys. Soc. Japan*, 1984, **53**(5), 1728–36.
6. Gershuni, G.Z. and E.M. Zhukhovitsky. On the stability of convective flow in vibrational field with respect to three-dimensional disturbances, *Trans. Acad. Sci. SSSR (Izvestiya), Mech. Zhidk. Gasa*, 1988, **2**, 116–22.
7. Zavarykin, M.P., S.V. Zorin and G.F. Putin. Experimental study of vibrational convection, *Commun. Acad. Sci. SSSR (Doklady)*, 1985, V. 281, No. 4, 815-816.
8. Gershuni, G.Z. and E.M. Zhukhovitsky. Plane-parallel advective flows in vibrational field, *Engn.-Phys. J.*, 1989, **56**(2), 238–42.
9. Gershuni, G.Z., E.M. Zhukhovitsky and A.A. Nepomnyashchy. *The Stability of Convective Flows*, Nauka, Moscow, 1989, 320 pp.

10. Gershuni, G.Z., P. Laure, V.M. Myznikov, B. Roux and E.M. Zhukhovitsky. On the stability of plane-parallel advective flows in long horizontal layers, *Micrograv. Quart.*, 1992, **2**(3), 141–51.
11. Katanova, T.N. Linear stability of advective flow in high frequency vibrational field, in *Convective Flows*, Perm, 1991, pp. 50–4.
12. Birikh, R.V. On vibrational convection in a plane layer with longitudinal temperature gradient, *Trans. Acad. Sci. SSSR (Izvestiya), Mech. Zhidk. Gasa*, 1990, **4**, 12–15.
13. Gershuni, G.Z. and E.M. Zhukhovitsky. On free thermal convection in vibrational field under weightlessness conditions, *Commun. Acad. Sci. SSSR (Doklady)*, 1979, **249**(3), 580–4.
14. Gershuni, G.Z., E.M. Zhukhovitsky and Yu.S. Yurkov. On vibrational thermal convection under weightlessness conditions, in *Hydromechanics and Heat/Mass Transfer in Weightlessness*, Nauka, Moscow, 1982, pp. 90–8.
15. Gill, A.E. The boundary-layer regime for convection in a rectangular cavity, *J. Fluid Mech.*, 1966, **26**(3), 515–36.
16. Gershuni, G.Z., E.M. Zhukhovitsky and V.M. Shikhov. The stability of vibrational convective fluid flow in a plane layer, *Trans. Acad. Sci. SSSR (Izvestiya), Ser. Phys.*, 1985, **49**(4), 643–8.

3 Non-linear Problems

In this chapter some problems of vibrational convection in closed cavities and in an open volume are considered. When the vibrational Rayleigh number (or the Grashof number) is large enough, the phenomena are of an essentially non-linear character. In some problems considered the effects of equilibrium instability or of steady flow instability are demonstrated. Perhaps the only effective approach to solving such problems is the numerical one.

The first sections of this chapter are devoted to the convective phenomena of a purely vibrational nature, i.e. those that take place when the static gravity is absent (weightlessness conditions). In the last section we discuss the situation where the thermogravitational convection is superimposed with high-frequency vibrations.

1 Rectangular Cavity

Let us begin with the consideration of a vibrational convection in the cavity of the simplest geometry. A fluid fills the rectangular cavity with rigid boundaries (Fig. 60). The heating conditions are as follows. The horizontal parts of the boundaries $y = 0$ and $y = a$ are maintained at constant different temperatures $T = \Theta$ and $T = 0$, respectively. On the lateral boundaries $x = 0$ and $x = la$, the temperature varies linearly with the y coordinate: $T = \Theta(1 - y/a)$. The vibration axis is parallel to the x axis, so the unit vector $\mathbf{n}$ has the projections $\mathbf{n}(1, 0, 0)$. A two-dimensional problem is considered. The only non-zero components are v_x, v_y, w_x and w_y; they depend on x and y as well as on the time t.

In this case, the mechanical quasi-equilibrium is not possible since conditions (1.29) to (1.31) are not valid. (They are satisfied only in the limiting case of an infinitely long layer where $l \to \infty$ and the quasi-equilibrium field $\mathbf{w}_0$ has only the longitudinal component.)

The description of the thermovibrational convective flow is based on the general set of equations (1.17) to (1.20). Here the thermogravitational buoyancy force is not taken into account. Dimensionless variables are introduced adopting

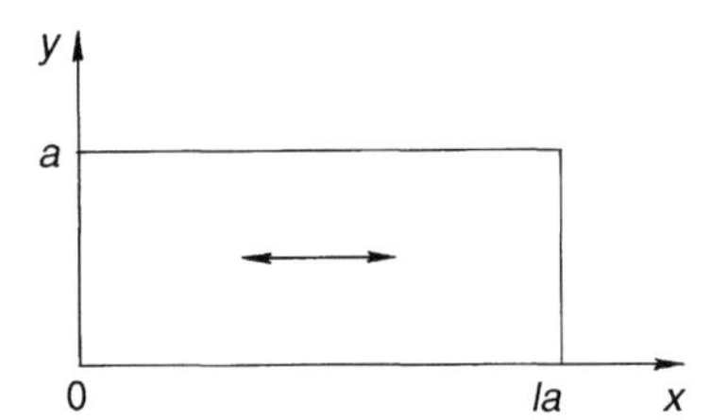

Fig. 60 Rectangular cavity; axes of coordinates

the following units: a for length, a^2/χ for time, χ/a for velocity, Θ for temperature and for the **w** field, and $\rho\nu\chi/a^2$ for pressure. We also introduce stream functions ψ and F of the two-dimensional fields **v** and **w** according to

$$v_x = -\frac{\partial \psi}{\partial y}, \qquad v_y = \frac{\partial \psi}{\partial x}; \qquad w_x = -\frac{\partial F}{\partial y}, \qquad w_y = \frac{\partial F}{\partial x} \tag{3.1}$$

(χ and Θa are taken as the units for ψ and F, respectively). After eliminating the pressure and substituting the stream functions one may obtain a set of equations in terms of ψ, φ and T:

$$\frac{1}{Pr}\left(\frac{\partial \Delta\psi}{\partial t} + \frac{\partial \psi}{\partial x}\frac{\partial \Delta\psi}{\partial y} - \frac{\partial \psi}{\partial y}\frac{\partial \Delta\psi}{\partial x}\right) = \Delta\Delta\psi + Ra_{\mathrm{v}}\left[\frac{\partial T}{\partial x}\frac{\partial}{\partial y}\left(\frac{\partial F}{\partial y}\right) - \frac{\partial T}{\partial y}\frac{\partial}{\partial x}\left(\frac{\partial F}{\partial y}\right)\right] \tag{3.2}$$

$$\frac{\partial T}{\partial t} + \frac{\partial \psi}{\partial x}\frac{\partial T}{\partial y} - \frac{\partial \psi}{\partial y}\frac{\partial T}{\partial x} = \Delta T \tag{3.3}$$

$$\Delta F = -\frac{\partial T}{\partial y} \tag{3.4}$$

Here Δ is the two-dimensional Laplacian and the vibrational Rayleigh number Ra_{v} is defined through the temperature difference Θ and the length of the rectangular cavity side a. The boundary conditions are:

$$\begin{aligned} &x = 0, \quad x = l: \quad \psi = 0, \quad \frac{\partial \psi}{\partial x} = 0, \quad T = 1 - y, \quad F = 0 \\ &y = 0: \quad \psi = 0, \quad \frac{\partial \psi}{\partial y} = 0, \quad T = 1, \quad F = 0 \\ &y = 1: \quad \psi = 0, \quad \frac{\partial \psi}{\partial y} = 0, \quad T = 0, \quad F = 0 \end{aligned} \tag{3.5}$$

This problem was solved by means of the finite difference technique in [1–3]. The additional variable — the vorticity ζ of the velocity field—was introduced as $\zeta = -\Delta\psi$. The grid used inside the integration domain was uniform across both spatial directions. All spatial derivatives were approximated by central difference ratios. Mainly, the implicit finite difference scheme of fractional steps was applied. The associated system of algebraic equations was solved by the ADI method. The Poisson equations for the stream functions were tackled at each time step by the iterative SOR procedure. The steady solutions were obtained as a result of a transient process. Some calculations were performed by means of the explicit finite difference scheme. The results worked out by both schemes practically coincide. For the square cavity ($l = 1$) the spatial grid steps varied from $1/30$ to $1/40$ and for rectangular cavities ($1 < l \leqslant 16$) the grid nodes were equidistant with a $1/20$ step.

By means of the finite difference technique one can determine numerically the instantaneous fields of velocity and temperature as well as the limiting ones corresponding to the settled (steady) flow regime. These distributions allow us to find such important characteristics of the flow as maximal and minimal values of the stream function ψ featuring the intensity of the mean flow, the total value Ψ of all discrete values ψ_{ik} in the grid nodes describing the total angular momentum in the cavity and also the dimensionless heat flux through the cavity—the Nusselt number—which may be obtained as

$$Nu = \frac{1}{2} \oint_{\Gamma} \left| \frac{\partial T}{\partial n} \right| \mathrm{d}l$$

Here $\partial T / \partial n$ is the normal component of the temperature gradient and integration is carried out along the whole boundary of the cavity. In the regime of no convection (pure heat diffusion) it is evident that $Nu = l$. The difference $Nu - l$ is the measure of the convective contribution to the heat transfer.

Let us discuss first the results of calculations concerning the square cavity ($l = 1$) at the fixed Prandtl number $Pr = 1$. As has already been mentioned, the equilibrium is not possible in the case of finite l and the flow exists at arbitrary small Ra_{v} values. In the range of small Ra_{v}, the flow has a weak intensity and the four-vortices structure is symmetrical with respect to the reflection in the planes $x = \frac{1}{2}$ and $y = \frac{1}{2}$ (see Fig. 61a). As Ra_{v} increases, the instability sets in at the critical value $Ra_{\mathrm{v}}^{(\mathrm{A})} \approx 15 \times 10^3$. The secondary stable regime possessing another symmetry bifurcates from the main four-vortices one. It forms through fusion of two diagonal vortices (see Fig. 61b and c). The variants $2b$ and $2c$ are equivalent, and the choice between them is determined by initial disturbances. The secondary flow may be regarded as a one-vortex regime, because it consists of one main vortex formed as a result of fusion of a diagonal pair of vortices and two relatively weak vortices with the opposite direction of circulation. Thus, the set of symmetry elements of the secondary stable regime includes inversion.

The appropriate characteristic monitor of the transition to the secondary regime in numerical experiments is Ψ—the sum of the stream function values

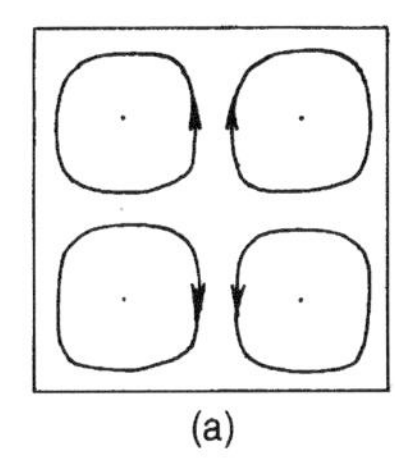
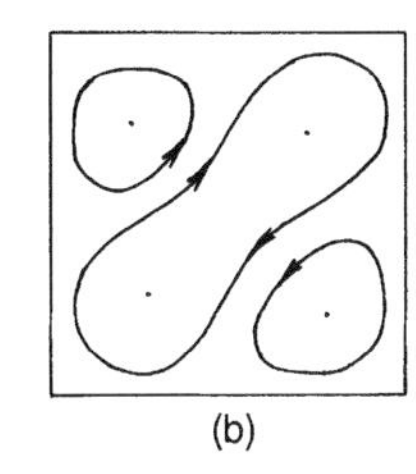
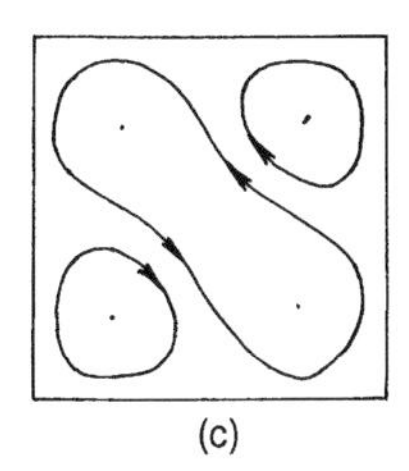

(a) (b) (c)

Fig. 61 Flow structures in the main and secondary regimes. (a) The main four-vortices regime; (b) and (c) two variants of the secondary one-vortex regime

in all the grid nodes. In the main four-vortices regime $\Psi = 0$. The formation of a one-vortex regime leads to destruction of symmetry and consequently to deviation of Ψ from zero. Thus the steady value of Ψ is a measure of the intensity of the secondary flow. The critical point $Ra_v^{(A)}$ may be found by extrapolation of the dependence $\Psi(Ra_v)$ to zero.

The existence of a four-vortices flow in the small Ra_v range is the result of the impossibility of mechanical equilibrium. The intensity of the flow at the critical point $Ra_v^{(A)}$ is small and the crisis is connected properly with the instability of 'deformed' equilibrium but not with instability of the flow. A similar situation ('imperfect bifurcation') has been described many times in the theory of thermogravitational convection (see [4–7]). The case under consideration is in a close analogy with [6], where the symmetry of the flow caused by perturbation of the equilibrium conditions does not coincide with the symmetry of the first critical perturbation.

In Fig. 62 the dependence of the dimensionless heat flux—the Nusselt number—on the vibrational Rayleigh number is presented for the two abovementioned modes. Perhaps the most interesting feature of the bifurcation pattern is the opportunity to observe the main four-vortices regime in the overcritical range $Ra_v > Ra_v^{(A)}$ (branch 1 m). In this range the four-vortices regime should be qualified as a metastable one. The states corresponding to this behavior are, strictly speaking, unstable, i.e. the transition from the metastable four-vortices regime to the stable one-vortex flow sets in after a lapse of time. However, the residence time for the metastable regime is large with respect to the time of transition to the one-vortex flow. It is especially long in the vicinity of the critical point (this is due to the fact that in this range the characteristic decrement responsible for the transition is small). The residence time decreases as Ra_v increases. There is some critical point on the line 1 m in which the residence time becomes compatible with the transition time. In this point the metastable regime becomes absolutely unstable and thus it may never be realized in numerical experiments. In the case under consideration this point is situated in the range of large Rayleigh numbers $Ra_v > 400 \times 10^3$.

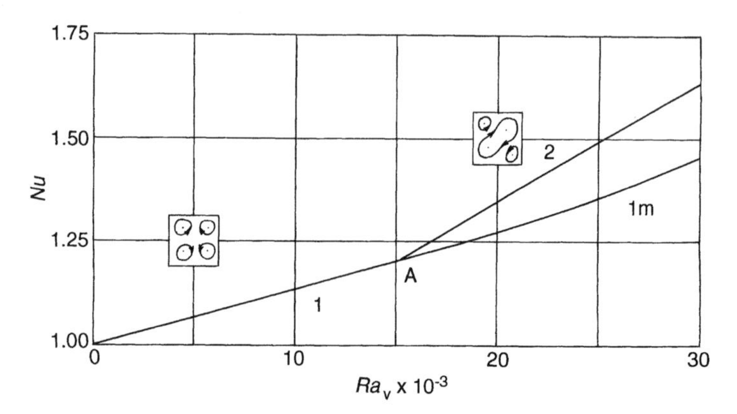

Fig. 62 Dimensionless heat flux as a function of the vibrational Rayleigh number; $Pr = 1$, $l = 1$: 1, stable four-vortices regime; 1 m, metastable four-vortices regime; 2, stable one-vortex regime

In the terminology of the dynamic system theory, one may consider the metastable regime to be a saddle one. Indeed, it is attractive for perturbations with the same symmetry as its own and it is repulsive for perturbations with another symmetry. In our case these are one-vortex perturbations admitting inversion. The presence of the metastable branch in the spectrum of steady regimes, as will be demonstrated in Section 5, affects considerably the transient processes.

In Fig. 63 the curves of heat flux $Nu(Ra_v)$ are shown for the square cavity ($l = 1$) at several values of the Prandtl number. As for the four-vortices metastable regime, it does not depend on Pr in the range $Pr \gg 1$. The only

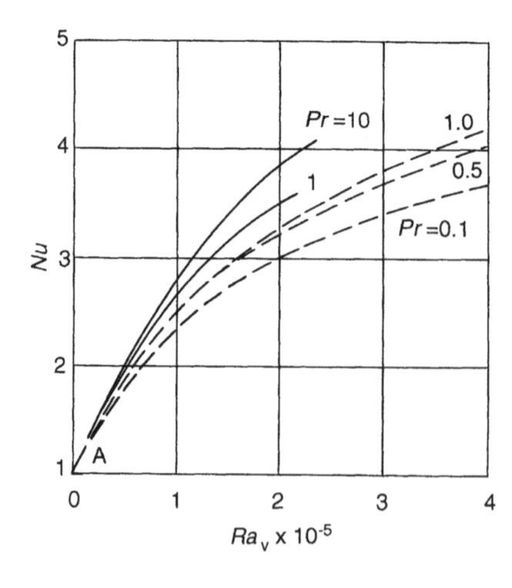

Fig. 63 Heat transfer curves for a square cavity and several values of the Prandtl number. Dashed lines, metastable branches

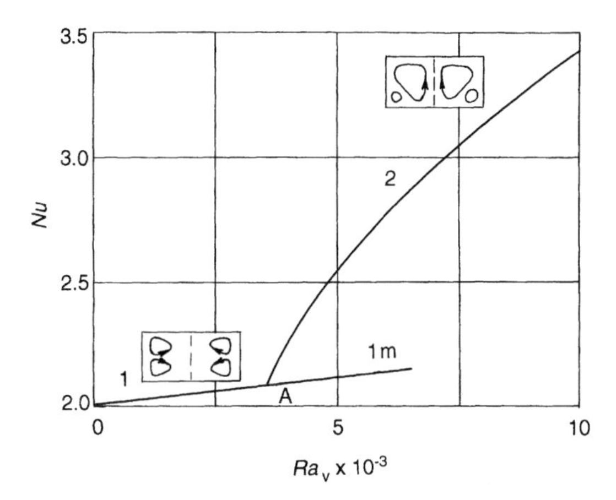

Fig. 64 Dimensionless heat flux as a function of the vibrational Rayleigh number; $Pr = 1$, $l = 2$: 1 and 1 m, stable and metastable main regimes; 2, stable secondary regime

parameter determining the heat transfer is Ra_v. For the stable one-vortex flow, the value $Pr \approx 10$ is the limiting one: when $Pr > 10$ the Nusselt number does not depend on the Prandtl number. The occurrence of some asymptotes not depending on Pr at large Pr could be understood. At large Pr one may neglect inertial terms in the equation of motion. Moreover, the boundary problem (3.2) to (3.5) will not include the parameter Pr at all. Hence, all the characteristics of the flow and heat transfer do not depend on Pr in this limiting case.

Now we give some results for the case of a stretched rectangular cavity ($l \geqslant 2$) for $Pr = 1$. When $l = 2$, the results are qualitatively similar to those for a square cavity. The dependence of the Nusselt number on the vibrational Rayleigh number is given in Fig. 64. The main state of the 'deformed' equilibrium is now related to the existence of four weak vortices. At $Ra_v^{(A)} \approx 3.6 \times 10^3$, due to instability, the secondary two-vortices flow bifurcates from the four-vortices one apparently 'softly'. In the range $Ra_v > Ra_v^{(A)}$, the two-vortices regime is stable. As for the main four-vortices mode, it is metastable in the overcritical range and its residence time decreases as Ra_v grows. At $Ra_v > 6.5 \times 10^3$ this regime becomes absolutely unstable and cannot be realized in a numerical experiment.

When stretching l along the vibration axis, the critical value $Ra_v^{(A)}$ decreases and the metastable part of line 1 shrinks. The situation then approaches the limiting case of an infinite layer where the basic state is a pure equilibrium and its instability leads to formation of a periodic set of cells (Section 3 in Chapter 1).

At $l = 4$ the development of instability may be illustrated by Fig. 65 where streamlines and isotherms are presented for three values of Ra_v. The extremal

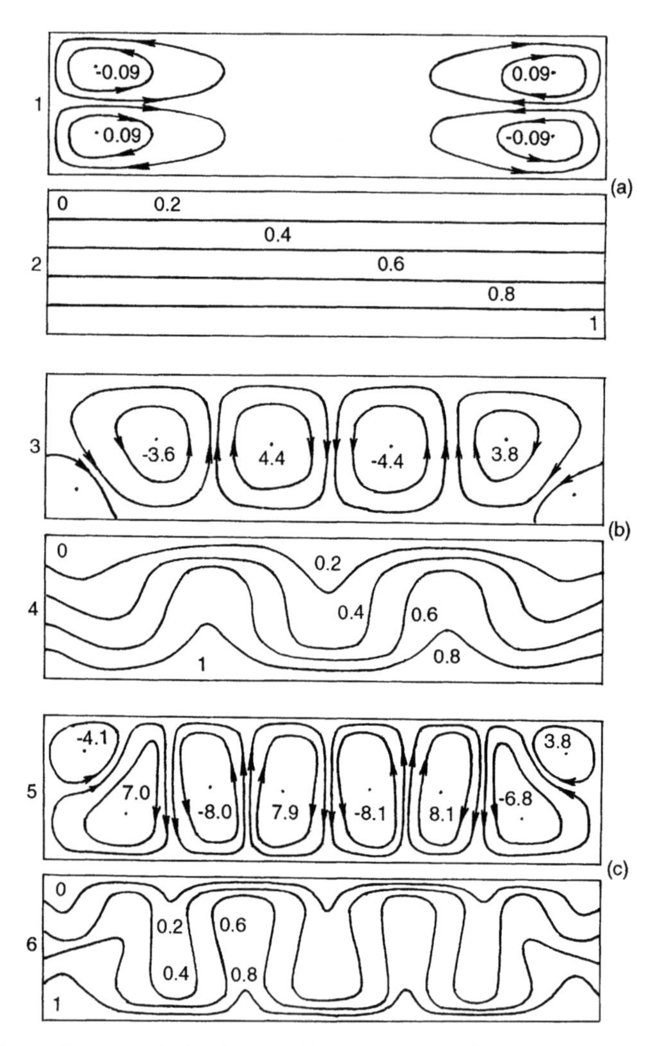

Fig. 65 Streamlines and isotherms for a rectangular cavity; $Pr = 1$, $l = 4$: (a) $Ra_v = 1.5 \times 10^3$, (b) $Ra_v = 10 \times 10^3$, (c) $Ra_v = 100 \times 10^3$

values of the stream function shown in the figure render the extent of intensity of the vortices. At the subcritical value $Ra_v = 1.5 \times 10^3$, a set of four weak vortices appears in the regions close to the ends of the cavity (weak deformation of equilibrium). The temperature field and the heat flux do not deviate practically from their quasi-equilibrium behavior. The developed set of convective cells depends on the supercritical values of Ra_v and the number of cells within the cavity increases with Ra_v.

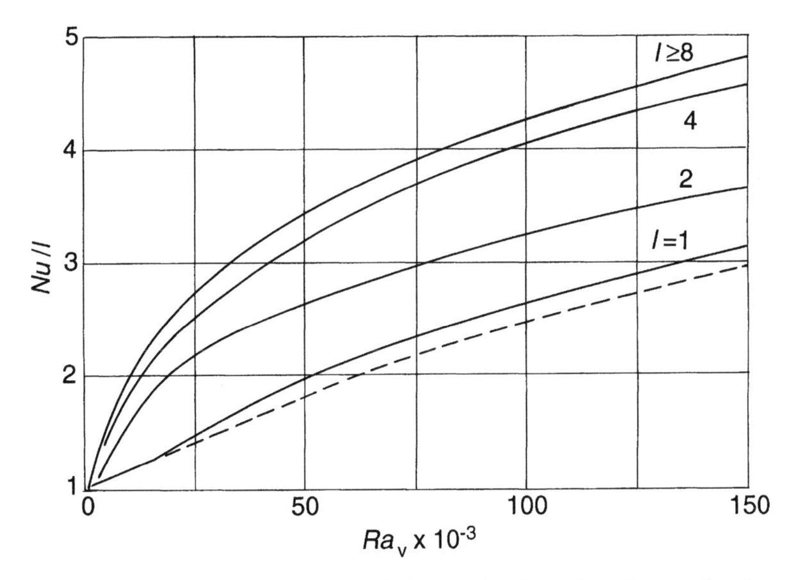

Fig. 66 Dimensionless heat flux per unit length for $Pr = 1$ and several values of l. Dashed line, metastable regime at $l = 1$

At large $l (l \geqslant 8)$ the situation resembles very much that of the infinitely long layer: the deformation of equilibrium is very weak and has a form of low-intensive vortices near the ends of the cavity. The critical point is approaching its position determined with the linear stability theory $Ra_v^{(A)} \approx 2129$ (Section 3 in Chapter 1). The regime of cells bifurcates softly from the basic one.

In Fig. 66 the dimensionless heat flux per unit length Nu/l is given as a function of Ra_v for different l. The curves for $l = 8$ and $l = 16$ do not differ from each other within the accuracy of the plot. The curve $l = 8$ may be considered the limiting one—it does not diverge practically from the heat transfer curve for an infinite layer. It can be seen that establishing a secondary regime of convective cells is connected with a significant intensification of the heat flux. Therefore, at $Ra_v = 100 \times 10^3$, the heat flux is several times its value in the equilibrium.

It is interesting to remark that at large $l (l \geqslant 8)$ different steady regimes can emerge. They differ in number of cells in the cavity, but they are very close from the viewpoint of stability and heat transfer. On the basis of computations, an approximate formula for the heat flux may be proposed in the parameter range $l \geqslant 8$, $Ra_v > 10^4$:

$$\frac{Nu}{l} = 0.101 \times Ra_v^{0.324}$$

According to calculations, at $Ra_v \geqslant 2 \times 10^5$ the steady regime becomes unstable: the transient process leads to convective oscillations.

In the above, a rectangular cavity elongated in the direction of the vibration axis in the presence of a transversal temperature gradient was considered. The opposite case of a rectangular cavity elongated in the direction of the temperature gradient and exposed to transversal vibrations was studied in [8]. The limit $l \to \infty$ corresponds to absolutely stable quasi-equilibrium (see Section 4 in Chapter 1). Calculations have shown that at $l \to \infty$ the convective motion decreases but not monotonically. Before going down, the convection intensity and longitudinal heat flux reach their maximal values at some aspect ratio l^*. For example, $l^* \approx 4$ at $Ra_v = 1 \times 10^5$.

We finish this section with a brief review of the results [9–11] where two problems of vibrational convection were solved for a rectangular cavity under different conditions of heating.

The first problem [9, 10] deals with a rectangular cavity with rigid boundaries and a longitudinal vibration axis (see Fig. 60). The boundary conditions are as follows. The ends $x = 0$ and $x = la$ are thermally insulated; the boundary $y = a$ is isothermal and maintained at the temperature $T = 0$; on the boundary $y = 0$ a constant longitudinal temperature gradient is imposed, $T = \Theta + Ax$. Thus, in the dimensionless form we have

$$x = 0, \quad x = l: \qquad \frac{\partial T}{\partial x} = 0$$

$$y = 0: \qquad T = 1 + \alpha x$$

$$y = 1: \qquad T = 0$$

Here $\alpha = Aa/\Theta$ is the dimensionless temperature gradient. Under the adopted conditions, equilibrium is impossible. The computations performed by the finite difference technique for $Pr = 1$, $l = 5$, $\alpha = 1$ show that at small values of the vibrational Rayleigh number the one-vortex flow takes place. As Ra_v increases, the transition sets in without bifurcation to a two-vortices structure and then to a three-vortices one. In addition, it has been found that at $Ra_v \geqslant 180$ there exists another steady convection regime. It differs from the basic one in number of vortices and is hardly excited. In the range $180 \leqslant Ra_v \leqslant 400$ both regimes coexist. For $Ra_v \geqslant 400$ regular oscillations appear. At high Ra_v ($Ra_v \geqslant 2100$), a steady regime with several vortices settles once again. The same transition takes place for different l and α.

The second problem [11] differs from the first one in the temperature condition on the $y = 1$ boundary, where now a constant gradient $T = \alpha x$ is imposed. Therefore, in addition to a constant transversal temperature difference, on the long boundaries there exists a constant temperature gradient in the direction of the vibration axis. Under those circumstances, a one-vortex flow is inherent in the range of small Ra_v as Ra_v enhances, forming developed boundary layers near the long sides of the cavity. Boundary layers seize the internal core region which is practically motionless. As l increases, the flow tends to a plane-parallel one described in Section 4 in Chapter 2.

In addition to the afore-mentioned, [12–15] are devoted to the bifurcation analysis of secondary vibrational convective regimes in a plane infinitely long layer. In [16, 17] the effect of vertical high-frequency vibrations on the non-linear convection in a plane horizontal fluid layer heated from below was studied. The Galerkin technique was applied using the approximation of the Lorentz triplet type. It was shown that as the vibrational parameter grows, the chaos threshold is brought down. Upon its further increase, the scenario of the transition to chaos changes.

2 Infinite Cylinder of a Circular Cross-section

In this section the problem is considered of vibrational convection in an infinite right cylinder of circular cross-section under lateral heating and transverse vibrations. In Section 3 in Chapter 1 this situation has been discussed from the viewpoint of feasibility of a quasi-equilibrium. It has been found that the quasi-equilibrium may exist only in two cases: (i) when the equilibrium temperature gradient and the vibration axis are parallel and (ii) when they are mutually perpendicular being as well perpendicular to the cylinder axis. In the parallel case, the equilibrium is absolutely stable. At mutual perpendicularity, the loss of equilibrium stability takes place at some critical value of the Rayleigh number that was evaluated in Section 3 in Chapter 1. The general situation is studied below in its non-linear formulation. We assume that the angle between the temperature gradient and the vibration axis is arbitrary. Varying this angle as a parameter allows one to keep watch on the transition from a true equilibrium situation to the general one where equilibrium is impossible and the convection sets in at an arbitrarily small temperature difference.

Let us turn to Fig. 67 where the normal cylinder cross-section is shown. In this plane two systems of coordinates are introduced: the Cartesian one (x, y) and the polar one (r, φ), where the polar angle φ counts from the x axis. The vibration axis coincides with the y axis. The temperature on the boundary does not depend on the longitudinal coordinate z and depends harmonically on the angle φ:

$$T_w = \Theta \sin(\varphi + \alpha) \tag{3.6}$$

In the absence of flow, this temperature distribution inside the cavity a constant temperature gradient of magnitude Θ/a. Its direction makes an angle α with the vibration axis and is described by the unit vector $\mathbf{m}$. Equilibrium situations are

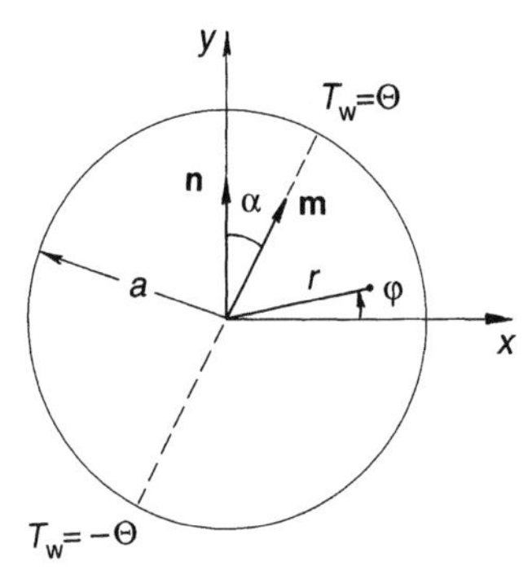

Fig. 67 Cross-section of a cylinder; coordinate axes

available at $\alpha = 0°$ ($\mathbf{m} \| \mathbf{n}$, absolute stability) and at $\alpha = 90°$ ($\mathbf{m} \perp \mathbf{n}$, the instability sets on at some critical value of Rayleigh number).

We consider a two-dimensional problem of vibrational convection assuming that vectors **v** and **w** have zero components along the cylinder axis z and all the fields do not depend on z. Stream functions ψ and F for the fields **v** and **w** are introduced by the following relations:

$$v_r = \frac{1}{r}\frac{\partial \psi}{\partial \varphi}, \qquad v_\varphi = -\frac{\partial \psi}{\partial r}; \qquad w_r = \frac{1}{r}\frac{\partial F}{\partial \varphi}, \qquad w_\varphi = -\frac{\partial F}{\partial r} \tag{3.7}$$

The set of dimensionless equations is written down for the stream functions ψ, F and temperature T choosing a, Θ, χ and Θa as units of length, temperature, ψ and F, respectively:

$$\frac{1}{Pr}\left[\frac{\partial \Delta\psi}{\partial t} + \mathscr{D}(\psi, \Delta\psi)\right] = \Delta\Delta\psi - Ra_{\mathrm{v}}\, \mathscr{D}\left(\frac{\partial F}{\partial x}, T\right)$$
$$\frac{\partial T}{\partial t} + \mathscr{D}(\psi, T) = \Delta T \tag{3.8}$$
$$\Delta F = -\frac{\partial T}{\partial x}$$

Here the Rayleigh number is $Ra_{\mathrm{v}} = (\beta b \Omega \Theta a)^2/(2\nu\chi)$, Δ denotes the two-dimensional Laplace operator and the following notations are used:

$$\mathscr{D}(f, g) = \frac{\partial f}{\partial y}\frac{\partial g}{\partial x} - \frac{\partial f}{\partial x}\frac{\partial g}{\partial y}$$
$$\frac{\partial}{\partial x} = \cos\varphi \frac{\partial}{\partial r} - \frac{\sin\varphi}{r}\frac{\partial}{\partial \varphi}$$
$$\frac{\partial}{\partial y} = \sin\varphi \frac{\partial}{\partial r} + \frac{\cos\varphi}{r}\frac{\partial}{\partial \varphi}$$

The boundary conditions on the circle $r = 1$ are

$$r = 1: \qquad \psi = 0, \qquad \frac{\partial \psi}{\partial r} = 0, \qquad F = 0, \qquad T = \sin(\varphi + \alpha) \tag{3.9}$$

In addition, the solution must be finite at the center $r = 0$.

In both limiting cases ($\alpha = 0°$ and $\alpha = 90°$), the problem under study has an equilibrium solution whose stability with respect to small perturbations is examined on the base of equations linearized near the equilibrium state (see Section 2 and 3 in Chapter 1). Here the computation results are presented on steady regimes of thermovibrational convection for arbitrary α. The calculations are performed by means of the finite difference method [18, 19] using an iterative scheme and equally spaced grid with 26×37 nodes along the radius and the angle. This ensures the accuracy of the order of a few percent when evaluating integral characteristics. The value of the Prandtl number is fixed: $Pr = 1$.

First, let us discuss the results corresponding to the equilibrium case $\alpha = 90°$ and to some closely related situations. The convection is excited as a result of the equilibrium stability loss and sets on when the Rayleigh number exceeds the

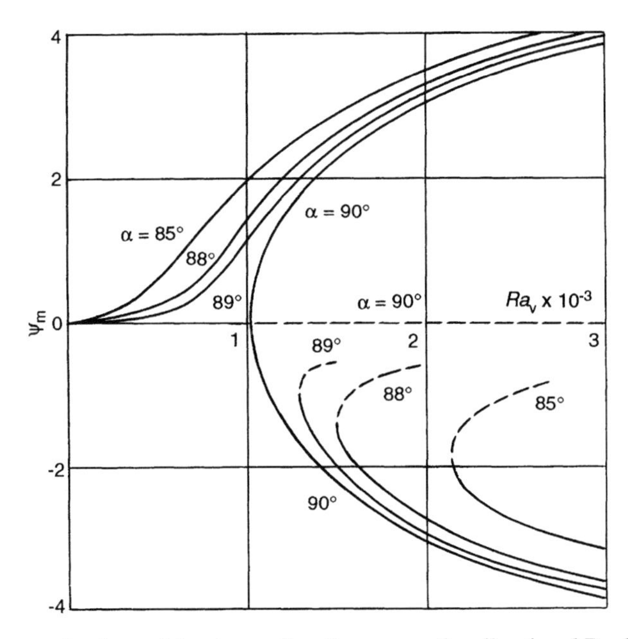

Fig. 68 Extremal values of the stream function versus the vibrational Rayleigh number for different angles in the range $\alpha \leqslant 90°$. Dashed lines, unstable parts of the amplitude curves

critical value $Ra_{vc} = 1029$. As the calculations show, in the subcritical region $Ra_v < Ra_{vc}$ there exists only one steady (equilibrium) solution. In the critical point $Ra_v = Ra_{vc}$, two steady solutions of a one-vortex structure bifurcate from the equilibrium. Both flows are stable and differ from each other just by the direction of circulation. The flow amplitude increases monotonically with overcriticality. In Fig. 68 the dependence of the extremal value ψ_m of the stream function on Ra_v is presented. The ψ_m value may be considered as a measure of the convective flow intensity. Both branches bifurcating from the equilibrium are equivalent and each of them can be realized in numerical experiment by an appropriate choice of sign of the initial perturbations. Extrapolation of the relationship $\psi_m(Ra_v)$ to the value $\psi_m = 0$ gives the bifurcation point $Ra_{vc} = 1020$, in good agreement with the result of the linear stability theory. Close to the threshold, the root-like dependence $\psi_m \approx \pm\sqrt{Ra_v - Ra_{vc}}$ takes place. This fact testifies that the convection excitation is of a soft character. The equilibrium solution in the range $Ra_v > Ra_{vc}$ is unstable.

If the angle α differs from 90° even a little, the equilibrium is not possible and the flow sets in at an arbitrarily small Rayleigh number Ra_v. If the angle α is close to 90° the branch of the equilibrium conditions is weak, which leads to a specific pattern of the convection excitation as a result of a 'non-ideal bifurcation'. This situation is presented in Fig. 68.

Let, for example, $\alpha = 89^\circ$. As the corresponding curve shows, there exists the main branch of solution coming from the origin of the coordinates. This solution describes the fluid flow with a 'positive' circulation. (According to the definition of the stream function adopted in this section, the positive values of ψ_m correspond to a counter-clockwise circulation.) At small Ra_v, the flow intensity is small because the conditions of heating are close to the equilibrium ones. The intensity increases with Ra_v. In the neighborhood of the critical value Ra_{vc}, a sharp growth of intensity takes place. The branch is stable within the studied range. This steady solution is unique in the range of small Ra_v. At some point $Ra_{v*} > Ra_{vc}$ (according to our calculations $Ra_{v*} = 1260$), another two branches emerge associated with a flow of negative ('incorrect') circulation. One of these branches corresponding to the maximal in modulus ψ_m values is stable; it is shown by the solid line in the figure. Another branch approaching the Ra_v axis is shown by the dashed line and is unstable. Thus Ra_{v*} is the end-point for the branches with the 'incorrect' circulation. At this point both branches emerge. In computations, using the iterative scheme, the unstable branch may be distinguished if one takes care that the residual is minimal when Ra_v and α are varied in small steps.

The amplitude curves for other α close to 90° are similar. As α increases, i.e. as the situation withdraws from the equilibrium, the point Ra_{v*} rapidly shifts to the range of large Ra_v. The stable branch with the 'incorrect' circulation becomes more and more excited, and its numerical realization often becomes rather troublesome.

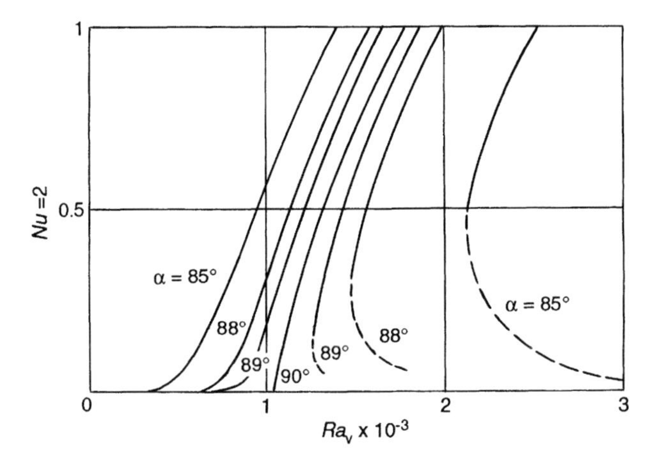

Fig. 69 Dimensionless heat flux as a function of the vibrational Rayleigh number; $\alpha \leqslant 90^\circ$

The peculiarities of the convection excitation for α close to 90° are reflected in the heat transfer curves. In Fig. 69 the dependence of the dimensionless heat flux, the Nusselt number, on Ra_{v} is presented for several α ($Nu = 2$ corresponds to the purely conductive regime). As in Fig. 68, solid lines correspond to the stable branches and dashed lines to the unstable ones.

The bifurcation pattern has been considered in the range $\alpha \leqslant 90^\circ$. If $\alpha \geqslant 90^\circ$, the described properties of the solutions do not change except for one thing. The main branch corresponds now to the flow with the negative circulation and two branches emerging at the point $Ra_{\mathrm{v}*}$ to the positive one. The associated amplitude curves may be obtained from those in Fig. 68 by replacing ψ_{m} with $-\psi_{\mathrm{m}}$.

Thus the excitation conditions for vibrational convection in the range close to the critical Ra_{v} value, where the perturbations of equilibrium are small, are similar to those studied in the theory of thermogravitational convection in the static gravity field. For the situation under consideration, the symmetry of the basic flow coincides with the symmetry of the secondary flow appearing in the first critical point. Therefore, the pattern is similar to the one described in [5] for the case of thermogravitational convection excited in a closed cavity heated from below in the presence of a small lateral temperature gradient.

We have considered the bifurcation structure close to the point at 90°. The intensity of the thermovibrational convective flow and heat transfer at an arbitrarily oriented external temperature gradient may be inferred from Figs. 70 and 71. In Fig. 70 the extremal value ψ_{m} of the stream function is presented as a function of the angle α between the directions of the external temperature gradient (vector **m**) and the vibration axis (vector **n**) for different values of the

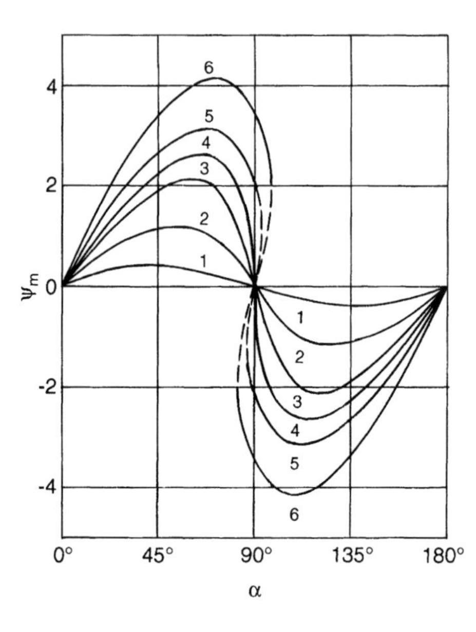

Fig. 70 Extremal values of the stream function depending on the angle α. Curves 1–6 correspond to Ra_v values of 100, 350, 700, 1000, 1400, 2500, respectively

vibrational Rayleigh number Ra_v. The curves 1–4 correspond to subcritical Ra_v values; thus at $\alpha \to 90°$ the stable equilibrium $\psi_m \to 0$ exists. The values $Ra_v = 1400$ and $Ra_v = 2500$ (5 and 6 are their respective curves) are supercritical. For them at $\alpha \to 90°$ the finite ψ_m values are obtained associated with the stable supercritical solutions. At these Ra_v values in the range $\alpha \approx 90°$, the non-single-valuedness takes place, as has been remarked in connection with Fig. 68. The unstable parts of curves 5 and 6 are shown by dashed lines. It can be seen that the extrema of the vibrational convection intensity are located in the intermediate range of angles. When Ra_v increases, the extrema shifts to the

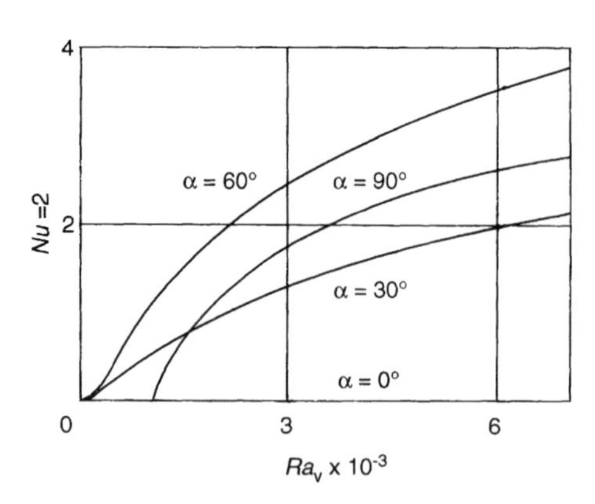

Fig. 71 Dimensionless heat flux as a function of the vibrational Rayleigh number for several values of α

point $\alpha = 90°$ whereas when Ra_v decreases, the extrema move towards $\alpha = 45°$ and $\alpha = 135°$, which can be proved analytically by the creeping flow method [19].

In Fig. 71 the dependence of the dimensionless heat flux on Ra_v is displayed for several values of α. At intermediate Ra_v, the vibrational convective part of the heat flux is 2–3 times greater than that in the purely conductive regime.

3 External Problem of Vibrational Convection

Let us consider now an example of an external problem of thermovibrational convection. It formulates as follows [20, 21]. A rigid infinite cylinder of a circular cross-section with the radius a is surrounded by an infinite fluid. The system as a whole vibrates harmonically along the direction perpendicular to the cylinder axis. On the cylinder surface the constant temperature Θ is maintained. Far from the cylinder, the temperature is taken to be zero. Two coordinate systems are introduced. The Cartesian one has its origin at the center of the cross-section, with the x and y axes in the plane of the cross-section and the z axis coinciding with the cylinder axis. The polar system (r, φ) is set on the plane of the cross-section. Let the axis of vibration coincide with the y axis and the polar angle be counted from the x axis. Conversely to the problem of the preceding section, here the external problem is studied where the fluid occupies the domain $r > a$. A two-dimensional solution is sought, so the stream functions ψ and F for the fields $\mathbf{v}$ and $\mathbf{w}$ are introduced in the same manner as (3.7). The dimensionless equations for ψ, F and T coincide with (3.8). On the surface of the cylinder, the average velocity and the normal component of pulsation velocity equal zero and the temperature is fixed:

$$r = 1: \qquad \psi = 0, \qquad \frac{\partial \psi}{\partial r} = 0, \qquad F = 0, \qquad T = 1 \tag{3.10}$$

At infinity, the flow ceases and the temperature tends to zero:

$$r \to \infty: \qquad \psi \to 0, \qquad \frac{\partial \psi}{\partial r} \to 0, \qquad F \to 0, \qquad T \to 0 \tag{3.11}$$

This boundary-value problem has been solved numerically by the method of finite differences in the finite domain $1 \leqslant r \leqslant a_\infty$, where a_∞ is large enough. The transposition of the boundary conditions from infinity to the finite distance a_∞ calls for some clarification. This transposition has been performed in the following way. At the boundary $r = a_\infty$, the transversal velocity and the transversal viscous stress are nil. As for the temperature, a recommendation [22] has been used that concerns the external problem of thermogravitational convection in a fluid surrounding a heated circular cylinder with an isothermal surface. The conditions for the temperature are chosen depending on the sign of the radial velocity: if the fluid flows into the region of integration from infinity, i.e. $v_r(a_\infty) < 0$, then $T(a_\infty) = 0$. If the fluid flows out of the heating zone, i.e. $v_r(a_\infty) > 0$, then the soft condition is maintained for the heat flux. Thus at $r = a_\infty$ the following model boundary conditions are assigned:

$$\frac{\partial \psi}{\partial r} = 0, \qquad \frac{\partial^2 \psi}{\partial r^2} = 0, \qquad \frac{\partial F}{\partial r} = 0$$

$$\text{if } \frac{\partial \psi}{\partial \varphi} < 0: \qquad T = 0 \tag{3.12}$$

$$\text{if } \frac{\partial \psi}{\partial \varphi} > 0: \qquad \frac{\partial^2 T}{\partial r^2} = 0$$

To solve the problem numerically, an equally spaced grid has been introduced along the radius and the angle. Both the iterative scheme and the alternating direction implicit scheme have been applied. The preliminary calculations in the whole region $\{1 \leqslant r \leqslant a_\infty, 0 \leqslant \varphi \leqslant 2\pi\}$ have shown that the solution is symmetrical with respect to reflection in the planes $x = 0$ and $y = 0$. Therefore the major part of the computation has been performed for the domain in between two rays $\varphi = 0$ and $\varphi = \pi/2$, with the appropriate reflection conditions on each of the rays. The grid was 100×32 nodes in the quadrant along the radius and the angle, respectively. A set of numerical experiments has been carried out for the fixed value of Prandtl number $Pr = 1$ in the range $Ra_\mathrm{v} \leqslant 5 \times 10^4$, the value a_∞ being changed from 20 to 80 depending on Ra_v.

Below we present some of the numerical results. In Fig. 72 the stream lines and isotherms are displayed for three values of the vibrational Rayleigh number. $Ra_\mathrm{v} = 1$ corresponds to the symmetrical four-vortices flow of a very small intensity. The isotherms are close to circles like the temperature field in a practically motionless fluid. It is interesting to note that the flow structure is similar to that near the circular cylinder, being surrounded by an isothermal fluid and vibrating in the transversal direction. The latter solution has been found by Schlichting [23]. As Ra_v grows, the velocity and temperature fields deform. The main perturbation takes the form of two jets transverse to the vibration axis along the rays $\varphi = 0$ and $\varphi = \pi$. Accordingly, the isotherms elongate in these directions. Therefore, the velocity and temperature fields acquire the features of a developed boundary layer.

To characterize the dimensionless local heat transfer through the cylinder surface, the heat flux density is introduced as $q_\mathrm{w} = -(\partial T/\partial r)|_{r=1}$. The constant value $q_\mathrm{w} = 1$ corresponds to the quiescent state. In a convection regime, q_w depends on φ and on Ra_v as a parameter. The polar diagrams $q_\mathrm{w}(\varphi)$ are given in Fig. 73 for two values of Ra_v. They agree well with the map of isotherms in Fig. 72.

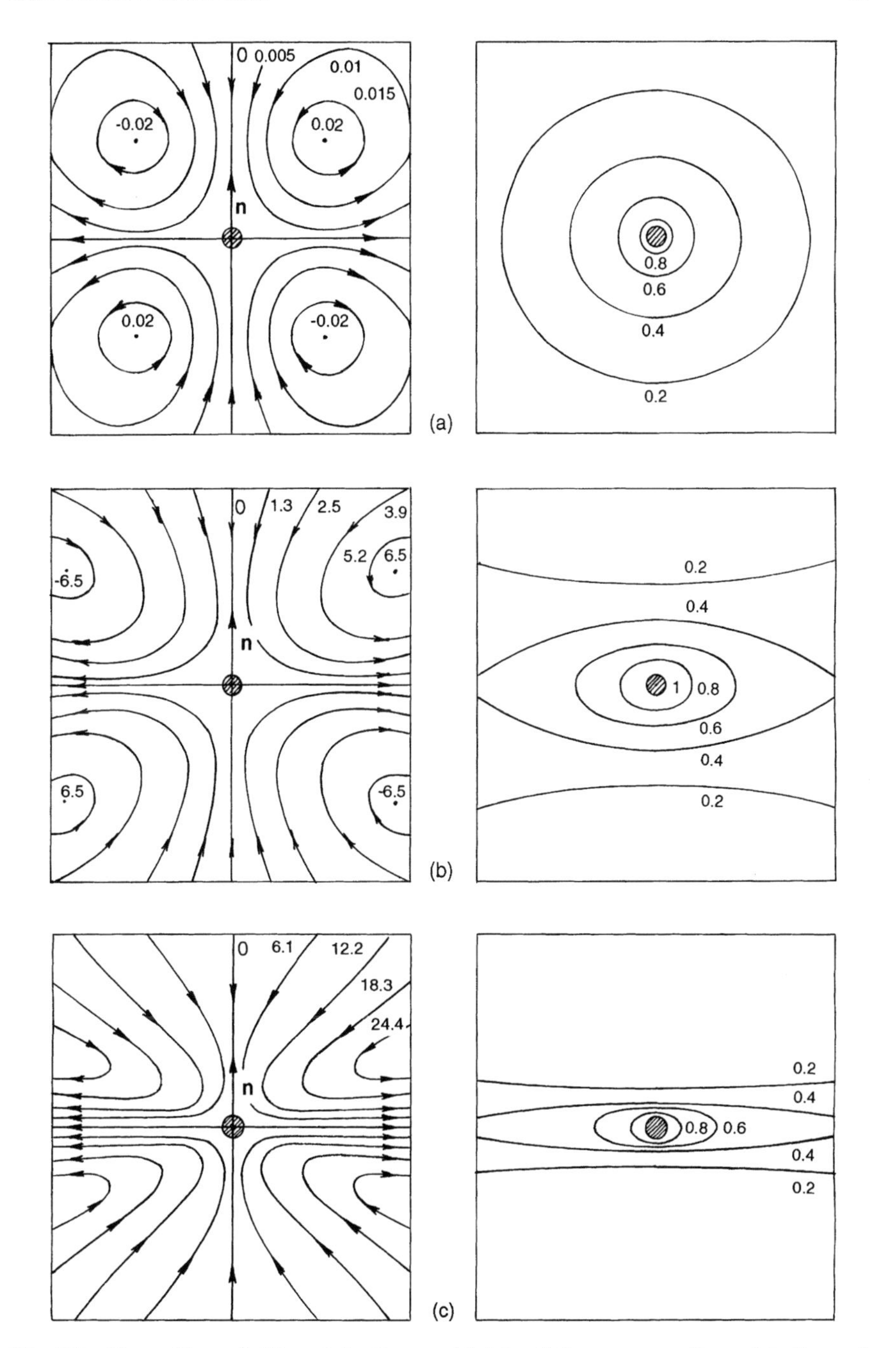

Fig. 72 Streamlines (left) and isotherms (right) of the average flow: (a) $Ra_v = 1$, (b) $Ra_v = 2.5 \times 10^2$, (c) $Ra_v = 5 \times 10^4$

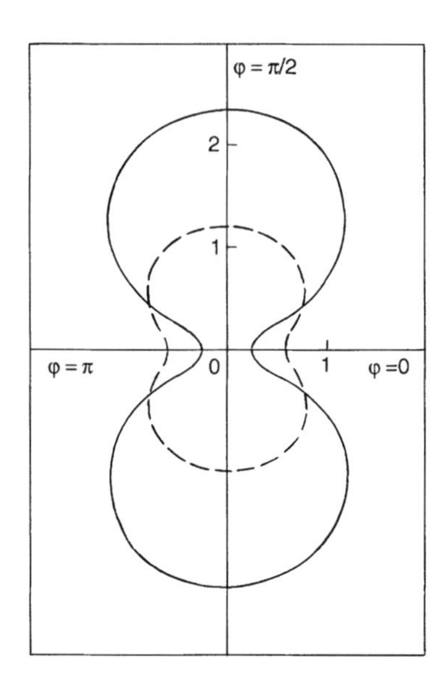

Fig. 73 Polar diagram of the heat flux density on the cylinder surface. Dashed line, $Ra_v = 10^2$; solid line, $Ra_v = 2 \times 10^4$

The integral dimensionless heat flux through the cylinder surface per unit distance along the cylinder axis may be described by the Nusselt number which is a function of Ra_v. The dependence $Nu(Ra_v)$ may be determined from the numerical data. In the limiting case of two developed transversal jets, this dependence has the following asymptotes:

$$Nu = 0.242 Ra_v^{1/6} \tag{3.13}$$

We shall revisit this problem in the next section after a discussion concerning the properties of a vibrational convective boundary layer. Also a comparison will be given of the numerical results with those obtained by means of the boundary layer approximation.

Chernatynsky [24] considered one more problem with a cylindrical symmetry. He studied vibrational convective flows in a cylindrical layer between two coaxial cylinders of different temperatures in the presence of transversal vibrations. The dimensionless parameters of the flow are the vibrational Rayleigh number and the ratio of radii of the cylinders. It has been found that in the vicinity of the poles the convection intensity is small since the conditions there are close to those in equilibrium. In the equatorial region the intensity is rather high at high Ra_v. The flow consists of several vortices with different circulations.

In [25] and [26] an experimental study of thermogravitational and thermovibrational convection in the cylindrical layer between horizontal coaxial cylinders was undertaken.

4 Vibrational Convective Boundary Layer

4.1 The Boundary Layer Approximation

In the previous section it has been shown that as the Rayleigh number increases, the flow around a circular cylinder vibrating in a transversal direction acquires the character of a boundary layer. In this section the structure of a vibrational convective boundary layer near the heat transfer surface is investigated on the basis of an approximate set of equations derived from the general system with the aid of the boundary layer ideology [21].

Let us consider a semi-infinite heated plate embedded in a fluid and performing transversal vibrations together with the fluid. The origin of the coordinate system is positioned at the edge of the plate, the x axis is directed normal to the plate and the y axis points along the plate. The unit vector $\mathbf{n}$ has the components $\mathbf{n}(1,0,0)$. A two-dimensional problem is studied in the plane xy. The set of equations for a steady flow has the form

$$v_x \frac{\partial v_x}{\partial x} + v_y \frac{\partial v_x}{\partial y} = -\frac{1}{\rho}\frac{\partial p}{\partial x} + \nu \Delta v_x + \varepsilon \left[w_x \frac{\partial}{\partial x}(T - w_x) + w_y \frac{\partial}{\partial y}(T - w_x) \right] \tag{3.14}$$

$$v_x \frac{\partial v_y}{\partial x} + v_y \frac{\partial v_y}{\partial y} = -\frac{1}{\rho}\frac{\partial p}{\partial y} + \nu \Delta v_y - \varepsilon \left(w_x \frac{\partial w_y}{\partial x} + w_y \frac{\partial w_y}{\partial y} \right) \tag{3.15}$$

$$v_x \frac{\partial T}{\partial x} + v_y \frac{\partial T}{\partial y} = \chi \Delta T \tag{3.16}$$

$$\frac{\partial v_x}{\partial x} + \frac{\partial v_y}{\partial y} = 0 \tag{3.17}$$

$$\frac{\partial w_x}{\partial x} + \frac{\partial w_y}{\partial y} = 0 \tag{3.18}$$

$$\frac{\partial w_y}{\partial x} - \frac{\partial w_x}{\partial y} = -\frac{\partial T}{\partial y} \tag{3.19}$$

Here $\varepsilon = 0.5(\beta b \Omega)^2$ is the dimensional vibrational parameter and Δ is the two-dimensional Laplace operator. The boundary conditions will be formulated later.

Let us write down equations (3.14) to (3.19) using the boundary layer approximation. In accordance with the assumptions of the boundary layer theory let us suppose that $v_y \gg v_x$, $w_y \gg w_x$. In addition, $\delta \ll l$, where δ is the boundary layer thickness and l is the reference longitudinal scale. Let v_0 and w_0 be the reference values of the longitudinal average and pulsation velocity components and Θ the reference temperature difference. Then $v_y \sim v_0$, $w_y \sim w_0$, $T \sim \Theta$.

From the continuity equations (3.17) and (3.18) one can obtain $v_x \sim \delta/lv_0$, $w_x \sim \delta/lw_0$. In equation (3.19) the second term on the left-hand side may be neglected and the following relation between the reference scales w_0 and Θ may be derived: $w_0 \sim \delta/l\Theta$. Let us estimate now the vibrational convective force. Due to the known simplifications, it is clear that the component of the force f_x may be approximately expressed as

$$f_x = \varepsilon\left(w_x \frac{\partial T}{\partial x} + w_y \frac{\partial T}{\partial y}\right)$$

which is of the order

$$f_x \sim \frac{\varepsilon w_0 \Theta}{l} \sim \frac{\varepsilon w_0^2}{\delta}$$

The component f_y of the force does not simplify in the boundary layer approximation and has the order

$$f_y \sim \frac{\varepsilon w_0^2}{l}$$

Thus, the transversal force f_x is much greater than the longitudinal one f_y. This is the essential distinction between the vibrational convective boundary layer and the classic Polhausen layer which is used in the theory of natural thermo-gravitational convection near a vertical heated plate.

Consider equation (3.14) governing the motion in the transversal direction. It is natural to suppose that, under the action of a large vibrational force f_x, a large transversal temperature gradient is induced that is of the same order as f_x. This yields the estimation for the transversal pressure difference $p_0 \sim \rho\varepsilon w_0^2$. Now let us take a look at the projection (3.15) of the equation of motion on the longitudinal axis y. We assume, as usual, that the inertial and viscous terms are of the same order as the longitudinal pressure and vibrational force. Thus one gets $w_0 \sim v_0/\sqrt{\varepsilon}$ and the estimation for the boundary layer thickness:

$$\delta \sim \left(\frac{\nu l^2}{\sqrt{\varepsilon}\Theta}\right)^{1/3}$$

The last relationship may be rewritten in the form $\delta/l \sim Gr_\mathrm{v}^{-1/6}$, where $Gr_\mathrm{v} = \varepsilon\Theta^2 l^2/\nu^2$ is the vibrational analog of the Grashof number determined through the temperature difference Θ and the longitudinal scale l. To recall equation (3.14), one can conclude that the inertial and viscous terms may be neglected. In equation (3.15), the vibrational force must be retained since it is of the same order as the other terms.

Therefore, the set of boundary layer equations can be written as follows:

$$-\frac{1}{\rho}\frac{\partial p}{\partial x} - \varepsilon\left(w_x\frac{\partial T}{\partial x} + w_y\frac{\partial T}{\partial y}\right) = 0$$

$$v_x\frac{\partial v_y}{\partial x} + v_y\frac{\partial v_y}{\partial y} = -\frac{1}{\rho}\frac{\partial p}{\partial y} + \nu\frac{\partial^2 v_y}{\partial x^2} - \varepsilon\left(w_x\frac{\partial w_y}{\partial x} + w_y\frac{\partial w_y}{\partial y}\right)$$

$$v_x\frac{\partial T}{\partial x} + v_y\frac{\partial T}{\partial y} = \chi\frac{\partial^2 T}{\partial x^2}$$

$$\frac{\partial v_x}{\partial x} + \frac{\partial v_y}{\partial y} = 0 \tag{3.20}$$

$$\frac{\partial w_x}{\partial x} + \frac{\partial w_y}{\partial y} = 0$$

$$\frac{\partial w_y}{\partial x} = -\frac{\partial T}{\partial y}$$

The boundary conditions are taken where on the surface of the plate the temperature profile T_w is maintained and far from the plate the flow ceases and the temperature and the pressure tend to zero:

$$\begin{aligned} &x = 0: \quad v_x = 0, \quad v_y = 0, \quad w_x = 0, \quad T = T_w(y) \\ &x \to \infty: \quad v_y \to 0, \quad w_y \to 0, \quad T \to 0, \quad p \to 0 \end{aligned} \tag{3.21}$$

We note a common feature of the problem under consideration to that of the thermogravitational boundary layer near a horizontal plate: a large longitudinal pressure gradient appears as a result of a large transversal force (see [27]). However, in our case the longitudinal component of the mass force takes part in the boundary layer formation.

4.2 Relations of Similarity

Consider a situation where the temperature at the surface varies with the longitudinal coordinate according to the power law

$$T_w(y) = Cy^n \tag{3.22}$$

In this case a similarity transformation is possible. Let us introduce the similarity variable

$$\eta = \left(\frac{\varepsilon C^2}{\nu^2}\right)^{1/6} xy^{(n-2)/3} \tag{3.23}$$

and also new functions $u(\eta)$, $v(\eta)$, $f(\eta)$, $g(\eta)$, $\tau(\eta)$ and $q(\eta)$ connected with v_x, v_y, w_x, w_y, T and p by the relations

$$v_x = \left(\nu^2 C\sqrt{\varepsilon}\right)^{1/3} y^{(n-2)/3} u(\eta), \qquad v_y = \left(\nu C^2 \varepsilon\right)^{1/3} y^{(2n-1)/3} v(\eta)$$

$$w_x = \left(\frac{\nu^2 C}{\varepsilon}\right)^{1/3} y^{(n-2)/3} f(\eta), \qquad w_y = \left(\frac{\nu C^2}{\sqrt{\varepsilon}}\right)^{1/3} y^{(2n-1)/3} g(\eta) \tag{3.24}$$

$$T = Cy^n \tau(\eta), \qquad p = \rho\left(\varepsilon C^2 \nu\right)^{2/3} y^{(4n-2)/3} q(\eta)$$

Substituting (3.24) into (3.20) and (3.21), one obtains the following set of ordinary differential equations (the prime denotes differentiation with respect to the similarity variable η):

$$\begin{aligned} q' &= f\tau' - gg' \\ v'' &= uv' - u'v - \frac{2}{3}(2n-1)g - n\tau f + \frac{2n-1}{3}g^2 \\ \tau'' &= Pr(u\tau' - g'v) \\ g' &= -\left(n\tau - \frac{n-2}{3}\eta\tau'\right) \\ u' &= -\tfrac{1}{3}[(2n-1)v + (n-2)\eta v'] \\ f' &= -\tfrac{1}{3}[(2n-1)g + (n-2)\eta g'] \end{aligned} \tag{3.25}$$

with the boundary conditions:

$$\begin{aligned} &\eta = 0: \quad u = 0, \quad v = 0, \quad \tau = 1, \quad f = 0 \\ &\eta \to \infty: \quad v \to 0, \quad g \to 0, \quad \tau \to 0, \quad q \to 0 \end{aligned} \tag{3.26}$$

The parameters of the problem are the power index n and the Prandtl number Pr. From the similarity relation it follows that the thickness of the boundary layer depends on the longitudinal coordinate as

$$\delta \sim y^{(2-n)/3} \tag{3.27}$$

The question of reality of the solution and its physical interpretation needs a special consideration for each particular n. Requiring integrability along the plate of either the viscous stress $\left(\sigma_{xy}\right)_{\mathrm{w}}$ or the heat flux density q_{w}, one obtains, respectively, the restrictions $n > 0$ and $n > -\frac{1}{4}$. Hence, for $n > 0$ both requirements are satisfied, ensuring finite values of the viscous force and integral heat. This means that the case of the isothermal plate ($n = 0$) does not meet the demands, but the condition of a constant heat flux at the plate ($n = \frac{1}{2}$) does.

4.3 Some Examples of Similar Solutions

Let us discuss first the interesting particular case of a boundary layer with a constant thickness. In accordance with (3.27), the corresponding power exponent is $n = 2$. The characteristics of the flow depend on the longitudinal coordinate y according to the laws $v_x =$ costant, $v_y \sim y$, $w_x \sim$ costant, $w_y \sim y$, $T \sim y^2$, $p \sim y^2$. After integrating the equation for pressure, the function $q(\eta)$ may be eliminated and the problem written as follows:

$$\psi''' = \psi\psi'' - (\psi')^2 + \tfrac{1}{2}g^2$$
$$\tau'' = Pr(\psi\tau' - 2\psi'\tau)$$
$$g' = -2\tau, \qquad f' = -g \tag{3.28}$$
$$\eta = 0: \quad \psi = 0, \quad \psi' = 0, \quad \tau = 1, \quad f = 0$$
$$\eta \to \infty: \quad \psi' \to 0, \quad g \to 0, \quad \tau \to 0$$

Here ψ is a new variable connected with the stream function for the average velocity. The components of the velocity are:

$$u = \psi, \qquad v = -\psi' \tag{3.29}$$

The problem (3.28) has been integrated numerically by the Runge–Kutta method using the three-parameter shooting procedure. In Figs. 74 and 75 the profiles are presented of the longitudinal velocity component $v(\eta) = -\psi'(\eta)$

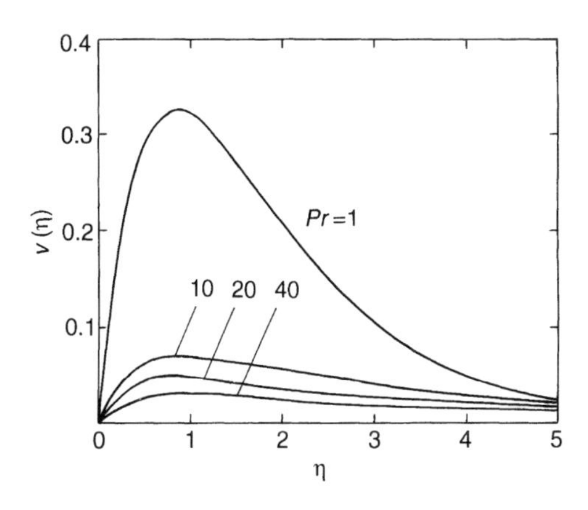

Fig. 74 Profiles of the longitudinal average velocity for the layer of a constant thickness; $n = 2$

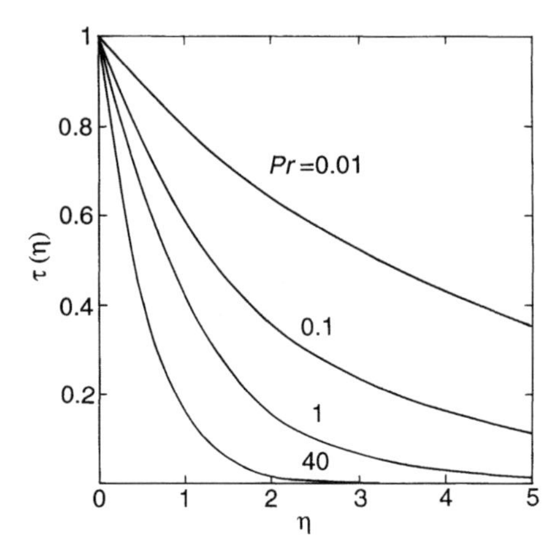

Fig. 75 Temperature profiles for the layer of a constant thickness; $n = 2$

Table 2 Characteristics of a boundary layer with a constant thickness

Pr	$-\psi''(0)$	$-\tau'(0)$	$g(0)$	$-\psi'_{\mathrm{m}}$	δ_{v}	δ_{T}
0.1	4.5924	0.4911	4.3349	1.4707	4.46	11.86
1	1.0599	0.7452	2.2555	0.3265	4.52	5.30
10	0.2888	1.0863	1.3937	0.0707	8.80	2.74
20	0.2009	1.2201	1.2271	0.0452	10.64	2.36
30	0.1628	1.3059	1.1411	0.0348	11.80	2.18
40	0.1405	1.3705	1.0844	0.0288	12.60	2.06

and the temperature $\tau(\eta)$ within the boundary layer for some values of the Prandtl number.

Some characteristics of the velocity and temperature distributions are summarized in Table 2. They have the following meanings: $-\psi''(0)$ yields the viscous stress on the rigid surface, $-\tau'(0)$ gives the local heat transfer density, $g(0)$ is the amplitude of the longitudinal oscillatory velocity on the plate and $-\psi'_{\mathrm{m}}$ denotes the maximal value of the longitudinal velocity in the boundary layer. The first three characteristics have been used as the shooting parameters in the computation procedure. The table also includes the values of the boundary layer thicknesses, dynamic δ_{v} and thermal δ_{T}, determined from the numerical solution in the following way. The dynamic thickness is defined by the relation $\psi'(\delta_{\mathrm{v}}) = 0.1\psi'_{\mathrm{m}}$, the thermal one by $\tau(\delta_{\mathrm{T}}) = 0.01$.

One may see that as the Prandtl number increases, the flow intensity decreases and the dimensionless heat flux grows. The dynamic and thermal boundary layer thicknesses vary monotonically with Pr. In the limiting case of large Prandtl numbers, the method of combined asymptotic expansions [28] can

Table 3 Boundary layer characteristics for two n values; $Pr = 1$

n	$-v'(0)$	$-\tau'(0)$	$g(0)$	$q(0)$	v_m	η_m	δ_v	δ_T
1.5	0.8729	0.6367	2.3516	−4.0521	0.3178	0.98	3.84	7.84
1.75	0.9517	0.6872	2.2704	−3.8327	0.3154	0.92	4.40	5.94

be applied. The asymptotic formulas at $Pr \to \infty$ are:

$$
\begin{aligned}
-\psi''(0) &= 0.8754\, Pr^{-1/2} \\
-\tau'(0) &= 0.7421\, Pr^{1/6} \\
g(0) &= -1.983\, Pr^{-1/6}
\end{aligned}
$$

Computations for the other values of the power exponent n in (3.22) have been performed for the fixed value $Pr = 1$. The boundary-value problem (3.25) and (3.26) has been solved by the shooting procedure with four parameters. The profiles of the longitudinal velocity and temperature are qualitatively close to those in Figs. 74 and 75. Some characteristics of the velocity and temperature distributions for two values of n are given in Table 3. Here v_m is the maximal velocity, η_m is the value of the similarity variable which provides the maximal velocity and the thicknesses δ_v and δ_T are determined as in the case of the boundary layer with a constant thickness.

4.4 Vibrational Convective Plume

The case $n = -\frac{1}{4}$ needs a special consideration. The longitudinal convective heat flux in such a boundary layer does not depend on the longitudinal coordinate. Actually, this flux per unit length along the z axis may be written in the form†

$$
\rho c_p \int_0^\infty v_y T \, dx = \rho c_p \left(\nu^2 C^4 \sqrt{\varepsilon}\right)^{1/3} y^{(4n+1)/3} \int_0^\infty v(\eta)\tau(\eta)\, d\eta \tag{3.30}
$$

Here ρ is the density and c_p is the heat capacity. At $n = -\frac{1}{4}$ this flux does not depend on y and equals

$$
\rho c_p \left(\nu^2 C^4 \sqrt{\varepsilon}\right)^{1/3} I(Pr), \qquad \text{where } I(Pr) = \int_0^\infty v(\eta)\tau(\eta)\, d\eta \tag{3.31}
$$

† The conductive part of the heat flux in a developed boundary layer may be neglected.

The flow under study may be interpreted as a free vibrational convective plume created by a linear heat source along the z coordinate perpendicular to the xy plane. Now there is no plate and $x = 0$ is the plane of the reflection symmetry. With allowance for this symmetry, the total heat flux along the plume is

$$Q = 2\rho c_p \left(\nu^2 C^4 \sqrt{\varepsilon}\right)^{1/3} I(Pr) \tag{3.32}$$

The characteristics of the plume vary along the longitudinal coordinate according to the laws: $\delta \sim y^{3/4}$, $v_y \sim y^{-1/2}$, $v_x \sim y^{-3/4}$, $T \sim y^{-1/4}$, $w_y \sim y^{-1/2}$, $w_x \sim y^{-3/4}$, $p \sim y^{-1}$. The linear heat source performing transversal vibrations, together with the fluid, gives rise to two symmetrical jets directed perpendicularly to the vibration axis.

To describe this self-similar plume, it is necessary to use the set of equations for the similarity functions (3.25) and set $n = -\frac{1}{4}$. The conditions at infinity stay the same while the conditions on the rigid plate must be replaced by the symmetry ones:

$$x = 0: \qquad u = 0, \qquad v' = 0, \qquad \tau' = 0, \qquad g = 0 \tag{3.33}$$

The condition for the longitudinal oscillatory velocity can be derived from the symmetry condition $f'(0) = 0$ using the continuity equation for the vector $\mathbf{w}$, the last one in the set (3.26).

The problem may be reformulated in a compact form by introducing new variables ψ and φ related to the stream functions of the average and oscillatory parts of the velocity:

$$\begin{aligned} u &= \tfrac{1}{4}(\psi - 3\eta\psi'), \qquad & v &= -\psi' \\ f &= \tfrac{1}{4}(\varphi - 3\eta\varphi'), \qquad & g &= -\varphi' \end{aligned} \tag{3.34}$$

Substituting (3.34) into the general set, eliminating pressure and integrating the heat transfer equation, one can obtain the following boundary-value problem:

$$\begin{gathered} \psi''' = \tfrac{1}{4}\psi\psi'' + \tfrac{3}{16}\tau(\eta\varphi)' + \tfrac{1}{2}(\psi'^2 + \varphi'^2) \\ \tau' = \tfrac{1}{4} Pr\,\tau\psi \\ \varphi'' = -\tfrac{1}{4}\left(1 + \tfrac{3}{4} Pr\,\eta\psi\right)\tau \\ \eta = 0: \qquad \psi = 0, \qquad \psi'' = 0, \qquad \tau = 1, \qquad \varphi' = 0 \\ \eta \to \infty: \qquad \psi' \to 0, \qquad \varphi' \to 0 \end{gathered} \tag{3.35}$$

The conditions $\tau'(0) = 0$ and $\tau(\infty) = 0$ are satisfied automatically, as in the case of the boundary layer plume solution in thermogravitational convection (see [27]).

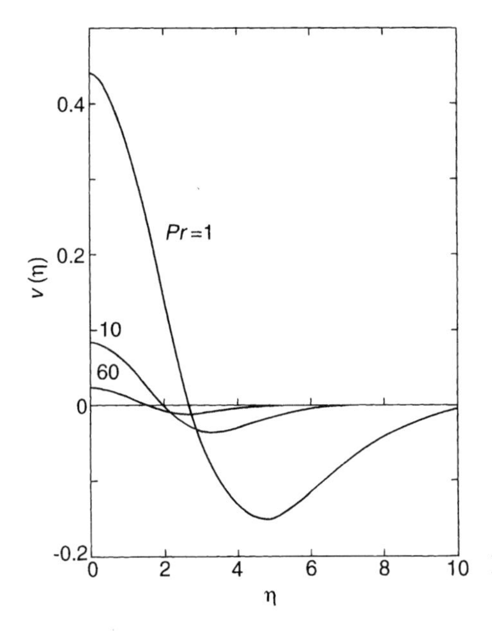

Fig. 76 Profiles of the longitudinal average velocity for the plume; $n = -\frac{1}{4}$

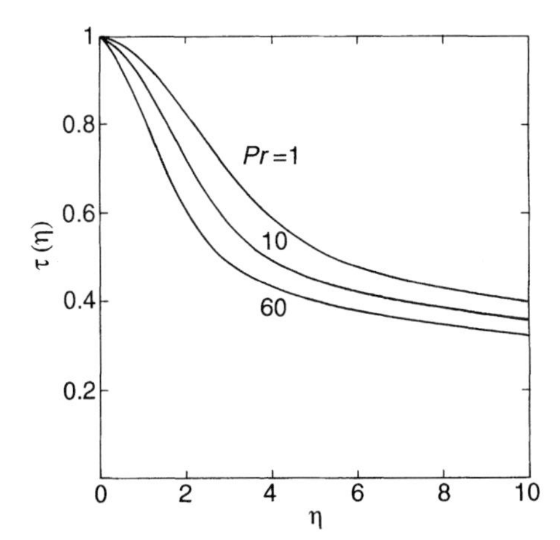

Fig. 77 Temperature profiles for the plume; $n = -\frac{1}{4}$

Some results of a numerical solution of the problem are displayed in Figs. 76 to 78. In Fig. 76 the profiles of longitudinal average velocity within the plume are given for several values of the Prandtl number. It can be seen that in the range $\eta_0 < \eta < \infty$ with η_0 depending on Pr, there exists the zone of a counter-flow:

$$v(\eta_0) = 0, \qquad v(\eta) < 0 \text{ if } \eta > \eta_0$$

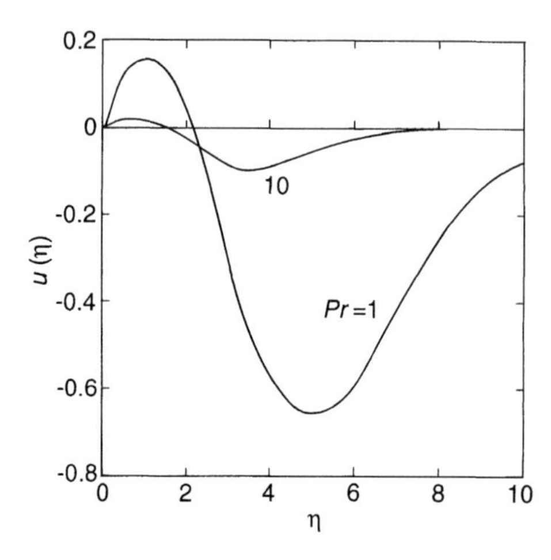

Fig. 78 Profiles of the transversal average velocity for the plume; $n = -\frac{1}{4}$

Calculations show that $\psi \to 0$ if $\eta \to \infty$ and thus the flow is closed:

$$\int_0^\infty v(\eta)\,\mathrm{d}\eta = 0$$

Because of this fact, the flow under study differs from its thermogravitational analog. The flow-closeness condition, when checked for the transversal average velocity distribution, does not hold. Indeed, $u \to 0$ at $\eta \to \infty$, i.e. the inflow to the plume is absent (Fig. 78). The longitudinal pulsation velocity component $g(\eta)$ is the odd function of η but the transversal one is even; thus $f(0) \neq 0$. The latter is due to the fact that the vibration direction is transverse with respect to the axis of the plume.

In Table 4 the following characteristics of the free vibrational convective plume are presented: the extremal value $-\psi'(0)$ of the average longitudinal velocity on the axis and its value at the point of minimum $\psi'_{\mathrm{m}} = \psi'(\eta_{\mathrm{m}})$, the values η_0 and η_{m} of the similarity variable, the value $f(0)$ of the transverse oscillatory velocity component on the axis and the thickness of the dynamic boundary layer δ_{v} determined from the relation $|\psi'(\delta_{\mathrm{v}})| = 0.1|\psi'(0)|$ for

Table 4 Characteristics of a self-similar vibrational convective plume at $n = -\frac{1}{4}$

Pr	$-\psi'(0)$	ψ'_{m}	η_0	η_{m}	f_0	δ_{v}
1	0.44071	0.14979	2.7	4.8	0.6239	7.8
10	0.08403	0.03234	1.9	3.4	0.3460	5.5
20	0.05262	0.02039	1.7	3.1	0.2768	4.9
40	0.03305	0.01283	1.5	2.8	0.2205	4.4
60	0.02519	0.00981	1.45	2.6	0.1929	4.0
100	0.01791	0.00698	1.4	2.4	0.1629	3.7

$\delta_v > \eta_m$. In the computations $\psi'(0)$ and $f(0)$ played the part of the shooting parameters.

The solution of the problem under consideration is closely connected with the problem of an external flow around a circular cylinder vibrating in the transversal direction, which was described in the preceding section. As we have shown in Section 3, at large Ra_v two symmetrical jets are formed with the axes perpendicular to the vibration axis. One may expect that, far from the cylinder, the structure of the jets obeys the same self-similarity as the one discussed in this section. A comparison of results confirms this. The distributions of longitudinal and transversal components of the average and pulsation velocities, as well as that of the temperature obtained by direct modeling, are the same as those yielded by the self-similar solution. In addition, numerical results are in a good agreement with the estimations deduced from the approximate boundary layer equations at the beginning of this section.

A quantitative comparison of the results of numerical modeling with those of the self-similar solution is also possible. Let us compare the most important characteristics—distributions of the longitudinal velocity and temperature at the axis of the plume. For this purpose we derive these distributions from the dimensionless self-similar equations. The self-similar solution does not include the reference length scale, so we take the radius of cylinder a as a unit of length, the ratio ν/a as a unit of velocity and the temperature of the cylinder surface $T|_{y=a} = \Theta$ as a unit of temperature. Then setting $n = -\frac{1}{4}$ and using dimensionless variables one finds from (3.24) that

$$\widetilde{v_y} = Gr_v^{1/3} v(0) \widetilde{y}^{-1/2}, \qquad \widetilde{T} = \widetilde{y}^{-1/4} \tag{3.36}$$

Here $\widetilde{v_y}$ and $\widetilde{T}$ are, respectively, the dimensionless longitudinal velocity and the temperature on the axis of plume, $\widetilde{y}$ is the dimensionless longitudinal coordinate and Gr_v the vibrational Grashof number determined as above. A comparison may be done for $Pr = 1$. In this case, $v(0) = 0.4407$ (see Table 4) and $Gr_v = Ra_v$. In Fig. 79 the temperature distributions are presented. It can be seen that there is good agreement between the numerical and self-similar solutions in the limiting case of large Gr_v and $\widetilde{y}$.

As for the longitudinal velocity on the axis of the plume, there is no entire quantitative agreement (see Fig. 80). As Gr and $\widetilde{y}$ increase, the numerical results tend to the self-similar limit ($\widetilde{v_y} \sim \widetilde{y}^{-1/2}$). However, there exists a considerable quantitative discrepancy, which may be due to the fact that the values of Gr_v and $\widetilde{y}$ are not large enough. Another circumstance that may have some effect is that the numerical solution has been constructed inside a finite domain.

Finally, we compare the data on the heat flux along the plume. The numerical solution at large Ra_v provides the Nusselt number defined by (3.13), with replacement of the Grashof number by the Rayleigh number since $Pr = 1$. The

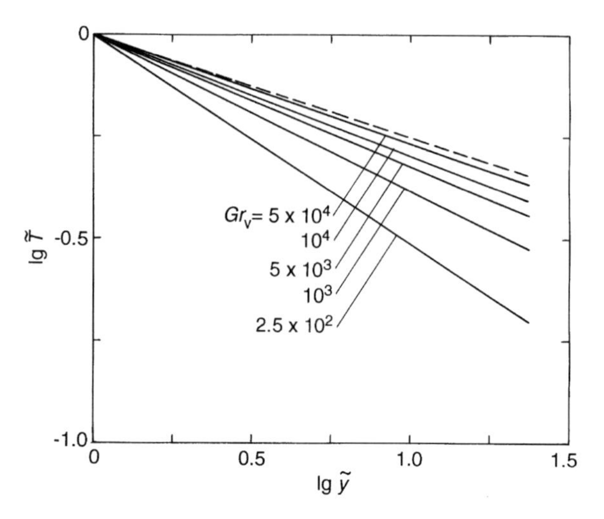

Fig. 79 Temperature on the axis of the plume as a function of the longitudinal coordinate. Solid lines, numerical solution; dashed straight line, self-similar solution

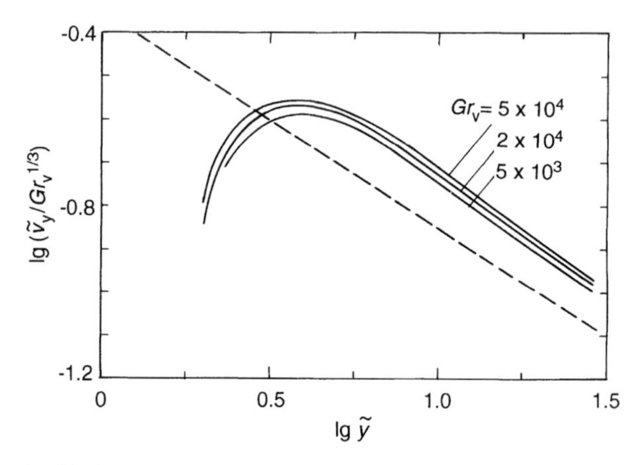

Fig. 80 Longitudinal averaged velocity on the axis of the plume as a function of the longitudinal coordinate. Solid lines, numerical solution; dashed straight line, self-similar solution

self-similar formula (3.32) gives

$$Nu = \tfrac{2}{5} I(Pr) Pr\, Gr_v^{1/6} \tag{3.37}$$

i.e. it yields the same limiting law $Nu \sim Gr_v^{1/6}$. At $Pr = 1$, it may be proved that $I(Pr) = 0.3298$ and the value of the numerical coefficient in (3.37) is 0.210

instead of 0.242 in the numerical solution. We consider this agreement as quite satisfactory.

In conclusion, we remark that the same consideration may be worked out for a point heat source. In this case, an axisymmetrical fan-like jet is formed in the equatorial plane and the symmetry axis coincides with the axis of vibrations. The longitudinal (radial) velocity, the temperature and the boundary layer thickness vary with r as $v_r \sim r^{-1}$, $T \sim r^{-1}$, $\delta \sim r$. Numerical solution of the corresponding external problem for a vibrating sphere [29] shows that as Gr_v increases a transition to the self-similar structure takes place.

5 Transient Regime of Thermovibrational Convection

5.1 Statement of the Problem

In this chapter mainly the steady regimes of flow and heat transfer have been considered. Now we give an example of a non-steady problem of thermovibrational convection under weightlessness and analyze a transient regime in a closed cavity. The problem is formulated as follows. A closed two-dimensional cavity has a form of a square of side a (Fig. 81). The imposed vibration has its axis parallel to the x axis, i.e. the unit vector $\mathbf{n}$ has the components $\mathbf{n}(1,0,0)$. The boundaries are assumed to be rigid; the average velocity vector as well as the normal component of the oscillatory velocity vanish at the boundaries. The side walls $x = 0$ and $x = a$ are thermally insulated. The upper boundary $y = a$ is isothermal and its temperature is taken as the reference point: $T|_{y=a} = 0$. The lower boundary $y = 0$, up to some initial moment $t = 0$ has the same temperature $T|_{y=0} = 0$. For $t < 0$, the fluid in the cavity is motionless and has a temperature coinciding with that of the boundaries. At $t = 0$, the temperature of the lower boundary rises abruptly up to the value Θ and further on is kept constant: $T|_{y=0} = \Theta$ at $t \geqslant 0$. In this way a transient process is initiated. It is easy to

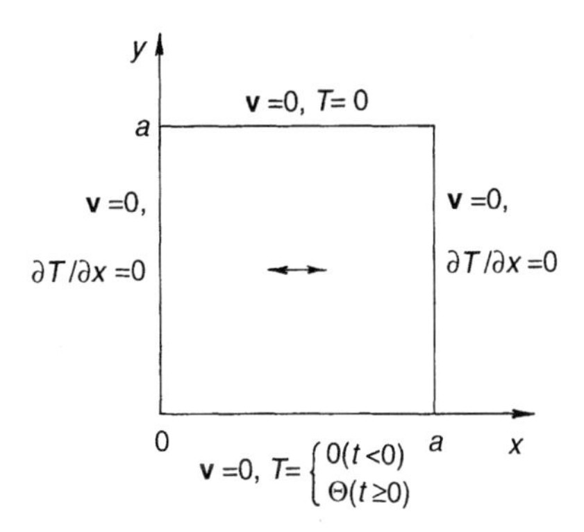

Fig. 81 Coordinate system and the boundary conditions

understand that under the described conditions a thermovibrational flow would necessarily be induced.

The equations of convection written in terms of temperature and stream functions ψ, F for the average velocity $\mathbf{v}$ and the vector $\mathbf{w}$ coincide with (3.2) to (3.4). The Rayleigh number is defined through the temperature difference Θ and the side length a. The dimensionless variables are based on the same units as in Section 1. The boundary and initial conditions for the dimensionless fields have the form

$$
\begin{aligned}
&x = 0, \quad x = 1: \quad \psi = 0, \quad \frac{\partial \psi}{\partial x} = 0, \quad \frac{\partial T}{\partial x} = 0, \quad F = 0 \\
&y = 1: \quad \psi = 0, \quad \frac{\partial \psi}{\partial y} = 0, \quad T = 0, \quad F = 0 \\
&y = 0, \quad t < 0: \quad \psi = 0, \quad \frac{\partial \psi}{\partial y} = 0, \quad T = 0, \quad F = 0 \\
&y = 0, \quad t \geqslant 0: \quad \psi = 0, \quad \frac{\partial \psi}{\partial y} = 0, \quad T = 1, \quad F = 0 \\
&0 \leqslant x \leqslant 1, \quad 0 \leqslant y \leqslant 1, \quad t < 0: \quad \psi = 0, \quad T = 0, \quad F = 0
\end{aligned}
\tag{3.38}
$$

These equations allow the flow and heat transfer in the transient regime to be monitored and the time evolution of all the fields to be observed. In particular, the main interest is to evaluate the reference time τ of the equilibrium settling which depends on the dimensionless parameters Ra_v and Pr.

5.2 Some Limiting Relations

Before proceeding to the numerical solution of the problem, it is useful to qualitatively discuss the situations that take place in some limiting cases.

First, let us consider the case of small vibrational Rayleigh numbers Ra_v. If $Ra_v = 0$, it is obvious that the transient process is not accompanied by convective phenomena and is determined by pure heat diffusion. The reference time of settling τ has the order a^2/χ. Evidently this estimation holds as long as the intensity of convection is low. It should be remembered that the value a^2/χ is taken as the time unit. Therefore, for a dimensionless settling time one gets

$$\tau = \text{constant} \qquad (Ra_v \ll 1) \tag{3.39}$$

where the constant does not depend on Ra_v and Pr.

Now, let us look at the limiting case of high values of the Prandtl number. Since the inertial terms in the equation of motion may be neglected, the Prandtl number does not enter the initial boundary-value problem. Thus the transient process itself as well as the settling time do not depend on Pr, i.e.

$$\tau = \tau(Ra_v) \qquad (Pr \gg 1) \tag{3.40}$$

In the opposite limiting case of small Pr ($Pr \ll 1$), the reference time of heat diffusion a^2/χ is small with respect to the reference viscous time a^2/ν. Then the settling time is determined mainly by the viscous time that in the dimensionless form gives

$$\tau \sim \frac{1}{Pr} \qquad (Pr \ll 1) \tag{3.41}$$

Finally, let us discuss the limiting situation $Ra_v \gg 1$, $Pr \gg 1$, where the temperature difference is large but the heat diffusion coefficient is small. After the temperature difference is switched on, a thermal wave travels slowly across the fluid-filled cavity, so the thickness of the layer within which the temperature distribution and stratification are non-uniform in the vibrational field increases with time according to the law $\delta(t) \sim \sqrt{\chi t}$. At some moment τ^*, the instability appears inside this layer. The value of τ^* may be estimated from the following 'quasi-static' considerations. The instability onset is defined as a moment of time when the vibrational Rayleigh number Ra_v (defined through $\delta(t)$ as a reference scale) exceeds the critical value (Fig. 82), i.e. from the relation

$$Ra_v = \frac{(\beta b \Omega \Theta)^2 t^*}{2\nu} = \text{constant}$$

where the constant does not, of course, equal the value 2129 corresponding to the case of an infinite layer with rigid isothermal boundaries. This leads to the estimation $\tau^* \sim \nu/(b\beta\Omega\Theta)^2$, or in the dimensionless form

$$\tau^* \sim \frac{1}{Ra_v} \qquad (Ra_v \gg 1, Pr \gg 1) \tag{3.42}$$

Note that τ^* is indeed the time of the instability onset ('the time of induction'), so τ^* is only a part of the whole transient time τ.

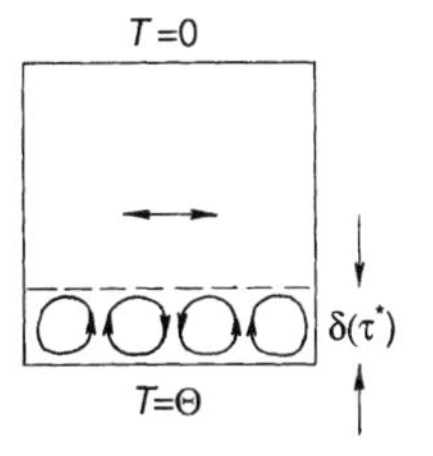

Fig. 82 Instability appearance at $t = \tau^*$ in the range $Pr \gg 1$, $Ra_v \gg 1$

5.3 Numerical Solution

To solve the problem numerically, the finite difference method was applied. The alternating direction implicit scheme of fractional steps was used in the framework of a two-field approach. The Poisson equation was solved by means of the Fourier transformation along one of the spatial coordinates [31]. Most of the computations were performed using a 21×21 grid.

To test the periodicity of transient process and to determine quantitatively the transient time, the following integral characteristics are introduced based on the instantaneous fields of velocity and temperature. The first of them is the sum of discrete stream function values over all the grid nodes:

$$\Psi(t) = \sum_{i,k} \psi_{ik} \tag{3.43}$$

which is proportional to the total angular momentum of the fluid. The second characteristic, the dimensionless heat flux (the Nusselt number), is defined as

$$Nu(t) = -\frac{1}{2}\int_0^1 \left(\left.\frac{\partial T}{\partial y}\right|_{y=0} + \left.\frac{\partial T}{\partial y}\right|_{y=1} \right) \mathrm{d}x \tag{3.44}$$

as well as the instantaneous ones, the maximal and minimal values of the stream function are found as measures of the intensity of vortices of both signs. To describe the duration of the transient process, the time τ is adopted and is determined in the following way. Let $f(t)$ be one of the functions presenting the field evolution in the transient process, e.g. $Nu(t)$ or $\Psi(t)$, and let f_∞ be its limiting value at $t \to \infty$ corresponding to a steady regime if it exists (Fig. 83). Then supposing that f_∞ is non-zero, t may be assumed to be greater than τ if $|f(t) - f_\infty| \leqslant 5$. In most of the numerical experiments $Nu(t)$ has been chosen as a 'tracking' function, and the maximal value of τ was taken as the transient time.

As with all the other characteristics of a transient process in a non-linear dynamic system, the transient time τ depends not only on parameters Ra_v and

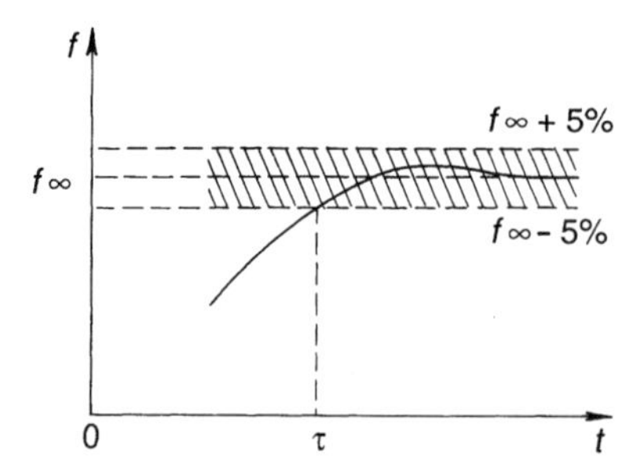

Fig. 83 On the definition of the transient time

Pr but also on the initial conditions. In the majority of the performed computations, the initial conditions resembled a motionless and isothermal fluid. In some of them, the initial condition was set in the form of a discrete vorticity distribution prepared by means of a noise generator. The comparison shows that both ways give the same periodicity of the transient process and practically the same values of the settling time if the amplitude of the initial random distribution is small enough.

5.4 Numerical Results for Pr = 1

The character of a transient process is essentially determined by the type of steady-state regime to which it leads. If the vibrational Rayleigh number is not too large, one may expect that the system possesses stable (or unstable) steady states. We begin the study by considering them.

If $Pr = 1$, the situation is qualitatively close to that described in Section 1, Fig. 60. In Fig. 84 the dependence of *Nu* is presented with the vibrational Rayleigh number for steady regimes. As in the case of side walls with high heat conductivity (Section 1, Fig. 62), in the range of small Ra_v there exists only one stable steady flow with a four-vortices structure, symmetrical with respect to reflection in the planes $x = \frac{1}{2}$ and $y = \frac{1}{2}$. Its intensity is weak and the Nusselt number is close to 1 (line 1). At the point $Ra_v^{(A)} \approx 8 \times 10^3$ this regime destabilizes and a new one bifurcates, possessing the inversion symmetry with respect to the central point. This flow (it may be conventionally called one-

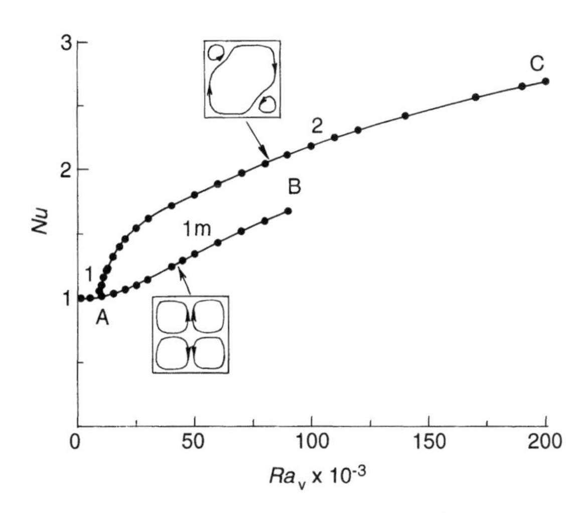

Fig. 84 The Nusselt number versus the vibrational Rayleigh number for steady regimes; $Pr = 1$

vortex) is stable in the range $Ra_v^{(A)} < Ra_v < Ra_v^{(C)}$, where $Ra_v^{(C)} \approx 200 \times 10^3$ (line 2). When $Ra_v > Ra_v^{(C)}$, the one-vortex regime is not steady any longer and transforms into non-damped oscillations. As for the main four-vortices flow, it is metastable in the range $Ra_v^{(A)} < Ra_v < Ra_v^{(B)}$ ($Ra_v^{(B)} \approx 90 \times 10^3$, line 1 m). This means that it is unstable with respect to small perturbations but is long-lived, i.e. its residence time is large as compared to the time of its formation as well as the time of the transition to a stable regime 1 m $\rightarrow$ 2. We remark that, in contrast to the situation considered in Section 1, the stability border of the four-vortices flow goes down. Now $Ra_v^{(A)} \approx 8 \times 10^3$ instead of $Ra_v^{(A)} \approx 15 \times 10^3$, as in the case of highly heat conductive walls, and the bifurcational pattern is pronounced more clearly. Both these distinctions are connected with the 'weakened' boundary conditions for the temperature on the side walls.

In Fig. 85 the sum of the stream function Ψ values in the nodes is presented as a function of Ra_v. The appearance of non-zero Ψ values is eventually related to the settling of the one-vortex regime which does not have reflection symmetry.

The structure of the spectrum of the non-linear steady regimes enables us to understand the qualitative features of the transient process. One has to remember that each numerical experiment has a motionless isothermal fluid as a departure point and a stable regime of thermovibrational convection (if it ever occurs) as its final state.

The transient time τ is shown in Fig. 86 as a function of the vibrational Rayleigh number Ra_v. The dependence is non-monotonic. In the range of

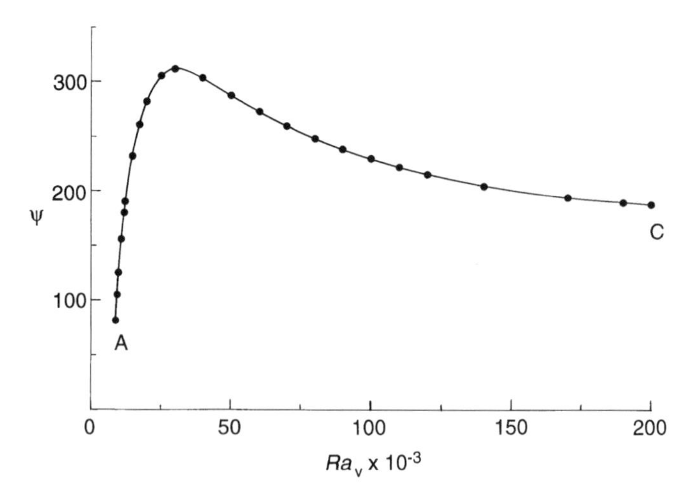

Fig. 85 The sum of the stream function values versus the vibrational Rayleigh number for the one-vortex regime; $Pr = 1$

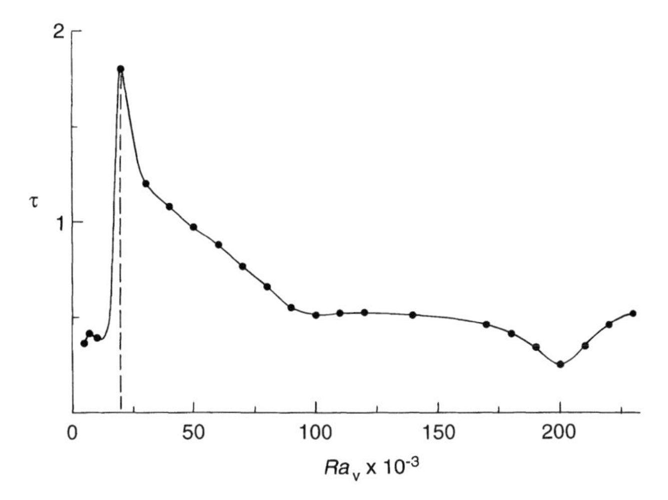

Fig. 86 Transient time as a function of the vibrational Rayleigh number; $Pr = 1$

small Ra_v only one stable steady four-vortices regime exists, formed as a result of the settling process. Its transient time does not in practice depend on Ra_v. The maximal value of $\tau(Ra_v)$ is attained in the vicinity of the point where the instability of the main flow sets in. Such a growth of the transient time growth might be understood with the aid of the following considerations. The transition time essentially depends on the characteristic mode increment. Near the critical point, this increment is small. The settling process in the range $Ra_v^{(A)} < Ra_v < Ra_v^{(B)}$ (where the metastable regime exists) always evolves from the motionless state to the stable one-vortex regime through the intermediate long-lived four-vortices flow. The reconstruction of the flow in the transient process is illustrated in Fig. 87 by the maps of streamlines and isotherms at the point $Ra_v \approx 20 \times 10^3$ (the dashed vertical in Fig. 86) and corresponds to the maximal value of $\tau(Ra_v)$. The evolution of two integral characteristics $Nu(t)$ and $\Psi(t)$ is given in Fig. 88. The curve $Nu(t)$ allows the periodization of the transient process to be understood. In it one can distinguish three different stages: the initial one associated with a transition to the metastable state, the residence time and the time of transition to the stable one-vortex regime. It can be seen that the residence time of the long-lived metastable state forms a significant part of the total transient time. Deviation of $\Psi(t)$ from zero becomes apparent only in the final stage and is due to formation of a non-symmetrical one-vortex flow. As Ra_v increases along the line 1 m, the residence time of the metastable regime decreases and at point B becomes comparable to the times of formation and destruction of the metastable flow. Hence, in the range $Ra_v > Ra_v^{(B)}$, the four-vortices regime is by all means unstable. The minimum on the curve $\tau(Ra_v)$ in the neighborhood of $Ra_v \approx 200 \times 10^3$ corresponds to the

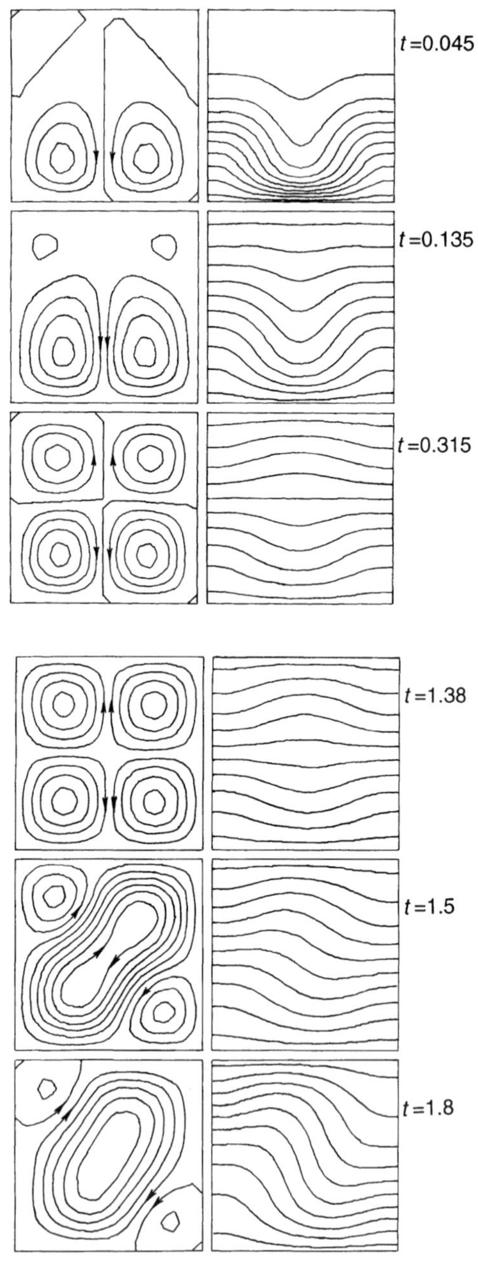

Fig. 87 Evolution of streamlines (left) and isotherms (right); $Pr = 1$; $Ra_v = 20 \times 10^3$

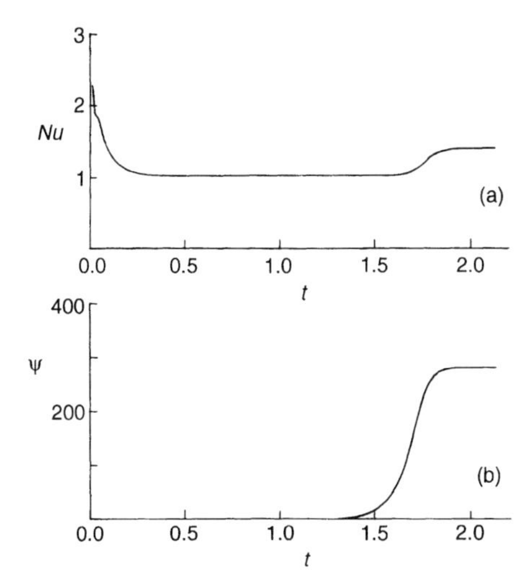

Fig. 88 The Nusselt number (a) and the sum of the stream function values (b) versus time; $Pr = 1$, $Ra_v = 20 \times 10^3$

appearance of non-damped oscillations.† If $Ra_v > Ra_v^{(C)}$, the amplitude of oscillations increases with Ra_v and in the range $Ra_v > 230 \times 10^3$ the oscillations are smaller than $\pm 5\%$ of the average value.

5.5 Effect of the Prandtl Number

The described structure of the spectrum of steady regimes and corresponding properties of the transient process are very sensitive to variations of the Prandtl number. In Fig. 89 the bifurcation pattern for steady regimes is displayed in the plane (Ra_v, Nu) at $Pr = 10$. Up to the graph accuracy, it coincides with that inherent to the limiting case $Pr \to \infty$. We note two distinctions from the situation with $Pr = 1$. First, in the main four-vortices regime, the metastable branch is absent, so at $Ra_v \geqslant Ra_v^{(A)}$ this flow is absolutely unstable. The critical point $Ra_v^{(A)}$ has practically the same value as for $Pr = 1$. The instability of the main state develops against the background of a slow flow and thus it is determined by the critical value of Ra_v independently of Pr, as in the case of equilibrium. Second, inside the overcritical range in addition to the one-vortex regime bifurcating the point $Ra_v^{(A)}$, there exists one more stable regime. It may be conventionally called a two-vortices one and possesses the reflection symmetry with respect to the plane $x = \frac{1}{2}$ (line 3). In the range $70 \times 10^3 < Ra_v < 100 \times 10^3$ this flow may be qualified as metastable, but at $Ra_v > 100 \times 10^3$ it becomes stable. Hence, in this range two stable regimes coexist,

† Recall that the analysis is based on the averaged approach to the description of the thermovibrational convection. This means that we imply slow oscillations whose frequencies are small with respect to the frequency of vibrations.

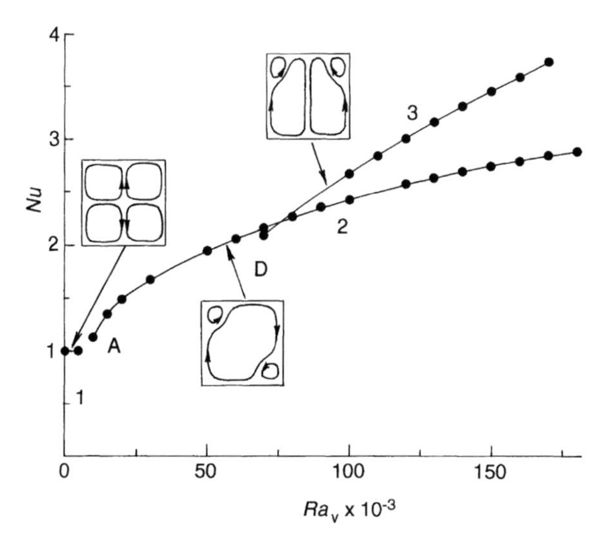

Fig. 89 The Nusselt number as a function of the vibrational Rayleigh number for steady regimes; $Pr = 10$ and $Pr \to \infty$

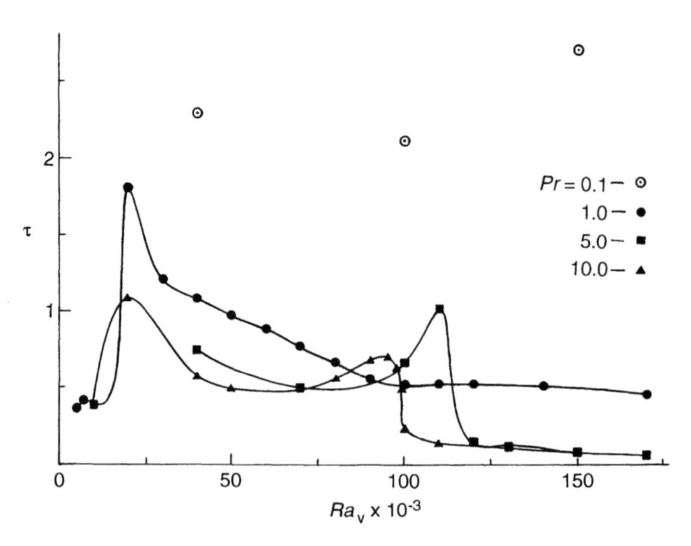

Fig. 90 Transient time as a function of the vibrational Rayleigh number for some values of the Prandtl number

each of them having in the phase space of the system an attraction area of its own.

The transition time as a function of Ra_v is plotted in Fig. 90 with the results obtained for different values of the Prandtl number. The curve $\tau(Ra_v)$ for $Pr = 10$ has two maxima. One of them resides close to the stability border and another to the intersection point of the two supercritical regimes. At small Ra_v values, the limiting regime has a four-vortices structure. In the range $15 \times 10^3 < Ra_v < 100 \times 10^3$ the limiting regime is the secondary one-vortex flow, and at $Ra_v > 110 \times 10^3$ a stable two-vortices flow is set as a result of the transient process. In the figure, the dependence $\tau(Ra_v)$ is presented for $Pr = 5$. It has also two maxima related to competition of two stable steady regimes with different symmetries. The points pertinent to $Pr = 3$ are not plotted, but computations show that such a kind of competition is absent because in the overcritical region only a stable steady one-vortex regime exists as in the case $Pr = 1$.

As a whole, the calculations already carried out confirm the estimations (3.39) to (3.42). In the range of small Ra_v, the transient time does not in practice depend on Ra_v and Pr. For estimations one may take $\tau \approx 0.4$. When $Ra_v \gg 1$ and $Pr \gg 1$, one would expect that the transient time is inversely proportional to Ra_v. This is justified numerically, and for $Ra_v > 110 \times 10^3$ one finds $\tau \approx 12 \times 10^3/Ra_v$. If the Prandtl number values are small, the one-vortex regime is the limiting one in the supercritical region. The threshold of oscillation onset is shifted towards smaller Ra_v. Therefore, for $Pr = 0.1$ (isolated points in Fig. 90), for $Ra_v = 150 \times 10^3$ the flow is at the threshold of the onset of oscillation. At the same time, for $Pr = 0.01$ the oscillations set in at $Ra_v = 50 \times 10^3$. At $Ra_v = 40 \times 10^3$ computations yield $\tau = 2.3$ and $\tau = 20$ for $Pr = 0.1$ and $Pr = 0.01$, respectively. Thus it is possible to adopt $\tau = 0.2/Pr$ as a rough approximation according to (3.41).

In conclusion, note that in the considered situation the vectors of the vibration axis and temperature gradient are mutually perpendicular. If both vectors are parallel (in other words, vector $\mathbf{n}$ is parallel to the y axis), the problem becomes trivial. In this case a non-steady mechanical equilibrium is possible and it is absolutely stable. The vibration damps the appearing perturbations and the transient process is determined by the pure molecular heat diffusion mechanism.

6 Secondary Thermogravitational Flows in the Presence of a Vertical Vibration

In the preceding sections of this chapter the problems have been considered of thermovibrational convection under purely weightlessness conditions, i.e. when the static gravity field is absent. In this section the effect of high-frequency vibrations on the thermogravitational convection in a closed cavity heated from below is investigated (after paper [32]). The problem is posed in a two-dimensional formulation for a square cavity. Its horizontal boundaries $y = 0$ and $y = a$ are maintained at constant temperatures $T = \Theta$ and $T = 0$, respectively. On the vertical boundaries the temperature is a linear function of the vertical coordinate $T = \Theta(1 - y/a)$. All the boundaries are assumed to be rigid, so the average velocity is thereby equal to zero together with the normal component of the pulsation velocity. The vibration axis is vertical and its unit vector **n** has the components $\mathbf{n}(0, 1, 0)$. The consideration follows paper [32].

First, we remark that, in contrast to the problems discussed in Sections 1 and 5 of this chapter, the problem under study admits mechanical equilibrium because in the basic state the temperature gradient is parallel to the vibration axis. Indeed, it is easy to verify that the temperature field $T_0 = \Theta(1 - y/a)$ and the field $\mathbf{w}_0 = 0$ satisfy the equilibrium conditions in the presence of a static field (1.100) and (1.101) under appropriate boundary conditions. Thus, convection may set in only as a result of a breach of the mechanical equilibrium. At the same time, it is clear that ∇T_0 and **n** are parallel in the basic state, and the vibrational convective mechanism is not operative. Because of this, one may expect that the effect of vibrations will be stabilizing (see Section 2 in Chapter 1). As for the secondary flow that bifurcates from the equilibrium, in general the temperature gradient is not precisely parallel to the vibration axis. Therefore, the thermovibrational mechanism of convection interferes, and the effect of vibration might be either stabilizing or destabilizing, depending on the structure of secondary flow and the actual values of the parameters.

To describe a two-dimensional flow, let us introduce the stream functions ψ and F for the average and pulsation velocity fields according to relations (3.1) and choose the dimensionless variables as in Section 1. The set of equations for ψ, F, T taking into account thermogravitational and thermovibrational forces has the form

$$
\begin{gathered}
\frac{1}{Pr}\left(\frac{\partial \Delta\psi}{\partial t} + \frac{\partial \psi}{\partial x}\frac{\partial \Delta\psi}{\partial y} - \frac{\partial \psi}{\partial y}\frac{\partial \Delta\psi}{\partial x}\right) \\
= \Delta\Delta\psi + Ra\frac{\partial T}{\partial x} - Ra_{\mathrm{v}}\left[\frac{\partial T}{\partial x}\frac{\partial}{\partial y}\left(\frac{\partial F}{\partial y}\right) - \frac{\partial T}{\partial y}\frac{\partial}{\partial x}\left(\frac{\partial F}{\partial y}\right)\right] \\
\frac{\partial T}{\partial t} + \frac{\partial \psi}{\partial x}\frac{\partial T}{\partial y} - \frac{\partial \psi}{\partial y}\frac{\partial T}{\partial x} = \Delta T \\
\Delta F = \frac{\partial T}{\partial x}
\end{gathered}
\tag{3.45}
$$

Here Ra, Ra_{v} are the usual thermogravitational Rayleigh number and its thermovibrational analog, respectively.

The boundary conditions for the set (3.45) are

$$x = 0, \qquad x = 1: \qquad \psi = 0, \qquad \frac{\partial \psi}{\partial x} = 0, \qquad T = 1 - y, \qquad F = 0$$

$$y = 0: \qquad \psi = 0, \qquad \frac{\partial \psi}{\partial y} = 0, \qquad T = 1, \qquad F = 0 \tag{3.46}$$

$$y = 1: \qquad \psi = 0, \qquad \frac{\partial \psi}{\partial y} = 0, \qquad T = 0, \qquad F = 0$$

To solve the boundary-value problem, the Galerkin method was used. The unknown variables ψ and F are represented in the form

$$\psi = \sum_{i,j=0}^{M} c_{ij}(t)\varphi_{ij}(x, y), \qquad F = \sum_{i,j=0}^{N} f_{ij}(t)q_{ij}(x, y) \tag{3.47}$$

Here $\varphi_{ij}(x, y)$ and $q_{ij}(x, y)$ are the parts of the complete sets of the basic functions and the coefficients $c_{ij}(t)$ and $f_{ij}(t)$ are time-dependent. The trial functions φ_{ij} satisfy the boundary conditions for ψ and q_{ij} for F. The temperature is presented in the form

$$T = (1 - y) + T'$$

The deviation of T' from the equilibrium distribution $(1 - y)$ satisfies the same boundary conditions as F. Thus, one may set

$$T = (1 - y) + \sum_{i,j=0}^{K} d_{ij}(t)q_{ij}(x, y) \tag{3.48}$$

Hence, we have two sets of trial functions φ_{ij} and q_{ij} for three variables, viz. ψ, F and T. These basic functions are constructed as linear combinations of the Chebyshev polynomials of the first and second order with respect to each coordinate x and y, so that the boundary conditions are satisfied [33]. According to the Galerkin technique, the orthogonality conditions lead to the set of non-linear ordinary differential equations for the coefficients c_{ij}, f_{ij} and d_{ij} depending on time:

$$\dot{X}_l(t) = a_{lj} X_j (T) + b_{ljk} X_j(t) X_k (t) \tag{3.49}$$

where X_l is either of the coefficients. In due course, a finite-dimensional dynamic system is obtained. Its solution approximates the behavior of the complete system (3.45) and (3.46). Steady regimes of convection correspond to the fixed points of (3.49). Such regimes have been found by the Newton method and the settling procedure. The stability of the fixed points has also been

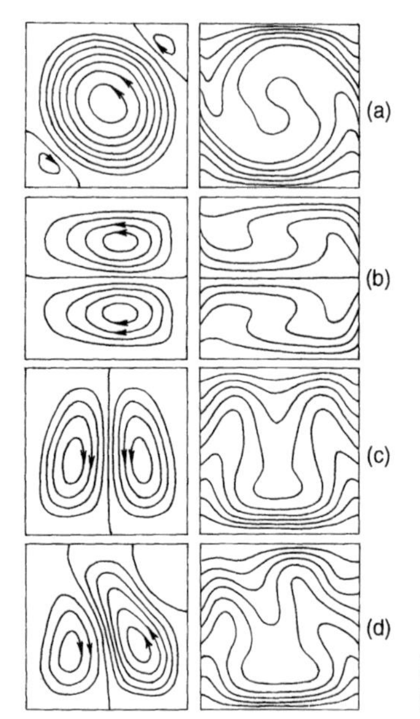

Fig. 91 Steady flow types. Streamlines (left); isotherms (right)

investigated by means of (3.49) using linearization around the fixed point and analysis of the characteristic increment spectrum. Certainly, non-linear oscillations, transient processes, etc., may be studied with the aid of the set (3.49). In numerical calculations, approximations (3.47) and (3.48) contain six trial functions each ($M = N = K = 6$).

Further, we discuss some numerical results. The computations have been performed for three values of the Prandtl number $Pr = 0.02$, $Pr = 1$ and $Pr = 15$. In the examined parameter range Ra, Ra_v and Pr, the existence of four steady regimes has been found in addition to the equilibrium. The pertinent fields of stream function and temperature are demonstrated in Fig. 91. It can be seen that the regimes have different symmetry. The flow (a) possesses the inversion symmetry whereas two regimes ((b) and (c)) have the reflection symmetry with respect to the planes $y = \frac{1}{2}$ and $x = \frac{1}{2}$, respectively. The flow (d) is unsymmetrical.

The ranges of stability for the three mentioned values of the Prandtl number are presented in Figs. 92 to 94. The neutral curves are plotted in the plane where the axes are the Rayleigh number Ra and the dimensionless parameter $\alpha = Ra_v/Ra = b^2\Omega^2\beta\Theta/(2ga)$. Solid lines correspond to the stability border of the mechanical equilibrium. It is known (see [34]) that in the absence of vibrations the critical value of the Rayleigh number does not depend on Pr and equals $Ra^* \approx 5030$. The instability in this case is related to the lowest critical mode of the spectrum possessing the inversion symmetry. In the critical point,

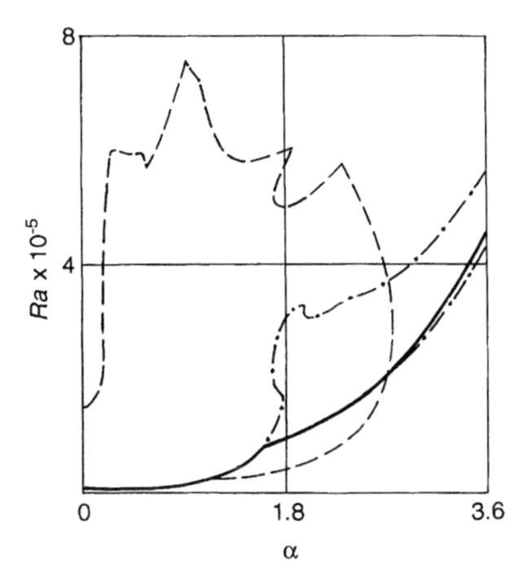

Fig. 92 Neutral curves; $Pr = 0.02$

the steady flow of structure (a) bifurcates straightly. As might have been expected, the increase in the vibration intensity elevates the stability border. Calculations reveal a break in the neutral curve corresponding to equilibrium. Its position $\alpha = 1.6$ does not depend on Pr. At $\alpha < 1.6$ the instability is caused by perturbations of the (a) type, and at $\alpha > 1.6$ by those of the (b) type.

First, let us consider the stability borders for steady flows at $Pr = 0.02$ given in Fig. 92. In accordance with computations, three stable steady regimes are distinguished: the equilibrium and two steady flows of the types (a) and (b). The domain of the equilibrium stability lies below the solid line, the steady flow of the inversion symmetry is stable in the area bounded by the dashed curve and the flow of the reflection symmetry (relative to the plane $y = \frac{1}{2}$) has the stability domain bordered by the dot-and-dash line. One sees that the stability domains of different types overlap, i.e. non-uniqueness of the solution takes place. At the same values of parameters, the flows with different symmetry are possible and their realization depends on the initial conditions. This means that as the parameter varies, the flow structure reorganizes and is accompanied by hysteresis phenomena. In particular, the stability domains for the flows of (a) and (b) types penetrate the domain which, with respect to the equilibrium neutral curve, is subcritical. Thus, for $\alpha < 1.14$, as the Rayleigh number increases, a soft excitation of flow (a) takes place as a result of the equilibrium stability loss. In the range $1.14 < \alpha < 2.7$, as Ra grows, the subcritical equilibrium loses its stability with respect to perturbations of the (a) type. For $\alpha > 2.7$ the hard instability is present relative to perturbations of the (b) type. On the upper border of the stability domain of the (a)-type flow, an oscillatory instability occurs. The breaks on the dashed curve reflect the changes in the form of the critical oscillatory perturbations.

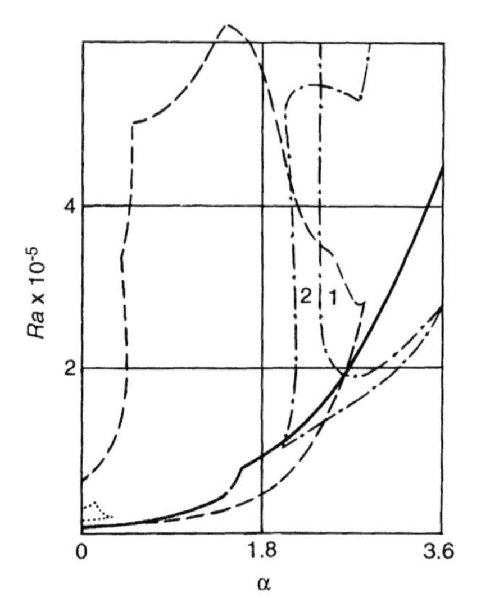

Fig. 93 Neutral curves; $Pr = 1$

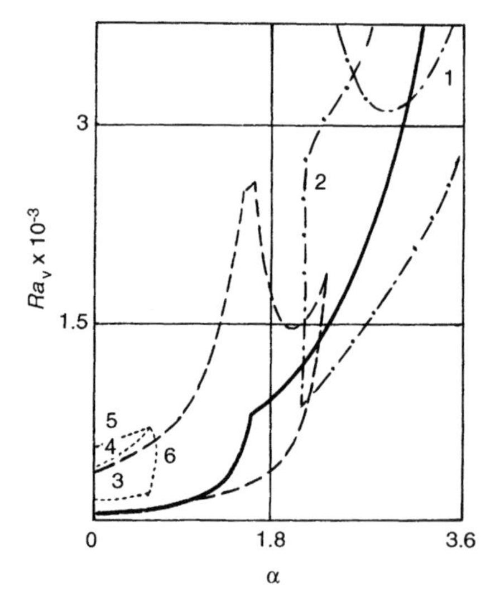

Fig. 94 Neutral curves; $Pr = 15$

The picture alters substantially as the Prandtl number grows (see Figs. 93 and 94). As before, the dashed lines bound the stability domains for the (a)-type flows and the dot-and-dash lines 1 bound those for the flows of the (b) type. In addition to them, as the Prandtl number increases the domains of stability appear for the flows of (c) and (d) types. Those domains for non-symmetrical (d) flows are indicated in Figs. 93 and 94 by the dot-and-dash curves 2. When

$Pr = 15$, there exist the stability domains for the flow of reflection symmetry with respect to the vertical $x = \frac{1}{2}$ (dashed lines 3, 4 and 6) and for an asymmetrical flow (dashed lines 4 and 5). As Pr decreases, the area bounded by the curves 5, 6 and 3 tends to shrink but it still persists at $Pr = 1$ (Fig. 93).

The analysis of oscillatory perturbations and the estimations given in [32] lead to the conclusion that the characteristic neutral frequencies are of the order of 10^3 in terms of χ/a^2. Hence, the reference times for oscillatory perturbations are much greater than the period of vibrations and the consideration in the framework of the averaged approach is physically justified.

References

1. Gershuni, G.Z., E.M. Zhukhovitsky and Yu.S. Yurkov. On vibrational thermal convection in rectangular cavity, in *Problems of Viscous Flows*, Novosibirsk, SB AS USSR, 1981, pp. 19–23.
2. Gershuni, G.Z., E.M. Zhukhovitsky and Yu.S. Yurkov. On vibrational thermal convection in weightlessness conditions, in *Hydromechanics and Heat/Mass Transfer in Weightlessness*, Nauka, Moscow, 1982, pp. 90–8.
3. Gershuni, G.Z., E.M. Zhukhovitsky and Yu.S. Yurkov. Vibrational thermal convection in rectangular cavity, *Trans. Acad. Sci. SSSR (Izvestiya), Mech. Zhidk. Gaza*, 1982, **4**, 94–9.
4. Yanenko, N.N. *Method of Fractional Steps for Solving Many-Dimensional Problems in Mathematical Physics*, Nauka, Novosibirsk, 1967, 196 pp.
5. Chernatynsky, V.I. and M.I. Shliomis. Convection in proximity to critical Rayleigh number at almost vertical temperature gradient, *Trans. Acad. Sci. SSSR (Izvestiya), Mech. Zhidk. Gaza*, 1973, **1**, 64–70.
6. Tarunin, E.L. Convection in closed cavity heated from below at the infraction of equilibrium conditions, *Trans. Acad. Sci. SSSR (Izvestiya), Mech. Zhidk. Gaza*, 1977, **2**, 203–7.
7. Hall, P. and I.C. Walton. Bènard convection in a finite box: secondary and imperfect bifurcations, *J. Fluid Mech.*, 1979, **90**(2), 377.
8. Khallouf, H., G.Z. Gershuni and A. Mojtabi. Numerical study of two-dimensional thermovibrational convection in rectangular cavity, *Numer. Heat Transfer*, 1995, **A27**, 297–305.
9. Siraev, R.R. On vibrational convection in rectangular cavity, in *Convective Flows*, Perm, 1987, pp. 55–61.
10. Siraev, R.R. Numerical study of steady and oscillatory regimes of vibrational convection in rectangular cavity with varying temperature in weightlessness, *J. Engn. Phys.*, 1989, **57**(1), 33–7.
11. Siraev, R.R. Vibrational convection in rectangular cavity in the boundary layer regime, in *Convective Flows*, Perm, 1989, pp. 68–73.
12. Markman, G.S. and A.L. Urintsev. On the effect of high frequency vibration on the appearance of secondary convective regimes, *Trans. Acad. Sci. SSSR (Izvestiya), Mech. Zhidk. Gaza*, 1976, **2**, 90–6.
13. Markman, G.S. and A.L. Urintsev. On appearance of finite-amplitude convection in vibrating fluid layer heated from above, *Trans. Acad. Sci. SSSR (Izvestiya), Mech. Zhidk. Gaza*, 1978, **1**, 27–35.

14. Zen'kovskaya, S.M. and S.N. Ovchinnikova. Thermovibrational convection in a fluid layer under weightlessness or reduced gravity, *J. Appl. Mech. and Techn. Phys.*, 1991, **2**, 84–90.
15. Braverman, L. and A. Oron. Weakly nonlinear analysis of the vibrational convective instability in a fluid layer, *Eur. J. Mech., B/Fluids*, 1994, **13**(5), 557–72.
16. Zaks, M.A., D.V. Lyubimov and V.I. Chernatynsky. On the effect of vibration on the regimes of supercritical convection, *Trans. Acad. Sci. SSSR (Izvestiya), Fiz. Atmosfery i Okeana*, 1983, **19**(3), pp. 312–14.
17. Lyubimov, D.V. and M.A. Zaks. Two mechanisms of the transition to chaos in finite-dimensional models of convection, *Physica D*, 1983, **9**, 52–64.
18. Gershuni, G.Z., E.M. Zhukhovitsky and A.N. Sharifulin. Vibrational thermal convection in cylindrical cavity, *Numer. Meth. Cont. Med. Mech., Novosibirsk*, 1983, **14**(4), 21–33.
19. Sharifulin, A.N. Vibrational convection in cylindrical cavity under arbitrary direction of heating, in *Convective Flows*, Perm, 1981, pp. 22–9.
20. Siraev, R.R. Vibrational thermal convection near the homogeneously heated cylinder, *Trans. Acad. Sci. SSSR (Izvestiya), Mech. Zhidk. Gaza*, 1989, **3**, 23–6.
21. Gershuni, G.Z., E.M. Zhukhovitsky and R.R. Siraev. A convective boundary layer in vibrating fluids at zero gravity, *Stab. and Appl. Anal. Cont. Media*, 1991, **1**(4), 349–71.
22. Kuehn, T.H. and R.J. Goldstein. Numerical solution to the Navier–Stokes equations for laminar natural convection about a horizontal isothermal circular cylinder, *Int. J. Heat and Mass Transf.*, 1980, **23**(7), 971–9.
23. Schlichting, H. *Theory of Boundary Layer*, Nauka, Moscow, 1969, 742 pp.
24. Chernatynsky, V.I. Numerical study of vibrational convection in cylindrical layer, in *Convective Flows*, Perm, 1989, pp. 32–7.
25. Ivanova, A.A. and V.G. Kozlov. Vibrational gravitational convection in horizontal cylindrical layer, in *Convective Flows*, Perm, 1985, pp. 45-57.
26. Ivanova, A.A. Stability of free convective flow in horizontal cylindrical layer in vibrational field, in *Convective Flows*, Perm, 1989, pp. 37-44.
27. Jaluria, Y. *Natural Convection*, Mir, Moscow, 1983, 399 pp.
28. Naife, A. *Methods of Perturbations*, Mir, Moscow, 1976, 455 pp.
29. Siraev, R.R. Vibrational thermal convection about homogeneously heated sphere, *Inst. Sci. Inform.*, 1989, **1836-B89**, 14 pp.
30. Chernatynsky, V.I., G.Z. Gershuni and R. Monti. Transient regimes of thermovibrational convection in a closed cavity, *Micrograv. Quart.*, 1993, **3**(1), 55–67.
31. Hockney, R.V. The potential calculation and some applications, in *Methods in Computational Physics*, Academic Press, New York, London, 1970, No. **9**, p. 35.
32. Gelfgat, A.Yu. Development and instability of steady convective flows in square cavity heated from below in the field of vertical vibrations, *Trans. Acad. Sci. SSSR (Izvestiya), Mech. Zhidk. Gaza*, 1991, **2**, 9–18.
33. Gelfgat, A.Yu. Variational method of solving the problems of viscous incompressible fluid dynamics in rectangular cavities, in *Applied Problems of Math. Phys.*, Riga, 1987, pp. 14–24.
34. Gershuni, G.Z. and E.M. Zhukhovitsky, *Convective Stability of Incompressible Fluid*, Nauka, Moscow, 1972, 392 pp.

4 Internal Heat Sources

In this chapter the free convection arising in a fluid with internal heat generation as a result of joint action of both thermogravitational and thermovibrational mechanisms is considered. The situations are discussed where internal heat sources are distributed uniformly over the fluid volume and when the internal heating is connected with an exothermic reaction taking place in a fluid. For both cases the linear stability of the mechanical equilibrium is studied. The finite amplitude non-linear regimes developing after the loss of stability are considered for a plane horizontal layer with longitudinal vibrations.

1 Basic Equations: Mechanical Equilibrium

Let us consider a plane horizontal fluid layer bounded by the parallel planes $z = 0$ and $z = h$ (Fig. 95). In the fluid, internal heat generation takes place with the intensity Q defined as the amount of heat released into the unit volume per unit time. In this section, as well as in two following ones, we assume that the heat generation is spatially uniform, i.e. $Q = \text{constant}$. The internal heat generation leads to a non-uniform temperature distribution whose actual form depends on the thermal conditions on the boundaries. (This point will be discussed later on.) The vibration axis is oriented arbitrarily with respect to the layer. The unit vector along the vibration axis $\mathbf{n}$ is oriented at the angle α to the horizontal axis; its components are $\mathbf{n}(\cos\alpha, 0, \sin\alpha)$. In the static gravity and vibrational fields, the non-uniform temperature distribution causes initiation of the combined thermogravitational and thermovibrational convection. The set of equations to describe the phenomenon differs from (1.17) to (1.19) by the form of the heat transfer equation, where now the internal heat sources must be added:

$$\frac{\partial T}{\partial t} + \mathbf{v}\nabla T = \chi \Delta T + \frac{Q}{\rho c_p} \tag{4.1}$$

Let us introduce dimensionless variables. The following reference values are taken as the units of length, time, velocity and pressure, respectively: h, h^2/ν,

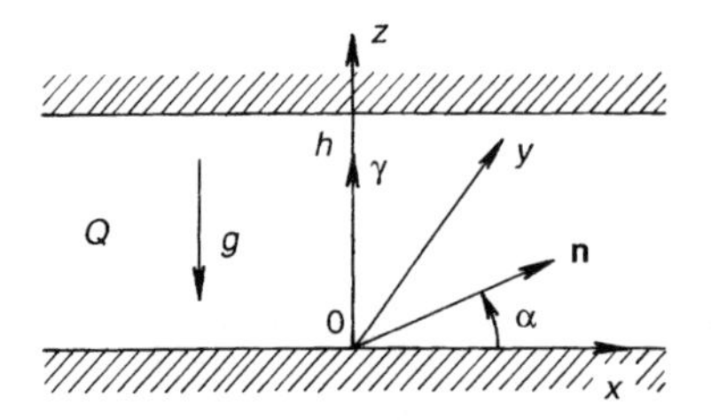

Fig. 95 Coordinate system

χ/h and $\rho\nu\chi/h^2$. It is convenient to choose $Qh^2/\rho c_p$ as the unit of temperature and of **w**. Thus the dimensionless system of equations takes the form

$$\frac{\partial \mathbf{v}}{\partial t} + \frac{1}{Pr}(\mathbf{v}\nabla)\mathbf{v} = -\nabla p + \Delta \mathbf{v} + Ra_{\mathrm{q}}\, T\boldsymbol{\gamma} + Ra_{\mathrm{v}}(\mathbf{w}\nabla)(T\mathbf{n} - \mathbf{w})$$

$$Pr\frac{\partial T}{\partial t} + \mathbf{v}\nabla T = \Delta T + 1 \tag{4.2}$$

$$\operatorname{div}\mathbf{v} = 0, \qquad \operatorname{div}\mathbf{w} = 0, \qquad \operatorname{curl}\mathbf{w} = \nabla T \times \mathbf{n}$$

The set (4.2) includes three dimensionless parameters, viz. the Rayleigh number Ra_{q}, the vibrational Rayleigh number Ra_{v} defined through the internal heat generation power and the Prandtl number Pr:

$$Ra_{\mathrm{q}} = \frac{g\beta Q h^5}{\rho c_p \nu \chi^2}, \qquad Ra_{\mathrm{v}} = \frac{\varepsilon}{\nu\chi}\left(\frac{Qh^3}{\rho c_p \chi}\right)^2 = \frac{1}{2\nu\chi^3}\left(\frac{\beta b \Omega Q h^3}{\rho c_p}\right)^2, \qquad Pr = \frac{\nu}{\chi} \tag{4.3}$$

Now let us formulate the boundary conditions. We shall consider both boundaries to be rigid. The non-slip condition gives

$$z = 0 \text{ and } z = 1: \qquad \mathbf{v} = 0, \qquad w_z = 0 \tag{4.4}$$

The conditions of closeness for the average and oscillating flows must be satisfied:

$$\int_0^1 v_x \,\mathrm{d}z = \int_0^1 v_y \,\mathrm{d}z = 0 \tag{4.5}$$

$$\int_0^1 w_x \,\mathrm{d}z = \int_0^1 w_y \,\mathrm{d}z = 0 \tag{4.6}$$

The following three variants of the temperature boundary conditions will be considered:

Variant (a). Both boundaries are isothermal and maintained at the same temperature adopted as the zero point:

$$z = 0 \text{ and } z = 1: \qquad T = 0 \tag{4.7a}$$

Variant (b). One of the boundaries (e.g. the upper one) is isothermal and the other one is heat-insulated (i.e. the normal component of the heat flux is zero):

$$\begin{aligned} z = 0: &\qquad \frac{\partial T}{\partial z} = 0 \\ z = 1: &\qquad T = 0 \end{aligned} \tag{4.7b}$$

Variant (c). One of the boundaries (e.g. the upper one) is isothermal and on the other one the Newton linear heat exchange law holds:

$$z = 0: \qquad \kappa \frac{\partial T}{\partial z} = aT$$

Here all the values are dimensional, κ is the heat conductivity coefficient of the fluid and a is the heat exchange coefficient on the boundary. The dimensionless form of the boundary conditions is

$$\begin{aligned} z = 0: &\qquad \frac{\partial T}{\partial z} = BiT \\ z = 1: &\qquad T = 0 \end{aligned} \tag{4.7c}$$

Here the Biot number $Bi = ah/\kappa$ is the dimensionless parameter of the heat exchange on the boundary.

From the general set of equations (4.2) it is possible to obtain the necessary conditions of the mechanical quasi-equilibrium ($\mathbf{v} = 0$). The equilibrium fields T_0 and $\mathbf{w}_0$ have to satisfy the equations

$$\begin{gathered} Ra_q(\nabla T_0 \times \gamma) + Ra_v[\nabla(\mathbf{w}_0\mathbf{n}) \times \nabla T_0] = 0 \\ \Delta T_0 = -1 \\ \operatorname{div} \mathbf{w}_0 = 0, \qquad \operatorname{curl} \mathbf{w}_0 = \nabla T_0 \times \mathbf{n} \end{gathered} \tag{4.8}$$

with the appropriate boundary conditions.

For all three variants specified, it is easy to be prove that, independently of the orientation of the vibration axis, the mechanical equilibrium is possible. It has $T_0 = T_0(z)$ and the structure of the $\mathbf{w}$ field of the form $w_{0y} = w_{0z} = 0$,

$w_{0x} = w_0(z)$. These distributions are

$$\begin{cases} T_0 = \frac{1}{2}z(1-z) \\ w_0 = \frac{1}{12}(-1+6z-6z^2)\cos\alpha \end{cases} \tag{4.9a}$$

$$\begin{cases} T_0 = \frac{1}{2}(1-z^2) \\ w_0 = \frac{1}{6}(1-3z^2)\cos\alpha \end{cases} \tag{4.9b}$$

$$\begin{cases} T_0 = \dfrac{1}{2(1+Bi)}(1-z)[1+(1+Bi)z] \\ w_0 = \dfrac{\cos\alpha}{12(1+Bi)}\left[(z-Bi)+6Bi\,z-6(1-Bi)z^2\right] \end{cases} \tag{4.9c}$$

Of the considered ones, variant (c) is the most general: the limiting case $Bi \to \infty$ corresponds to isothermal boundaries and the opposite limiting case $Bi = 0$ corresponds to the thermally insulated lower boundary. Formulas (4.9c) then reduce to (4.9a) or (4.9b), respectively.

In Fig. 96, the equilibrium temperature profiles for different values of the Biot number are presented. It can be seen that for $Bi = 0$ the fluid layer is stratified potentially unstable (we mean stratification in the gravitational field). At $Bi \neq 0$ there exists a zone of unstable stratification abutting the upper

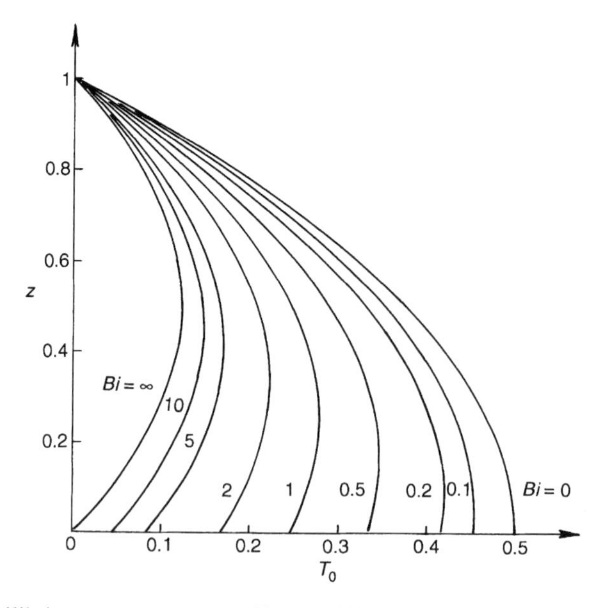

Fig. 96 Equilibrium temperature profiles at different Biot numbers

boundary. Its thickness decreases as Bi grows. The limiting case $Bi \to \infty$ corresponds to the situation of an unstably stratified upper half of the layer. At $Bi \neq 0$ there exists the zone of potentially stable stratification near the lower boundary, which becomes thicker as Bi increases.

The equilibrium temperature profiles in Fig. 96 correspond to the case of positive values of Ra_{q}, i.e. heat generation. At $Ra_{\mathrm{q}} < 0$ (heat sinks distributed in the fluid), the profiles $T_0(z)$ may be obtained from those presented in Fig. 96 by reflection with respect to the z axis. It is evident that in this case the upper part of the layer is stratified stably at $Bi \neq 0$. The cases $Ra_{\mathrm{q}} > 0$ and $Ra_{\mathrm{q}} < 0$ from the viewpoint of instability are not equivalent. It could be understood, for example, by considering the limiting case $Bi = 0$. At $Ra_{\mathrm{q}} > 0$, as was mentioned, the layer is unstably stratified, while at $Ra_{\mathrm{q}} < 0$ the stratification in the whole layer is stable. The Rayleigh (thermogravitational) instability mechanism is not active and for this case one may only expect an instability of the thermovibrational origin.

2 Equilibrium Stability

2.1 Spectral Amplitude Problem

Let us consider now the linear stability of the mechanical equilibrium determined by the profiles (4.9) [1–3]. Small perturbations of the temperature T', pressure p', oscillating velocity $\mathbf{w}'$ and slow average velocity $\mathbf{v}'$ are introduced. Substituting the perturbed fields into the general set (4.2), after linearization one obtains the set of equations for perturbations:

$$\frac{\partial \mathbf{v}'}{\partial t} = -\nabla p' + \Delta \mathbf{v}' + Ra_{\mathrm{q}} T' \boldsymbol{\gamma} + Ra_{\mathrm{v}}[(\mathbf{w}_0 \nabla)(T'\mathbf{n} - \mathbf{w}') + (\mathbf{w}'\nabla)(T_0 \mathbf{n} - \mathbf{w}_0)]$$

$$Pr \frac{\partial T'}{\partial t} + \mathbf{v}' \nabla T_0 = \Delta T' \tag{4.10}$$

$$\operatorname{div} \mathbf{v}' = 0, \qquad \operatorname{div} \mathbf{w}' = 0, \qquad \operatorname{curl} \mathbf{w}' = \nabla T' \times \mathbf{n}$$

On the boundaries of the layer $z = 0$ and $z = 1$ the average velocity $\mathbf{v}'$ and the normal component of the oscillatory velocity w_z' turn to zero. From (4.7) the temperature boundary conditions may be obtained for each variant under consideration.

By analogy with the stability problem for the layer with different temperatures of the boundaries (Section 5 in Chapter 1), we surmise that the most dangerous perturbations are two-dimensional ones of the form $\mathbf{v}'(v_x', 0, v_z')$, $\mathbf{w}'(w_x', 0, w_z')$. The temperature perturbations T' and p' also do not depend on y. Eliminating by the usual method the pressure perturbations p' and the horizontal components v_x', w_x', we introduce the normal modes

$$\left(v_z', w_z', t'\right) = (v, w, \theta) \exp\left(-\lambda t + \mathrm{i}kx\right)$$

where v, w, θ are the amplitudes depending on the transversal coordinate z. Then one obtains the set of linear homogeneous equations for the amplitudes:

$$-\lambda D v = D^2 v - k^2\, Ra_{\mathrm{q}} \theta - \ Ra_{\mathrm{v}}\ T_0'\left(\mathrm{i}kw' \cos\alpha + k^2 w \sin\alpha - k^2\theta \cos\alpha\right)$$

$$-\lambda\, Pr\,\theta = D\theta - T_0' v \tag{4.11}$$

$$Dw = -\left(\mathrm{i}k\theta' \cos\alpha + k^2\theta \sin\alpha\right)$$

Here as usual D stands for the operator $D = \mathrm{d}^2/\mathrm{d}z^2 - k^2$ and the prime denotes differentiation with respect to the transversal coordinate z.

At the boundaries of the layer, the amplitudes v and w satisfy the conditions

$$z = 0 \text{ and } z = 1: \qquad v = v' = 0, \qquad w = 0 \tag{4.12}$$

and the conditions for θ with respect to the above-mentioned three variants are

$$z = 0 \text{ and } z = 1: \qquad \theta = 0 \tag{4.13a}$$

$$z = 0: \qquad \theta' = 0; \qquad z = 1: \qquad \theta = 0 \tag{4.13b}$$

$$z = 0: \qquad \theta' = Bi\,\theta; \qquad z = 1: \qquad \theta = 0 \tag{4.13c}$$

The set of equations (4.11) with the boundary conditions (4.12) and one of the variants (4.13) forms the spectral amplitude problem with the eigenvalue λ depending on the parameters Ra_{q}, Ra_{v}, Pr, α and k. The boundary problem for the amplitudes was solved by the Runge–Kutta–Merson method in combination with the shooting procedure.

2.2 Computational Results

Variant (a). Let us turn to the results of the equilibrium stability study and consider first the variant of isothermal boundaries.

The limiting case $Ra_{\mathrm{v}} = 0$ corresponds to the absence of vibration. Here we obtain the problem of equilibrium stability for a plane horizontal fluid layer with uniformly distributed internal heat sources in the static gravity field. Instability is caused by the thermogravitational mechanism. It sets in at the critical value $Ra_{\mathrm{qm}} = 3.733 \times 10^4$ (the result is minimized with respect to k) and is related to the critical wavenumber $k_{\mathrm{m}} = 4.00$. These values completely coincide with the ones obtained earlier [4].

The opposite limiting case $Ra_{\mathrm{q}} = 0$ corresponds to the absence of the static gravity, i.e. to weightlessness. The instability in this case is solely due

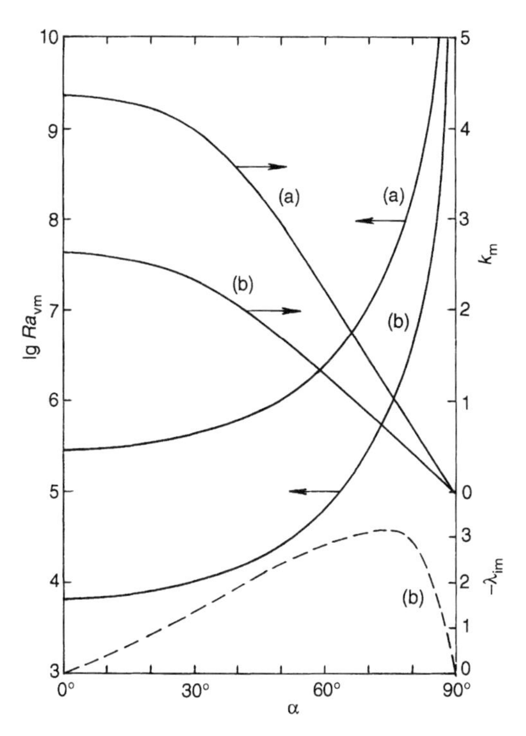

Fig. 97 Critical characteristics of thermovibrational instability in dependence on the angle of the vibration axis inclination (case of weightlessness, $Ra_q = 0$; variants (a) and (b))

to the thermovibrational mechanism and the development of monotonic perturbations ($\lambda_i = 0$). The stability border depends upon the orientation α of the vibration axis with respect to the layer boundaries. When α increases from zero the critical value Ra_{vm} increases monotonically and k_m decreases. For $\alpha \to 90°$, the critical value Ra_{vm} tends to infinity and k_m tends to zero (Fig. 97). In this case, for the vibration axis parallel to the temperature gradient, absolute stabilization takes place (see Section 2 in Chapter 1).

At arbitrary values of Ra_q and Ra_v, the instability of equilibrium is due to the joint action of both thermogravitational and thermovibrational mechanisms. The stability borders with respect to the most dangerous (in the k sense) perturbations are presented in Fig. 98 for different values of the angle; the corresponding values of k_m are displayed in Fig. 99.

At $\alpha = 0°$ and for relatively small inclination angles of the vibration axis to the horizontal ($\alpha < 53°$), one may see from Fig. 98 that both mechanisms act to mutual stabilization. At larger angles, the stabilizing role of vibrations is strong enough. In the limiting case $\alpha \to 90°$ and for large values of Ra_v,

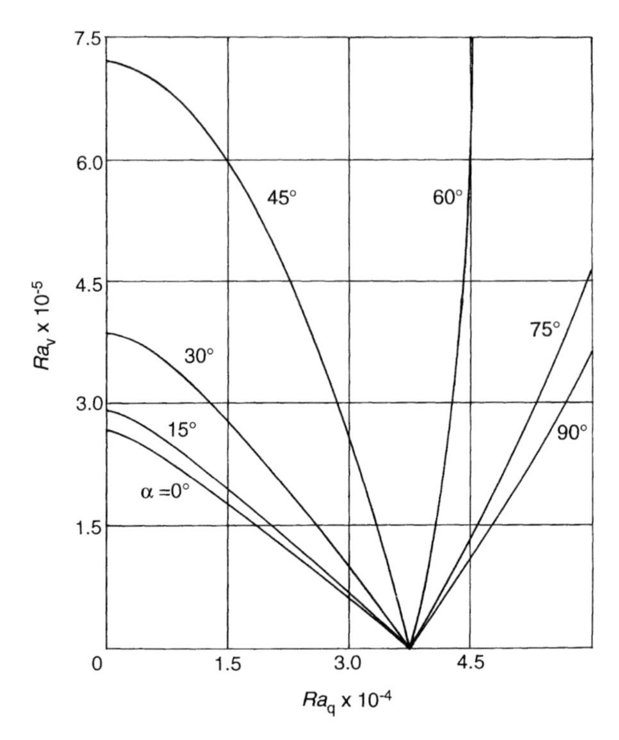

Fig. 98 Stability borders in the plane (Ra_q, Ra_v) for different α (variant (a))

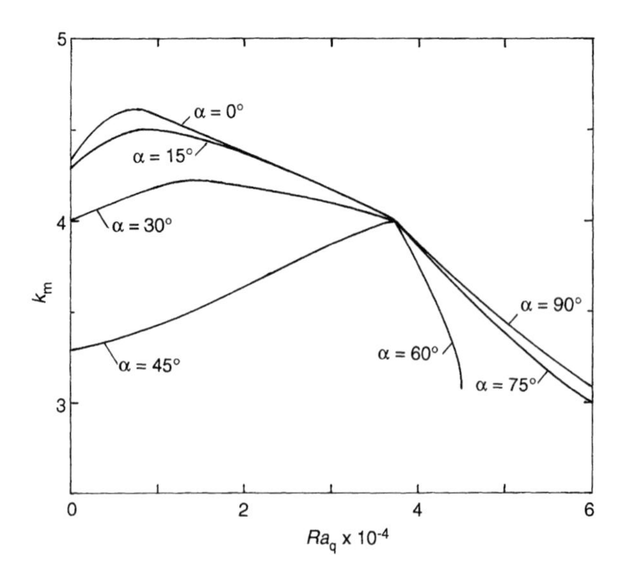

Fig. 99 Critical wavenumber k_m in dependence on Ra_q at different values of the angle α

the critical value Ra_{qm}, according to the calculation results, increases with Ra_v as $Ra_{qm} = 79.5\, Ra_v^{1/2}$. From that, using the definitions of Ra_q and Ra_v, one concludes that there exists the limiting value of the dimensional parameter of the vibration intensity $(b\Omega)^* = 0.01779gh^2/\sqrt{\nu\chi}$. If $b\Omega > (b\Omega)^*$, the equilibrium is absolutely stable at arbitrary values of the internal heat generation power Q.

As the calculations show, in contrast to the case of a horizontal layer with different temperatures of its boundaries (Section 5 in Chapter 1), in the situation under consideration the instability has, generally speaking, an oscillatory character. The monotonic instability takes place only at $\alpha = 0°$ (longitudinal vibration), at $\alpha = 90°$ (transverse vibration) and in the case where just one instability mechanism works, i.e. when $Ra_v = 0$ or $Ra_q = 0$. In all the other cases, at the stability border $\lambda_i \neq 0$, i.e. the instability has a form of two-dimensional rolls whose axes are parallel to the y axis. These rolls drift with the phase velocity $c = \lambda_i/k_m$ along the negative direction of the x axis since in the calculations $c < 0$.

In the case of an oscillatory instability, the critical parameters are, in general, the functions of the Prandtl number. However, the calculations show that at least in the range $0.1 \leqslant Pr \leqslant 10$ the critical values of Ra_q, Ra_v and k_m do not in practice depend on Pr. As for the phase velocity of neutral perturbations, in the said interval the product $c_m\, Pr$ identically does not depend on Pr. The values of this product for different α are presented in Fig. 100.

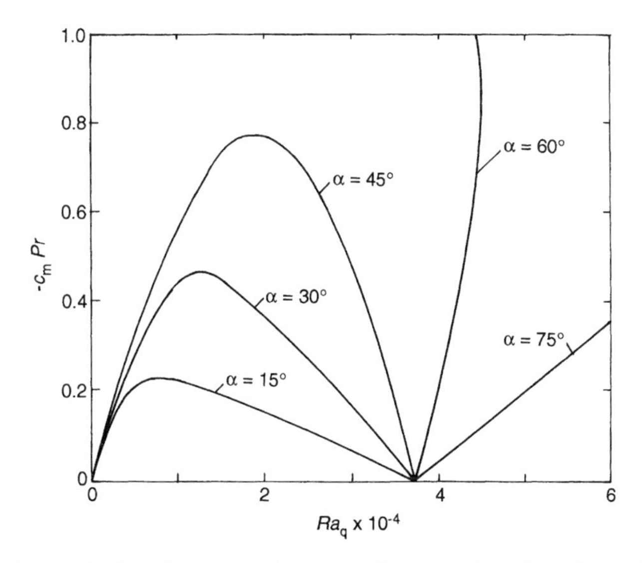

Fig. 100 Phase velocity of the neutral perturbations as a function of Ra_q for different angles α

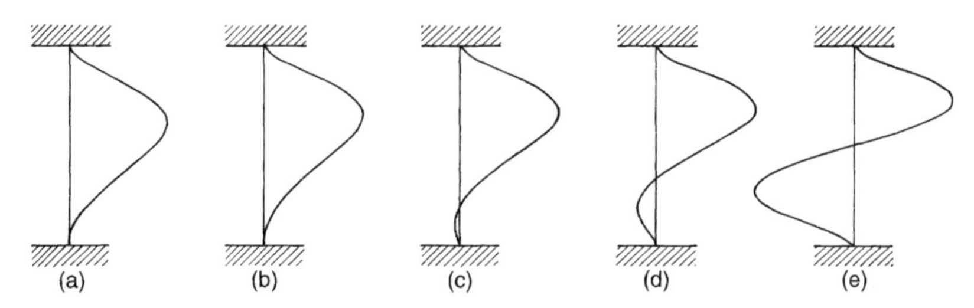

Fig. 101 Profiles of the transversal velocity on the stability border, $\alpha = 0$. (a) $Ra_q = 2.733 \times 10^4$, $Ra_v = 0$; (b) $Ra_q = 1 \times 10^4$, $Ra_v = 2.11 \times 10^5$; (c) $Ra_q = 4 \times 10^3$, $Ra_v = 2.53 \times 10^5$; (d) $Ra_q = 2 \times 10^3$, $Ra_v = 2.64 \times 10^5$; (e) $Ra_q = 0$, $Ra_v = 2.69 \times 10^5$

The form of critical perturbations may be determined as the eigenfunctions of the spectral amplitude problem (4.11) to (4.13). In Fig. 101 the profiles of the transversal component of the average velocity are presented at the stability border under longitudinal vibrations ($\alpha = 0$). At small values of Ra_v, when the gravitational instability mechanism predominates, the critical perturbation turns out to be a vertical one-vortex structure. These vortices emerge in the upper (unstably stratified) half of the layer and penetrate its lower half where the stratification is stable. When Ra_v increases and Ra_q decreases, the nature of the instability changes and in the limit $Ra_q \to 0$ it becomes purely thermovibrational. A vertical two-vortices structure forms with the opposing circulations of the vortices. In this case the upper and the lower parts of the layer are equivalent with respect to excitation of the thermovibrational convection.

Variant (*b*). Let us consider the equilibrium stability in the layer whose upper boundary is isothermal and the lower one is thermally insulated (the spectral amplitude problem has been formulated in (4.11), (4.12) and (4.13b)). In the absence of vibrations ($Ra_v = 0$) the critical parameters of instability have been obtained as $Ra_{qm} = 2.772 \times 10^3$ and $k_m = 2.63$, in good agreement with [5]. In the limiting case of weightlessness ($Ra_q = 0$), the critical parameters depend on the angle of inclination. These dependences are presented in Fig. 97 together with the corresponding results for the variant (a). In contrast to the case of the layer with isothermal boundaries, the instability is now caused by oscillatory perturbations with the exception of the limiting cases $\alpha = 0°$ and $\alpha = 90°$. The values of the neutral frequency λ_{im} are presented in Fig. 97 for $Pr = 1$. The critical parameters Ra_{vm} and k_m, similarly to variant (a), do not depend in practice on the Prandtl number.

The family of neutral curves in the plane (Ra_q, Ra_v) for different inclination angles is presented in Fig. 102, similarly to Fig. 98. As in variant (a), the stability is minimal at $\alpha = 0°$ (longitudinal vibration) when both mechanisms are destabilizing. As the contribution of the transversal

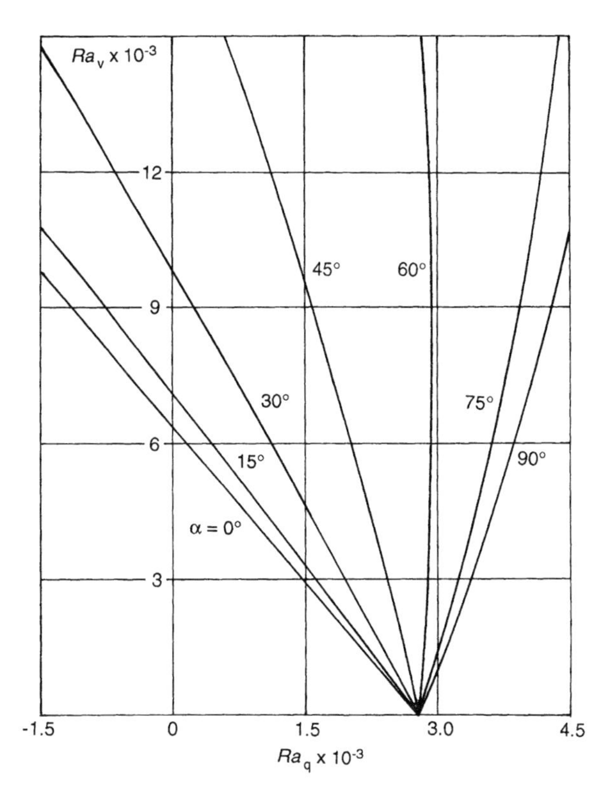

Fig. 102 Stability boundaries in the plane (Ra_q, Ra_v) for different α (variant b))

vibration increases, the stability enhances, and at $\alpha = 90°$ vibrations play a purely stabilizing role. In this case the asymptotics $Ra_{qm} \sim \sqrt{Ra_v}$ sets in for large Ra_v, so the absolute stabilization is reached at the threshold value of the vibration intensity parameter $(b\Omega)^* = 0.04874gh^2/\sqrt{\nu\chi}$. Contrary to variant (a), the neutral curves in the plane (Ra_q, Ra_v) are not symmetrical with respect to the axis Ra_v. Negative Ra_q corresponds to heat sinks (or to heat sources if the lower boundary is isothermal and the upper one is thermally insulated). The cases $Ra_q > 0$ and $Ra_q < 0$ correspond to different stratifications and, of course, are physically non-equivalent.

The location of the oscillatory instability borders in Fig. 102 does not depend on the Prandtl number in a wide interval. In contrast to this, the neutral frequency of oscillations depends on the Prandtl number rather strongly, as can be seen from Fig. 103, where the values λ_{im} along the neutral curve are given at $\alpha = 15°$.

Variant (c). Finally, let us consider the situation when one boundary of the layer is isothermal and on the other one the Newton heat exchange law holds.

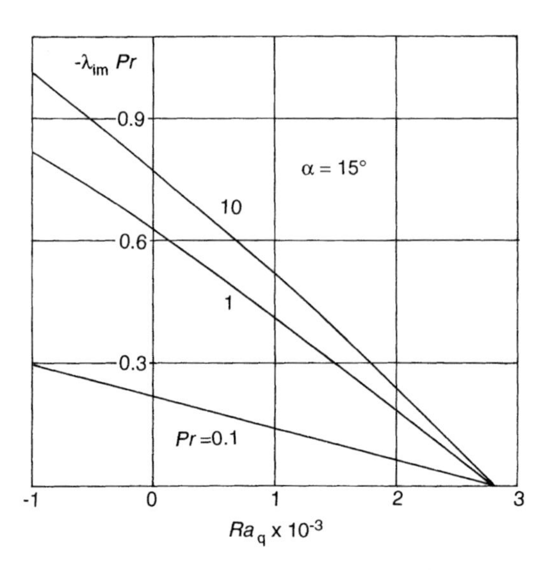

Fig. 103 Neutral frequency in dependence on Ra_q for different Pr

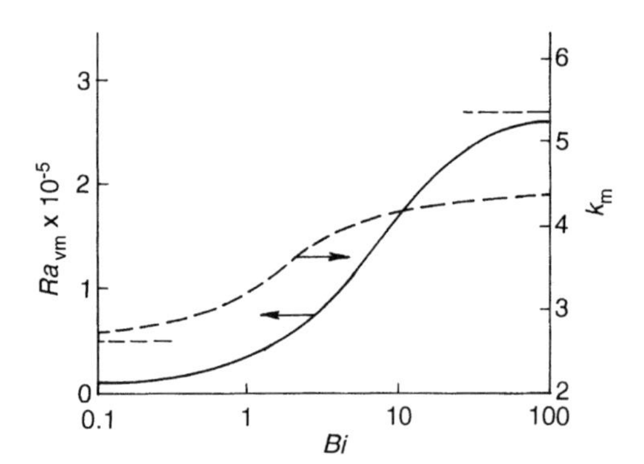

Fig. 104 Critical parameters in dependence on the Biot number ($Ra_q = 0$, weightlessness)

The stability of equilibrium is studied for the case of longitudinal orientation of the vibration axis ($\alpha = 0°$).

In the limiting case of weightlessness ($Ra_q = 0$), the critical values of the vibrational Rayleigh number Ra_{vm} and of the wavenumber k_m increase monotonically with the Biot number (Fig. 104). The stability curves in the plane (Ra_q, Ra_v) depend on Bi as the parameter. The family of those curves is presented in Fig. 105. Only in the case $Bi \to \infty$ (isothermal boundaries) is the curve symmetrical with respect to the Ra_v axis. For other values of Bi the symmetry is absent. It is interesting to note that for certain values of the parameter, the oscillatory instability exists in the range $Ra_q < 0$—the

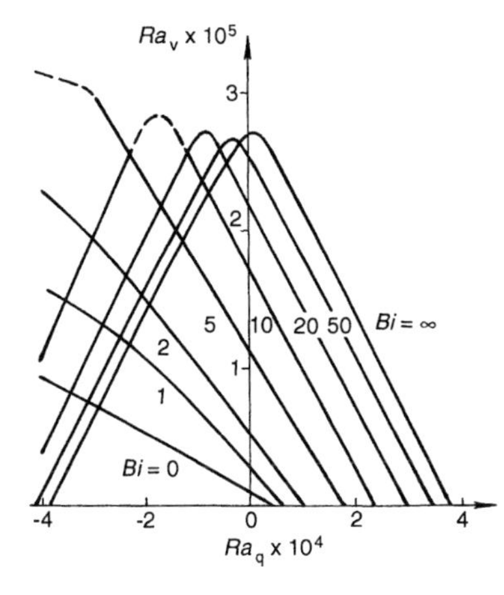

Fig. 105 Stability borders in the plane (Ra_q, Ra_v) for different values of Bi (variant (c)). Dashed parts of the curve, borders of oscillatory instability ($Pr = 1$)

corresponding parts of the curves for $Bi = 5$ and $Bi = 10$ are shown in Fig. 105 by dashed lines at $Pr = 1$. The origin of the oscillatory instability is due to interaction of the monotonic levels. One of them has a thermo-gravitational (Rayleigh) nature and develops in the thin lower layer that is unstably stratified, and the other one is due to the vibrational mechanism that acts in the stably stratified upper part of the layer. The situation is clarified by Fig. 106. Here the neutral curves are displayed for $Bi = 5$ and different Ra_q.

At $Ra_q = -2 \times 10^4$, there exist two monotonic neutral levels corresponding to both instability mechanisms. The analysis [3] proves that the eigenfunctions are localized in the appropriate parts of the layer. If $|Ra_q|$ increases, the levels approach each other and tend to cross. However, a 'simple' crossing of the neutral curves is forbidden. Instead, the interaction of the levels leads to their 'coalescence', with the formation of a wave-number interval where the instability has an oscillatory character. The

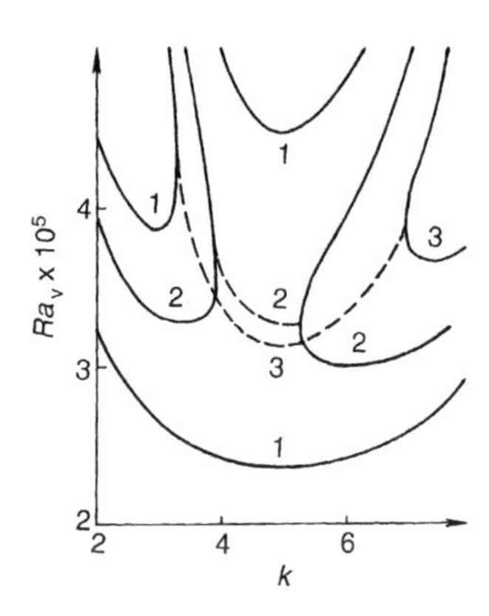

Fig. 106 Neutral curves for $Bi = 5$. The numeration of levels: 1, $Ra_q = -2 \times 10^4$; 2, $Ra_q = -3 \times 10^4$; 3, $Ra_q = -4 \times 10^4$. Dashed parts of the curves, borders of oscillatory instability ($Pr = 1$)

domains of the oscillatory instability are shown in Fig. 106 by dashed lines for $Pr = 1$. At $Ra_q = -3 \times 10^4$, the domain of the oscillatory instability also exists. However, it is not the most dangerous one since it does not correspond to the minimum of the neutral curve. Here the monotonic mode is the most dangerous. At $Ra_q = -4 \times 10^4$, the oscillatory mode is the most dangerous and a further increase of $|Ra_q|$ causes the interacting pair to split, yielding two monotonic levels.

A similar effect of the oscillatory instability formation as a result of interaction between two monotonic levels has been discovered [6] while studying convective instability of two-layer systems of immiscible fluids.

2.3 Comparison with Experiment

It is possible to compare different kinds of variant (b) where one boundary of the layer is isothermal whereas the other is thermally insulated. This exact case has been investigated experimentally in [7]. The equilibrium instability was observed in a horizontal layer cell 2.9 mm thick filled with an electrolite (1% $CaCl_2$ solution in ethanol) and heated by means of a variable electric current of industrial frequency. During the experiments, a longitudinal vibration ($\alpha = 0°$) was imposed with the amplitude up to 56 mm and three values of frequency, viz. 10, 16 and 25 Hz. The threshold value of the Rayleigh number was determined by observation of the heat transfer crisis. The results are presented in Fig. 107. Good agreement with the theoretical results is apparent. In Fig. 108, a picture of the convective structures is reproduced for $Ra_q = -5 \times 10^4$, $Ra_v = 15 \times 10^4$. In this range, the layer is stratified and is potentially stable ($Ra_q < 0$), and the

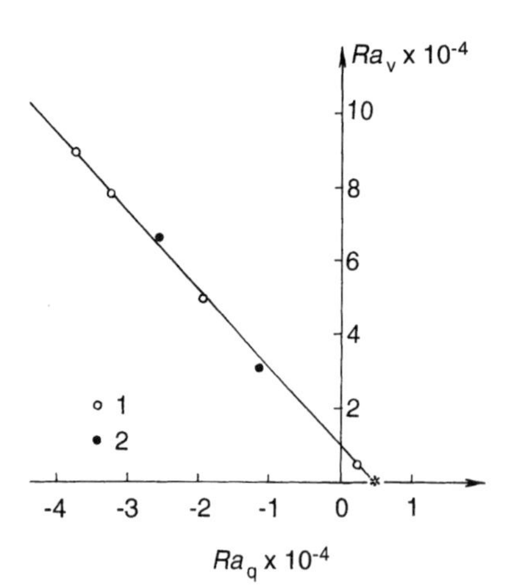

Fig. 107 Stability border. Solid line, theory; points, experiment. 1, $f = 10$ Hz; 2, $f = 16$ Hz

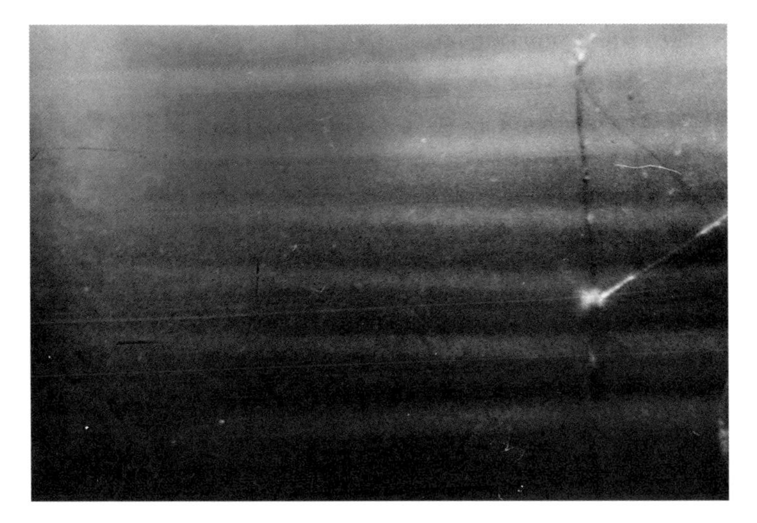

Fig. 108 Pictures of thermovibrational convection structures

equilibrium crisis is due to the thermovibrational mechanism. It can be seen that the convective structure resembles two-dimensional rolls whose axes are perpendicular to the vibration direction. The wavelength is in satisfactory accord with the theoretical predictions.

3 Non-linear Regimes

In this section we analyze the numerical results regarding steady and unsteady non-linear regimes of vibrational convection caused by instability of the state of mechanical equilibrium. To describe these regimes, it is necessary to turn back to the complete non-linear equations (4.2). Let us consider two-dimensional flows and write down the set of equations in terms of the stream functions φ and F for the fields of the average velocity $\mathbf{v}$ and the pulsation component $\mathbf{w}$, respectively:

$$\begin{aligned} v_x &= -\frac{\partial \varphi}{\partial z}, \qquad v_z = \frac{\partial \varphi}{\partial x} \\ w_x &= -\frac{\partial F}{\partial z}, \qquad v_z = \frac{\partial F}{\partial x} \end{aligned} \tag{4.14}$$

Eliminating the pressure from (4.2) and introducing the stream functions φ and

F, the set of equations takes the form

$$\frac{\partial \Delta\varphi}{\partial t} + \frac{1}{Pr}\left(\frac{\partial \varphi}{\partial x}\frac{\partial \Delta\varphi}{\partial z} - \frac{\partial \varphi}{\partial z}\frac{\partial \Delta\varphi}{\partial x}\right)$$
$$= \Delta\Delta\varphi + Ra_{\mathrm{q}}\frac{\partial T}{\partial x} + Ra_{\mathrm{v}}\left[\left(\frac{\partial^2 F}{\partial x^2}\frac{\partial T}{\partial z} - \frac{\partial^2 F}{\partial x \partial z}\frac{\partial T}{\partial x}\right)\sin\alpha\right.$$
$$\left. - \left(\frac{\partial^2 F}{\partial x \partial z}\frac{\partial T}{\partial z} - \frac{\partial^2 F}{\partial z^2}\frac{\partial T}{\partial x}\right)\cos\alpha\right] \quad (4.15)$$
$$Pr\frac{\partial T}{\partial t} + \left(\frac{\partial \varphi}{\partial x}\frac{\partial T}{\partial z} - \frac{\partial \varphi}{\partial z}\frac{\partial T}{\partial x}\right) = \Delta T + 1$$
$$\Delta F = \frac{\partial T}{\partial x}\sin\alpha - \frac{\partial T}{\partial z}\cos\alpha$$

Here $\Delta = \partial^2/\partial x^2 + \partial^2/\partial z^2$ is the two-dimensional Laplace operator.

The non-slip condition for the average flow velocity $\mathbf{v}$ and the non-overflow condition for the oscillatory component $\mathbf{w}$ yield

$$z = 0 \text{ and } z = 1: \qquad \varphi = \frac{\partial \varphi}{\partial z} = 0, \qquad F = 0 \qquad (4.16)$$

The temperature boundary conditions for the two afore-mentioned variants are

$$z = 0 \text{ and } z = 1: \qquad T = 0 \qquad (4.17a)$$

$$z = 0: \qquad \frac{\partial T}{\partial z} = 0$$
$$z = 1: \qquad T = 0 \qquad (4.17b)$$

It can be seen from the results of the previous section that, the stability border corresponds to the mode with quite a definite value of the wavenumber k_{m}. Thus, the solution of the non-linear problem has been sought in the rectangular domain on the xz plane: $\{0 \leqslant x \leqslant L, 0 \leqslant z \leqslant 1\}$, with the periodicity conditions on the lateral (vertical) boundaries:

$$f(0, z) = f(L, z) \qquad (4.18)$$

where f is either of the variables and L is the spatial period of the solution.

To solve the boundary problem (4.15) to (4.18), the finite difference technique was used. The computational procedure was based on the two-field approach requiring the Navier–Stokes equation for the average velocity to be put in terms of the stream function and the velocity vorticity. The implicit finite difference scheme and the alternative direction method were applied. At each time step, the stream function distributions were calculated by means of the iterative over-relaxation procedure.

Below, we present some numerical results on the non-linear regimes [2, 8, 9]. First, let us consider the symmetrical case of a layer with isothermal boundaries (variant (a)). The Prandtl number $Pr = 1$ is fixed and the spatial period $L = 1.5$ which corresponds to the wavenumber $k = 2\pi/L = 4.19$. As can be seen from Fig. 99, this value is close to the critical perturbation wavenumber determined by the linear stability theory for the case where the inclination angle of the vibration axis is not very large. Most of the calculations were performed using the homogeneous 31×21 grid; there were some control ones with a 46×31 grid. As the comparison shows, the discrepancies in the integral characteristics do not exceed a few percent.

Let $\alpha = 0°$, which corresponds to longitudinal vibrations. In accordance with the linear theory, the domain of the equilibrium stability in Fig. 98 is located below the curve $\alpha = 0°$. As the Rayleigh number Ra_q surmounts its critical value, the instability of a monotonic type sets in. In the case of pure weightlessness $Ra_q = 0$, the convection has a thermovibrational nature. In the integration area of size L, a four-vortices structure arises which is symmetrical with respect to the middle of the layer. The flow intensity may be characterized by the maximal (in modulus) value of the stream function $|\varphi_m|$. The dependence of $|\varphi_m|$ on Ra_v is presented in Fig. 109. From there it follows that the flow

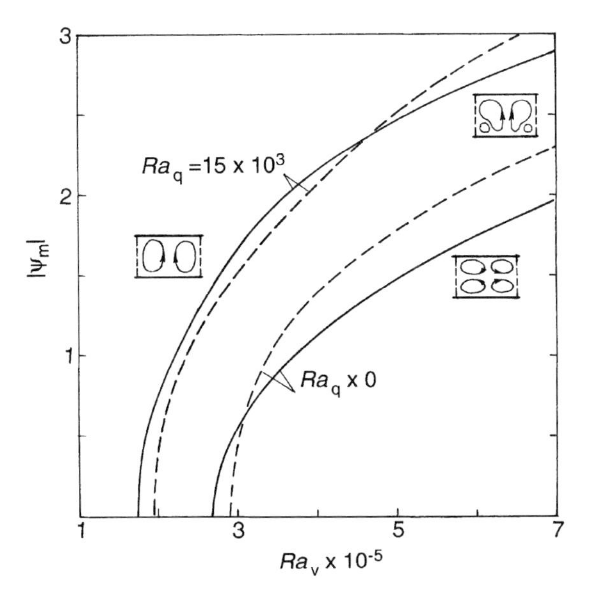

Fig. 109 Maximal value of the stream function versus Ra_v for two values of Ra_q. Solid lines, $\alpha = 0°$; dashed lines, $\alpha = 15°$

intensity increases monotonically with the overcriticality. Close to the bifurcation point, the following root law takes place: $|\varphi_{\rm m}| \sim (Ra_{\rm v} - Ra_{\rm vc})^{1/2}$. Here $Ra_{\rm vc}$ denotes the critical value of $Ra_{\rm v}$). According to the calculations, in the stability domain, all the initial perturbations decay. Thus the bifurcation to the stationary regime of thermovibrational convection is of a soft character. In the range of large overcritical values of $Ra_{\rm q}$ and $Ra_{\rm v}$, the transition to the oscillatory regime takes place (this transition was discovered in [10] for a purely gravitational convection).

The value $Ra_{\rm q} = 15 \times 10^3$ is high enough. Thus the convection setting in above the instability threshold has mainly a thermogravitational nature. Within the spatial cell of the period L, two vortices form whose centers are located in the domain of the unstable stratification. However, as $Ra_{\rm v}$ increases, the 'sign' of the convection changes gradually. The motion becomes predominantly thermovibrational and the characteristic four-vortices structure develops.

If the inclination of the vibration axis to the horizontal is small ($\alpha = 15°$, see Fig. 109), the shape of the bifurcation amplitude curves does not undergo qualitative changes. However, at $\alpha \neq 0°$, the instability has the oscillatory character and numerical calculations show that in the overcritical domain the cellular convective structure sets in, drifting along the layer with some constant velocity. Near the threshold, the drifting velocity is close to the phase velocity of neutral perturbations determined by the linear theory.

Let us consider two integral characteristics of the temperature field. The first one is the maximal temperature in the cavity $T_{\rm m}$. In the equilibrium regime, as one may see from (4.9a), $T_{\rm m} = 0.125$. The second reference value is the Nusselt number, defined as the heat flux through the upper boundary divided by its equilibrium value:

$$Nu = \frac{\displaystyle\int_0^L \left(\frac{\partial T}{\partial z}\right)_{z=1} {\rm d}x}{\displaystyle\int_0^L \left(\frac{\partial T_0}{\partial z}\right)_{z=1} {\rm d}x} \tag{4.19}$$

In the equilibrium regime $Nu = 1$. In the settled symmetrical four-vortices thermovibrational convection regime, evidently $Nu = 1$. If the convective motion has a thermogravitational component, the symmetry of the temperature field is disturbed because the vortices form in the upper (unstably stratified) part of the layer. As a result, the heat flux increases through the upper boundary and, accordingly, decreases through the lower boundary of the layer. Hence, at $Ra_{\rm q} \neq 0$, there exists a heat flux through the layer directed vertically upwards whose intensity is proportional to $Nu - 1$.

In Fig. 110, the dependences are presented of $T_{\rm m}$ and $Nu - 1$ on $Ra_{\rm v}$ for two values of $Ra_{\rm q}$ for the case of longitudinal vibrations and $\alpha = 15°$. It can be seen that in the overcritical range $T_{\rm m}(Ra_{\rm v})$ is non-monotonic. However, when the

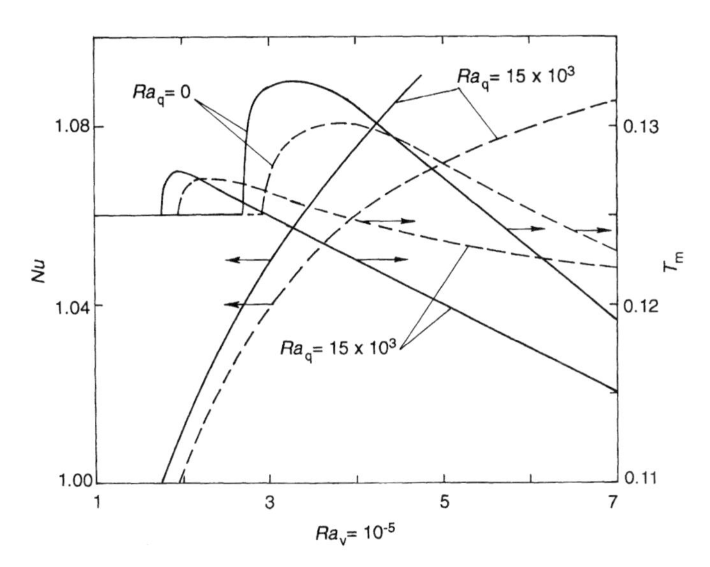

Fig. 110 Maximal temperature and the Nusselt number versus Ra_v for two values of Ra_q. Solid lines, $\alpha = 0°$; dashed lines, $\alpha = 15°$

overcriticality is high enough, T_m decreases with Ra_v growth in accordance with a nearly linear law. The heat flux through the layer increases monotonically with the overcriticality, and close to the critical point $Nu - 1 \sim (Ra_v - Ra_{vc})$.

If $\alpha \neq 0°$, the wave regime sets in as a result of the oscillatory instability of equilibrium. This regime manifests itself as a cellular structure drifting uniformly with the velocity U found from the non-linear calculations. Near the stability border, U is close to the value of the neutral phase velocity given by the linear stability theory. Figures 111 to 113 demonstrate the dependence of U on the parameters Ra_q, Ra_v and α in the overcritical region. It can be seen that both positive and negative values of U are possible. It is interesting to note that at some special values of the parameters, $U = 0$ and the regime is steady.† When the steady points are determined numerically, it is possible to obtain a map of regimes in the parameter space. In Fig. 114 such a map is shown in the plane (Ra_q, Ra_v) for $\alpha = 15°$. One may distinguish the domains of equilibrium as well as the 'left' and 'right' wave regimes. Variation of the border of domains $U < 0$ and $U > 0$ with the vibration axis inclination is shown in Fig. 115.

† From a dynamic point of view, the transition to a steady regime under variation of the parameters is evidently related to the symmetry of the steady regime with respect to an arbitrary translation along the layer.

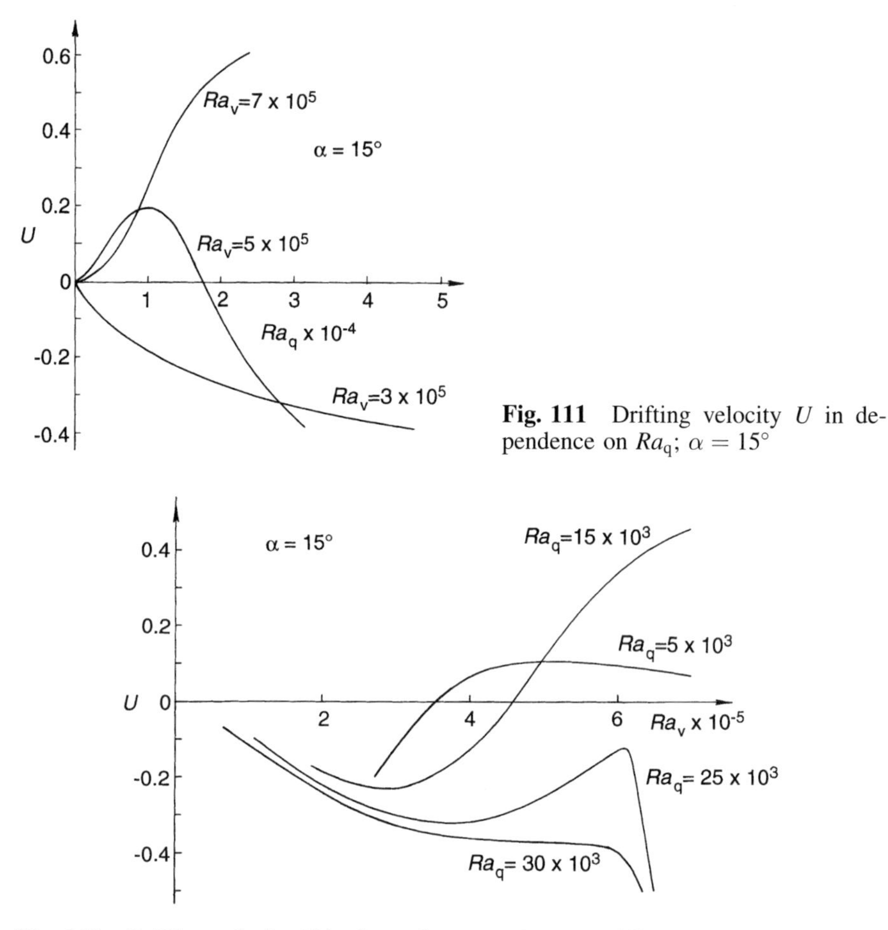

Fig. 111 Drifting velocity U in dependence on Ra_q; $\alpha = 15°$

Fig. 112 Drifting velocity U in dependence on Ra_v; $\alpha = 15°$

It is interesting to analyze the dependence of U on Ra_v at $\alpha = 20°$ in Fig. 113. It can be seen that $U > 0$ only in a narrow range of Ra_v. In addition, close to $Ra_v = 7 \times 10^5$ there exists a non-single-valuedness: three non-linear wave regimes with different values of the drift velocity U correspond to the same value Ra_v. One of these regimes is unstable. As Ra_v varies, a hard reorganization of the structure accompanied by a hysteresis takes place.

Now let us discuss some numerical results on non-linear regimes in the non-symmetrical case where one of the boundaries is isothermal and another one is thermally insulated (variant (b)). The calculations were been carried out for the space period $L = 2.5$, i.e. close to the critical wavenumber provided by the linear theory for small α. The 51×21 grid has been used.

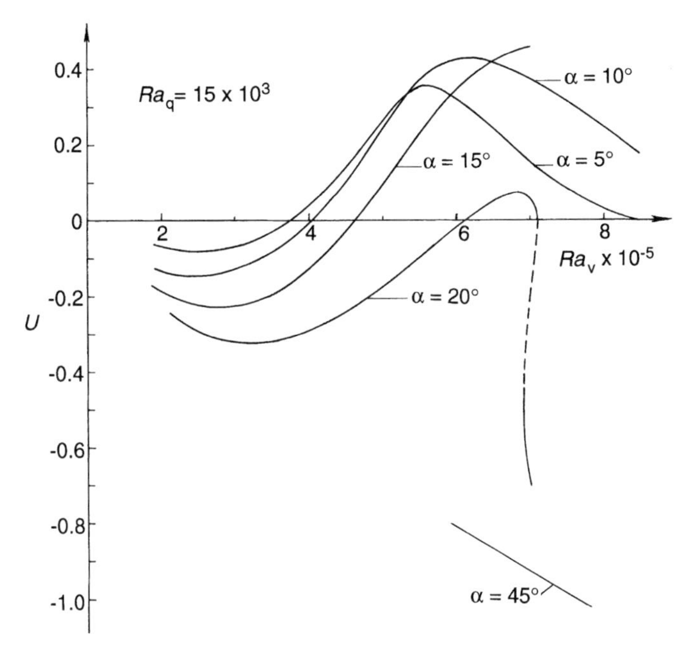

Fig. 113 Drifting velocity U in dependence on Ra_v for different α; $Ra_q = 15 \times 10^3$. Dashed part of the curve $\alpha = 20°$, the range of the wave instability

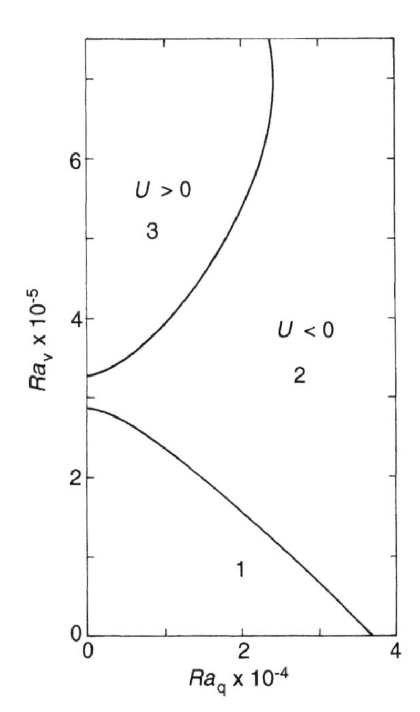

Fig. 114 Map of regimes in the plane (Ra_q, Ra_v) for $\alpha = 15°$: 1, the range of stable equilibrium; 2, the range of wave regimes with $U < 0$; 3, the range of wave regimes with $U > 0$

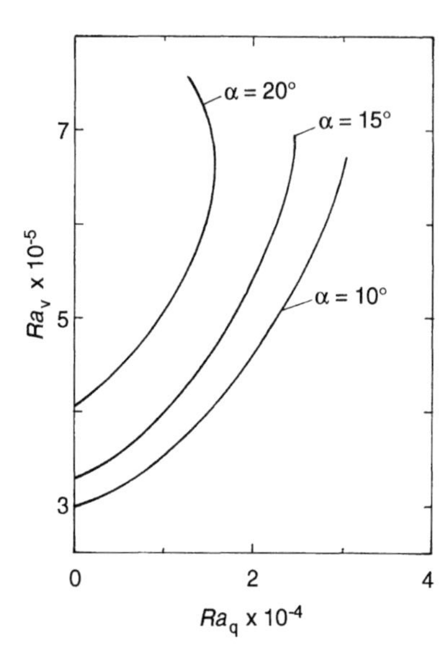

Fig. 115 Lines of steady regimes for three values of α

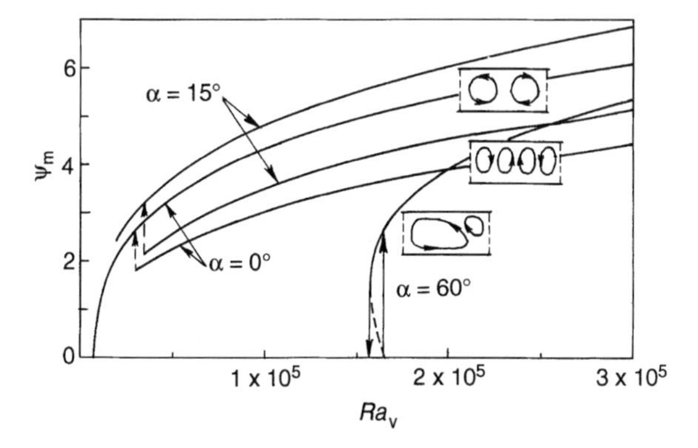

Fig. 116 Maximal value of the stream function versus Ra_v ($Ra_q = 0$, $Pr = 1$)

At $Ra_q = 0$ (weightlessness) and $\alpha = 0^\circ$ (longitudinal vibrations) the two-vortices steady regime bifurcates from the equilibrium in the critical point (see Fig. 116). However, in the wide Ra_v range the non-single-valuedness of the non-linear solution takes place. Due to that, the stable four-vortices regime coexists with the two-vortices one. Each of them may be attained by choosing an appropriate initial perturbation. The reorganization of the flow as Ra_v varies has a hard character and is accompanied by a hysteresis. Hence, as Ra_v decreases, a jump-like transition occurs from the four-vortices branch to the two-vortices one. The bifurcation from the equilibrium takes place in the point

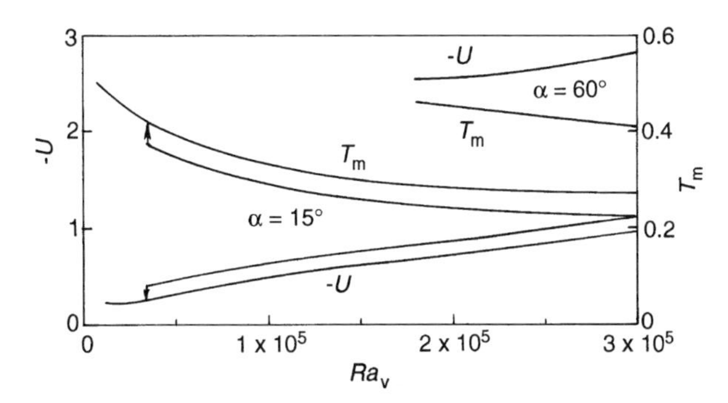

Fig. 117 Maximal temperature and drifting velocity versus Ra_v ($Ra_q = 0$, $Pr = 1$)

close to the critical value given by the linear theory. A similar pattern for excitation and reorganization of the non-linear regime was found at $\alpha = 15°$. However, here the wave regime (a cellular structure drifting along the layer) that bifurcates from the equilibrium is not steady. The critical point determined numerically is higher than the threshold pointed out by the linear theory. This is due to the fact that at $\alpha = 60°$, the critical wavelength is 4.8, which differs considerably from the value of the period $L = 2.5$ adopted in the calculations. Dependences of the maximal temperature T_m and the drift velocity U on Ra_v are shown in Fig. 117 for both branches. Coexistence of steady regimes of different forms is encountered also for other values of the Prandtl number (Fig. 118). When $Pr > 8$, only the four-vortices regime is stable.

In Fig. 119 the bifurcation characteristics are presented for the situation where both mechanisms of convective instability, i.e. the gravitational and vibrational, are operative. The figure corresponds to the case when $\alpha = 15°$ and $Ra_v = 10^5$. The point of the stability loss is $Ra_{qc} \sim -4 \times 10^4$. Here the stable four-vortices regime bifurcates from the equilibrium. At the point $Ra_q \sim 2 \times 10^4$, a hard transition to the two-vortices branch occurs. The hysteresis loop and the characteristics of the drifting regimes are given in the figure.

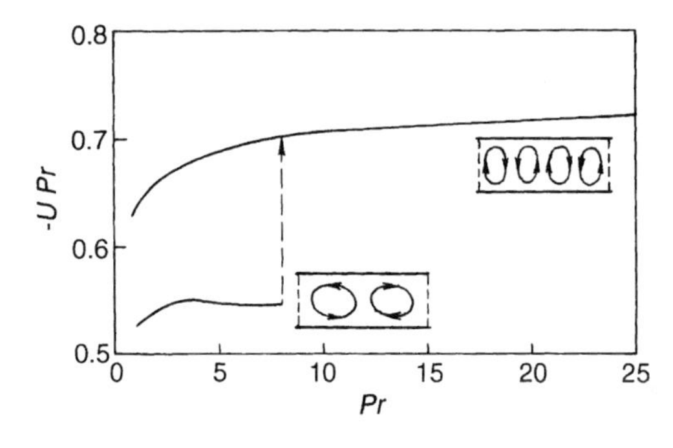

Fig. 118 Drifting velocity in dependence on the Prandtl number for different regimes for $Ra_q = 0$, $Ra_v = 10^5$, $\alpha = 15°$

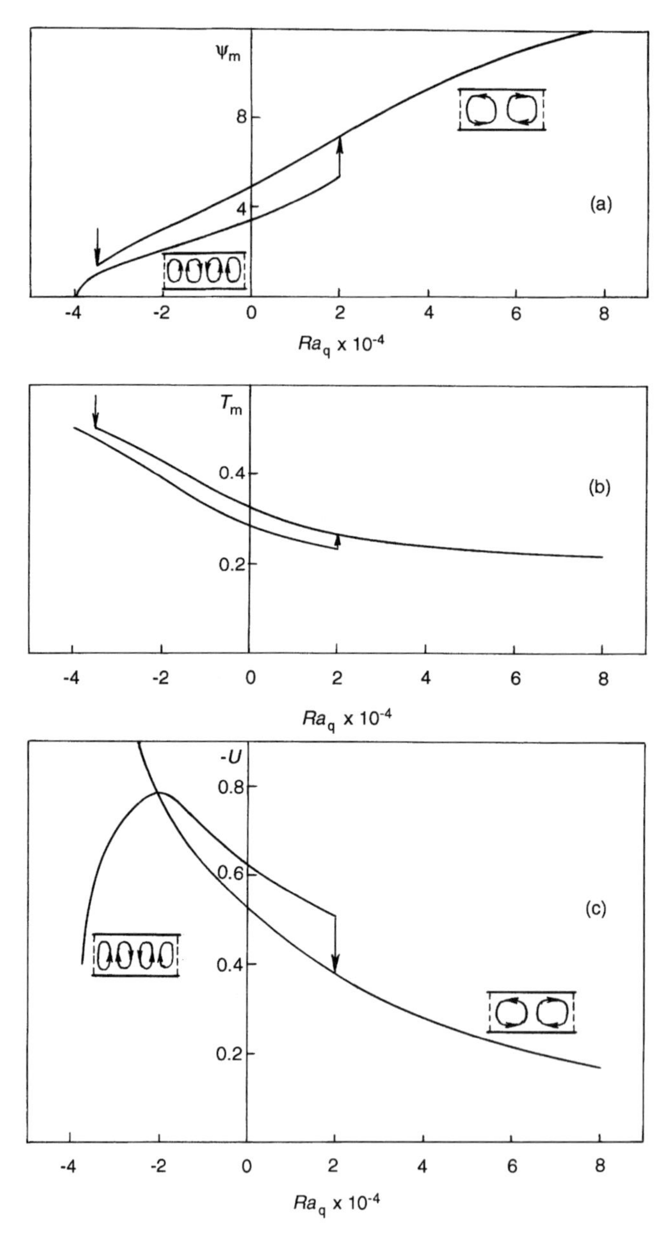

Fig. 119 Bifurcational characteristics for $Ra_v = 10^5$, $\alpha = 15°$, $Pr = 1$

In both variants of the boundary conditions considered—symmetrical (a) and non-symmetrical (b)—the structure of the wave regimes becomes more complicated in the range of large overcriticality (large Ra_q and Ra_v). The waves cannot be considered as a pure drift of cellular structures and oscillations with independent frequencies appear within one cell.

4 Chemically Active Liquid: Equilibrium Stability

We have considered a convection with uniformly distributed heat sources. Now let us turn to the case of a more complicated heat generation law. The fluid is supposed to be chemically active, so that an exothermic reaction of zeroth order takes place all over the volume. The heat effect of the reaction is high enough, so it is possible to neglect the dependence of heat generation on the reagent concentration. The convection is caused by the internal heat generation; the strength of the internal heat sources depends on temperature exponentially in accordance with the Arrenius law:

$$Q = Q_0 k_0 \exp\left(-\frac{E}{RT}\right) \tag{4.20}$$

Here T is the absolute temperature, E is the activation energy, R is the gaseous constant and Q_0 and k_0 are the thermal effect and the pre-exponent factor, respectively.

Suppose that the boundaries of the layer $z = 0$ and $z = h$ are isothermal and maintained at the same constant absolute temperature Θ. Generally speaking, a spatially non-uniform temperature distribution caused by the internal heat generation initiates convection in the gravity and vibrational fields. To describe the process, it is necessary to use the standard set of equations taking into account the existence of both forces: thermogravitational and thermovibrational. Hence, instead of (1.64) one must have

$$\frac{\partial T}{\partial t} + \mathbf{v}\nabla T = \chi \Delta T + \frac{Q_0 k_0}{\rho c_p} \exp\left[-\frac{E}{R(\Theta + T)}\right] \tag{4.21}$$

where now T is the temperature measured from its value at the boundaries.

We introduce the dimensionless variables based of the units used in Section 3 except for the temperature unit, which is now $R\Theta^2/E$. Assuming that the parameter $R\Theta/E$ is small, equation (4.21) takes the following dimensionless form:

$$Pr\frac{\partial T}{\partial t} + \mathbf{v}\nabla T = \Delta T + Fk \exp(T) \tag{4.22}$$

The rest of the equations of the set (4.2) remain the same. We have just changed the definitions for Ra_q and Ra_v and added one more parameter—the Frank-Kamenetsky number Fk. The complete set of parameters is

$$Pr = \frac{\nu}{\chi}, \qquad Ra_q = \frac{g\beta h^3 R\Theta^2}{\nu\chi E}, \qquad Ra_v = \frac{1}{2\nu\chi}\left(\frac{\beta b\Omega h R\Theta}{E}\right)^2$$
$$Fk = \frac{k_0 Q_0 E h^2}{\rho c_p \chi R\Theta^2}\exp\left(-\frac{E}{R\Theta}\right) \tag{4.23}$$

The boundary conditions (4.4) to (4.7a) are used in variant (a) with respect to temperature. Under these conditions, the mechanical equilibrium is available when the temperature depends solely on the transversal coordinate $T = T_0(z)$. Its profile may be determined from the following non-linear boundary problem:

$$\frac{d^2T_0}{dz^2} + Fk\exp(T_0) = 0$$
$$T_0(0) = T_0(1) = 0 \tag{4.24}$$

It is natural that this problem coincides with the classic one-dimensional Frank-Kamenetsky problem of a thermal explosion in a plane layer of a motionless reacting medium [11]. The steady solution of this problem exists within the interval $0 \leqslant Fk \leqslant Fk^*$, where the limiting value $Fk^* = 3.514$ determines the threshold of the thermal explosion. For $Fk > Fk^*$ there are no steady solutions of the problem (4.24) since the heat exchange through the boundaries cannot match the heat generation in the bulk that rises exponentially with temperature.

In the range $Fk < Fk^*$ there are two steady heat diffusive regimes. For each of them the temperature profile is an even function of the transversal coordinate with the maximal value T_{0m} reached in the middle point $z = \frac{1}{2}$. The values of T_{0m} as functions of Fk are given in Fig. 120. The high-temperature regime is unstable with respect to purely heat diffusive perturbations.

The field $\mathbf{w}_0$ in the equilibrium has the structure $\mathbf{w}_0(w_0, 0, 0)$ with the longitudinal component $w_0(z) = T_0(z)\cos\alpha + C$; here α is the inclination angle of the vibration axis and the constant C may be found from the condition of closeness of the oscillatory flow:

$$C = -\cos\alpha \int_0^1 T_0(z)\,dz$$

The low-temperature steady regime is stable in the range $Fk < Fk^*$ with respect to a proper heat diffusion. However, in a movable medium (fluid, gas) this regime may become unstable with respect to the convection onset. Convective thermogravitational instability of a plane horizontal layer of a reacting

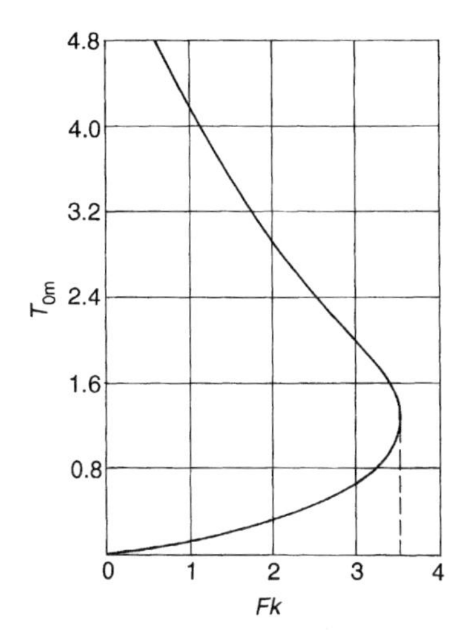

Fig. 120 Maximal temperature versus the Frank-Kamenetsky parameter

fluid was studied in [12] to [14]. Convective stability under static gravity and high-frequency vibration has been considered in [15].

Let us introduce small two-dimensional normal perturbations. As a result, the spectral problem for amplitudes of the transversal components v, w and temperature perturbation θ is obtained:

$$-\lambda Dv = D^2 v - k^2\, Ra_q\, \theta - Ra_v\, T_0'\big(ikw'\cos\alpha + k^2 w\sin\alpha - k^2\theta\cos\alpha\big)$$

$$-\lambda\, Pr\, \theta = D\theta - T_0' v + Fk \exp T_0 \theta \tag{4.25}$$

$$Dw = -\big(ik\theta'\cos\alpha + k^2\theta\sin\alpha\big)$$

$$z = 0, \quad z = 1: \qquad v = v' = 0, \qquad w = 0, \qquad \theta = 0 \tag{4.26}$$

This spectral problem has been integrated numerically by the Runge–Kutta–Merson method with the shooting procedure. At the same time, the shooting technique has been used for a numerical solution of the heat diffusion problem (4.24).

Let us discuss the results of calculations. According to [12] to [14], in the absence of vibrations ($Ra_v = 0$), the heat diffusion regime becomes convectively unstable when Ra_q exceeds the critical value depending on Fk. The critical value Ra_{qm} decreases monotonically as Fk increases, and it tends to a finite limit as $Fk \to Fk^*$. When $Fk > Fk^*$, the stability problem has no sense because in this range the equilibrium is not available.

The opposite limiting case corresponds to $Ra_q = 0$ (weightlessness). In this situation only the vibrational mechanism of convection excitation is in action. Two most interesting cases of the vibration axis orientation are considered

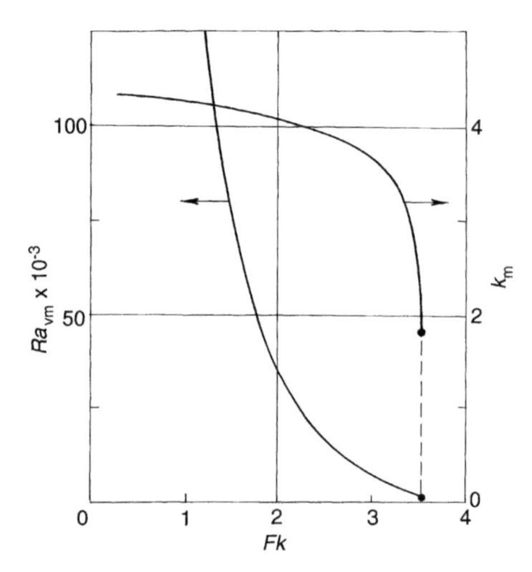

Fig. 121 Parameters of the vibrational instability in the case of weightlessness (longitudinal vibrations)

below: $\alpha = 0°$ (longitudinal vibrations) and $\alpha = 90°$ (transversal vibrations). As in the case of a uniform heat generation (see Section 2, variant (a)) the instability is related to the development of monotonic perturbations. To determine the stability border, it is necessary to set $\lambda = 0$ in the spectral amplitude problem. Thus the stability border does not depend on the Prandtl number.

Under longitudinal vibrations ($\alpha = 0$), the critical vibrational Rayleigh number depends on the wavenumber k and the Frank–Kamenetsky parameter Fk. The minimal value of Ra_v with respect to k and the corresponding wavenumber k_m of the most dangerous perturbation are presented in Fig. 121 as functions of Fk. It can be seen that the critical parameters decrease monotonically with Fk and for $Fk \to Fk^*$ one has $Ra_{vm} \to 1489$ and $k_m \to 1.8$. In the range of small Fk, the characteristic asymptote takes place. Indeed, at $Fk \to 0$ the internal heating is weak, and it is possible to set approximately $\exp(T_0) \sim 1$. Then the problem of the equilibrium heat diffusion regime (4.24) takes the form

$$T_0'' + Fk = 0, \qquad T_0(0) = T_0(1) = 0$$

i.e. the limiting transition occurs for the case of spatially uniform heat generation with intensity $Q = Fk\, R\Theta^2 \rho c_p \chi / (Eh^2)$. According to Section 2, in the case of uniform heat sources ($Ra_{\underset{\sim}{q}} = 0$, $\alpha = 0°$, variant (a)) the crisis sets in at the vibrational Rayleigh number $\widetilde{Ra}_v$ defined through Q: $\widetilde{Ra}_v = (\beta b \Omega h^3)^2 / [2\nu\chi(\rho c_p)^2] = 269.3 \times 10^3$. The most dangerous wavelength corresponds to $k_m = 4.35$. If we need to return to the vibrational Rayleigh number Ra_v used in this section, then the stability border may be presented as $Ra_v\, Fk^2 = 269.3 \times 10^3$. Thus in the limit $Fk \to 0$ one has $Ra_v = 269.3 \times 10^3 / Fk^2$ and $k \to 4.35$. The computational results agree well with this asymptotes (Fig. 121).

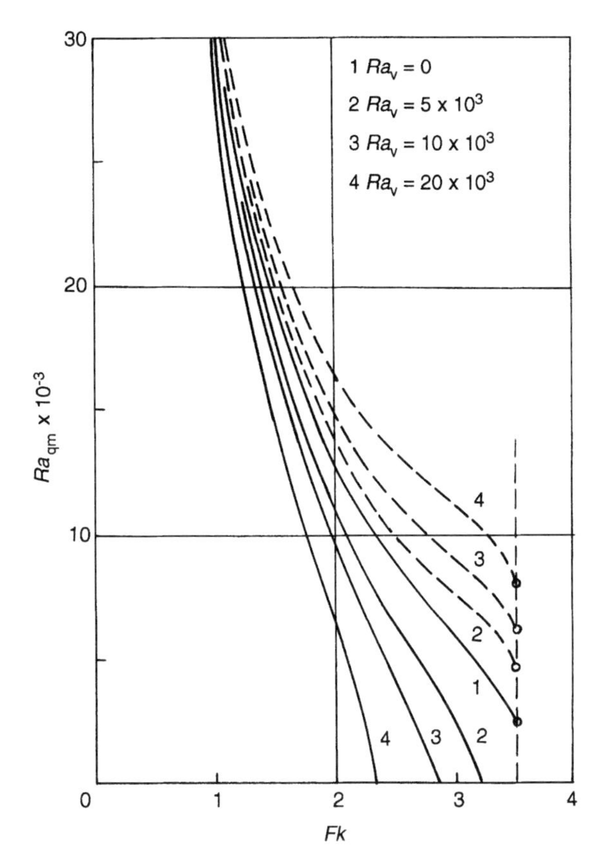

Fig. 122 Critical number Ra_{qm} versus Fk. Solid line, longitudinal vibrations; dashed line, transversal vibrations

In the case of transversal vibrations ($\alpha = 90°$), the thermovibrational mechanism does not work, making the situation like that of a uniform heat generation (Section 3). The effect of vibrations is absolutely stabilizing, and the heat diffusive regime is unconditionally stable at any finite value of Ra_v.

Let us now take in the joint action of both mechanisms. As in the case of weightlessness, instability is due to monotonic perturbations both for longitudinal ($\alpha = 0°$) and transversal ($\alpha = 90°$) vibrations. In Fig. 122 the critical value Ra_{qm} is presented as a function of the Frank-Kamenetsky parameter Fk for different Ra_v. If there are no vibrations ($Ra_v = 0$, curve 1), the location of the stability border entirely coincides with the data of [13] and [14]. It can be seen that the transversal vibrations act in a stabilizing manner; the stability border elevates with Ra_v. Longitudinal vibrations destabilize the equilibrium; the stability border goes down as Ra_v increases. For $Ra_v > 1489$ the neutral

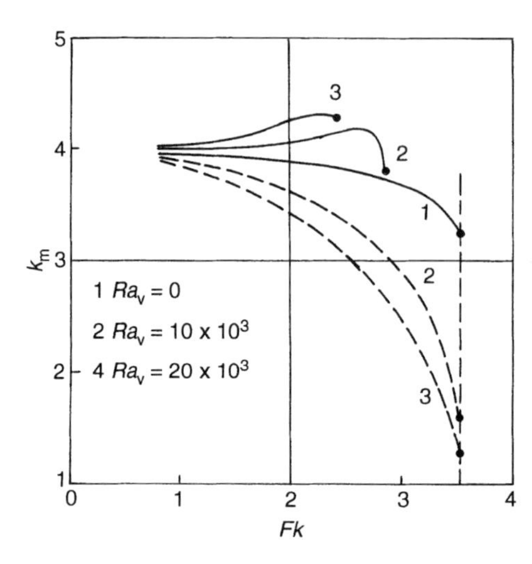

Fig. 123 Critical wavenumber versus *Fk*. Solid line, longitudinal vibrations; dashed line, transversal vibrations

curve intersects the Fk axis and cuts the interval of instability adjacent to the thermal explosion point Fk^*. This means that the steady heat diffusion regime under longitudinal vibrations becomes unstable in the range $Fk < Fk^*$ with respect to the onset of convection in the absence of the static gravity force ($Ra_{qm} = 0$).

The data on dependence of the critical wavenumber on the pertinent parameters is presented in Fig. 123. In Fig. 124 the stability borders in the plane (Ra_q, Ra_v) obtained by minimization with respect to the wavenumber k are given. Solid lines correspond to longitudinal vibrations and dashed lines to transversal ones. Transversal vibrations act in a stabilizing way. With regards to the longitudinal orientation of the vibration axis, the mutual destabilizing effect takes place when both mechanisms of the convection excitation operate simultaneously. Along each of the solid curves, the form of the instability changes as Ra_q and Ra_v vary. The inset in Fig. 124 shows the transversal velocity profiles for the points a, b and c of the curve corresponding to $Fk = 3$. Point a corresponds to the thermogravitational instability ($Ra_v = 0$) when a one-vortex structure appears in the unstably stratified part of the layer, and its center coincides with the center of the cell. As Ra_v increases, the relative role of the thermovibrational mechanism is enhanced, and at point c for weightlessness ($Ra_q = 0$) the instability has a two-vortices form of thermovibrational origin.

The stabilizing effect of transversal vibrations tempts the question of the possibility of stabilizing the equilibrium with respect to the high-temperature instability. The results presented in Fig. 125 partly deliver the answer. The neutral stability curves of the high-temperature regime are shown in the plane (k, Ra_q); the stability domains are those below the curves. The curve $Ra_v = 0$ describes the case of no vibrations. Even at $Ra_v = 0$, the interval of wave-

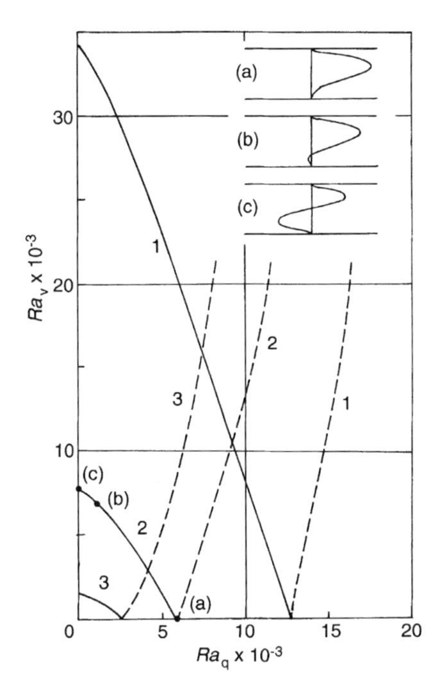

Fig. 124 Stability borders in the plane (Ra_q, Ra_v); 1, $Fk = 2$; 2, $Fk = 3$; 3, $Fk = 3.513$. Solid line, longitudinal vibrations; dashed line, transversal vibrations

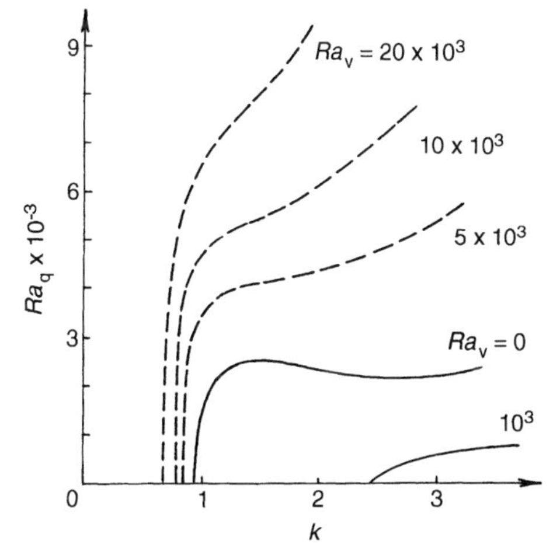

Fig. 125 Neutral curves for the high-temperature regime ($Fk = 3.5$). Solid line, longitudinal vibrations; dashed line, transversal vibrations

numbers $0 < k \leqslant 1$ exists where the long-wave heat diffusion instability occurs. As Ra_v increases, the instability range in the plane (k, Ra_q) contracts. Thus in a certain sense one may call it a stabilizing effect. However, it is necessary to emphasize that the long-wave instability (i.e. with $k \to 0$) sustains up to high enough Ra_v.

For the case of longitudinal vibrations, a destabilization takes place, and the instability region expands.

5 Chemically Active Liquid: Non-linear Regimes

The problem of non-linear regimes of convection developing after the mechanical equilibrium loses its stability is studied here in a way that is entirely similar to the one explained before (see Section 3 [16, 17]); namely, two-dimensional non-linear regimes are considered which are periodic along the longitudinal coordinate x. To describe them, the complete non-linear set of equations is rewritten in terms of φ, F, T, where φ, F are the stream functions for the average velocity and its oscillatory part, respectively. The set obtained differs from (4.15) by the form of the heat transfer equation. Consideration is restricted to the case of longitudinal vibrations with $\alpha = 0°$:

$$\begin{aligned} &\frac{\partial \Delta\varphi}{\partial t} + \frac{1}{Pr}\left(\frac{\partial \varphi}{\partial x}\frac{\partial \Delta\varphi}{\partial z} - \frac{\partial \varphi}{\partial z}\frac{\partial \Delta\varphi}{\partial x}\right) \\ &\quad = \Delta\Delta\varphi + Ra_q\frac{\partial T}{\partial x} + Ra_v\left(\frac{\partial^2 F}{\partial z^2}\frac{\partial T}{\partial x} - \frac{\partial^2 F}{\partial x \partial z}\frac{\partial T}{\partial z}\right) \\ &Pr\frac{\partial T}{\partial t} + \left(\frac{\partial \varphi}{\partial x}\frac{\partial T}{\partial z} - \frac{\partial \varphi}{\partial z}\frac{\partial T}{\partial x}\right) = \Delta T + Fk\exp T \\ &\Delta F = -\frac{\partial T}{\partial z} \end{aligned} \tag{4.27}$$

Here Δ is the two-dimensional Laplace operator. The conditions on the horizontal boundaries are as follows:

$$z = 0 \text{ and } z = 1: \qquad \varphi = \frac{\partial \varphi}{\partial z} = 0, \qquad F = 0, \qquad T = 0 \tag{4.28}$$

On the vertical boundaries of the domain of integration, the periodicity conditions are assumed for all the variables:

$$f(0, z) = f(L, z) \tag{4.29}$$

As in Section 3, the computation procedure is based on the finite difference technique. A 31×31 grid is used inside the cell of length $L = 1.5$ corresponding to the wavenumber $k = 4.19$.

First, let us discuss the case of purely thermovibrational convection in weightlessness ($Ra_q = 0$). In the range of stability $Fk < Fk^*$, $Ra_v < Ra_{vm}$ (see Fig. 121), any initial perturbation decays. There exists only one attracting

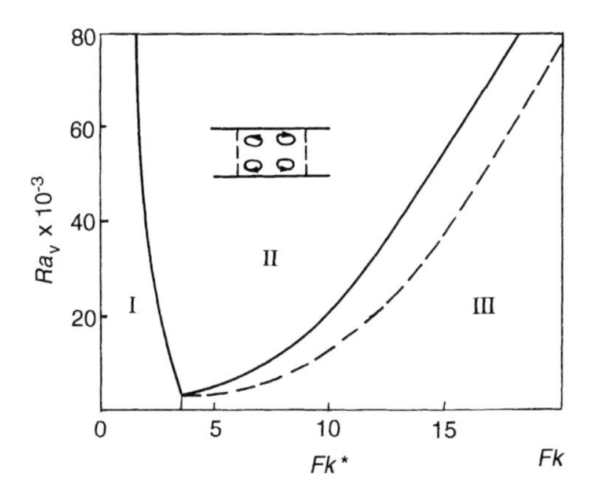

Fig. 126 Map of regimes in the plane (Fk, Ra_v), $Ra_q = 0$ (weightlessness). Solid lines, $Pr = 1$; dashed line, the border of thermal explosion for $Pr = 20$

regime in this region corresponding to equilibrium. If $Fk < Fk^*$ but $Ra_v > Ra_{vm}$, an initial small perturbation develops and the secondary steady regime of thermovibrational convection sets in. The range of its existence at fixed $Ra_v > Ra_{vm}$ spans over a certain interval along the Fk axis. The left boundary of this interval coincides in practice with the critical point determined by linear theory. At this point the secondary regime bifurcates from the equilibrium. If Fk increases at fixed Ra_v, the end-point of the steady regime existence is the one where the thermal explosion sets in. It is interesting to note that in the overcritical region, the steady flow is possible even at $Fk > Fk^*$.

Then, it is possible to construct a map of regimes in the plane (Fk, Ra_v). The result is shown in Fig. 126. The symbols I, II and III correspond, respectively, to the domains of stability, of steady flow and of thermal explosion. Some data in Fig. 126 testify that the position of the domain border depends on the Prandtl number. As for estimation of the dependence on the period length L, the calculations were performed with $L = 2.5$. There is no shift of the left boundary of the II domain because the line is almost vertical, and the critical wavenumber k_m depends very weakly on Fk. The period $L = 2.5$ accommodates two convective cells; the eight-vortices structure appears with a vertical plane of symmetry at the mid-point of the period. Thus this structure corresponds to two four-vortices ones each possessing $L = 1.25$. It is close to $L = 1.5$ and the change in L has practically no effect on the position of the right border of the II domain.

In the range of a steady convective regime, the flow has a symmetrical four-vortices structure per period. As a measure of flow intensity, the extremal value of the average flow stream function φ_m may be taken. The maximal temperature T_m is another interesting characteristic of a non-linear regime. In Fig. 127, T_m as a function Fk is presented for several values of Ra_v ($Ra_v = 0$ corresponds to the Frank-Kamenetsky heat diffusion regime). It can be seen that as Ra_v increases, i.e. the thermovibrational flow intensity strenghtens, the maximal

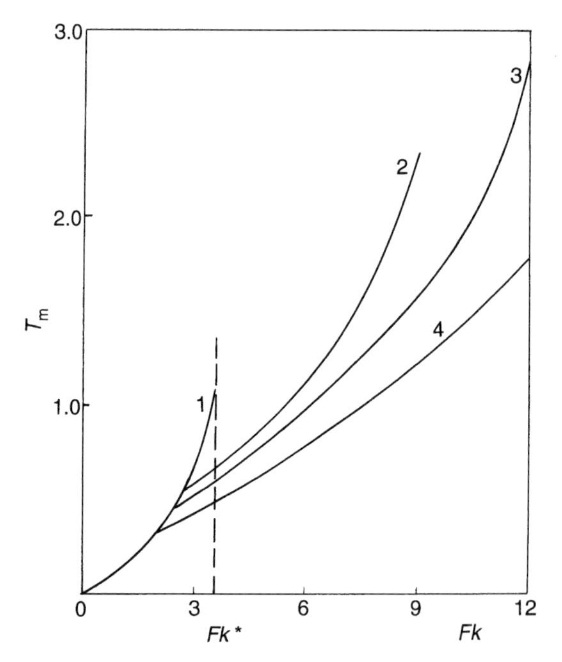

Fig. 127 Maximal temperature depending on *Fk*: 1, $Ra_v = 0$; 2, $Ra_v = 20 \times 10^3$; 3, $Ra_v = 40 \times 10^3$; 4, $Ra_v = 70 \times 10^3$

temperature decreases. The bifurcation from the Frank-Kamenetsky curve takes place at the point of the quasi-equilibrium stability loss.

The point of main interest here is the enhancement of the heat explosion threshold in a convective regime. From a physical point of view, the effect is quite understandable. The shift of the threshold is caused by the fact that a developed convection intensifies considerably the heat exchange in the layer in comparison with the heat diffusion regime. As see from Fig. 126, the explosion threshold grows with the Prandtl number.

Let us turn to the general case $Ra_q \neq 0$, $Ra_v \neq 0$ when both mechanisms of the thermal convection excitation, i.e. thermogravitational and thermovibrational, are present. Some numerical results are given here for the fixed values $Ra_v = 10 \times 10^3$, $Pr = 1$, $L = 1.5$. In Fig. 128 the map of regimes is presented in the plane (Fk, Ra_q). As in Fig. 126, the symbols I, II and III mark, respectively, the domains of stable equilibrium, of steady non-linear convective regimes and of thermal explosion. The left border of the II domain coincides with the stability border evaluated in the framework of linear theory (see Fig. 122). On its right border, any steady convective regime is destroyed due to the thermal explosion. The borders of this range for the case of no vibrations ($Ra_v = 0$ [18]) are presented for the sake of comparison and are drawn with dashed lines. Here the steady convective regime in the II region has a purely

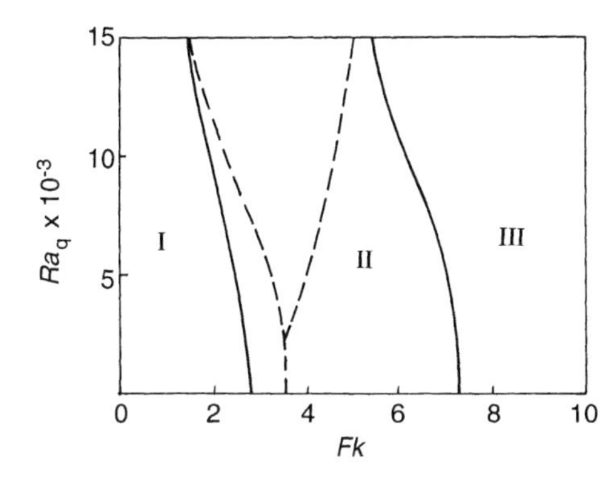

Fig. 128 Map of regimes in the plane (Fk, Ra_q); $Pr = 1$, $L = 1.5$. Solid lines, $Ra_v = 10 \times 10^3$; dashed lines, $Ra_v = 0$

thermovibrational nature. In Fig. 128 the effect of significant enhancement of the thermal explosion threshold due to both mechanisms of thermal convection excitation is well noticeable.

The structure of convective flow inside the II domain depends on the ratio of Ra_q to Ra_v. If Ra_q is small, the calculations show that convection has a symmetrical four-vortices shape related to its thermovibrational nature. As Ra_q increases, the structure changes gradually and the thermogravitational mechanism of instability excitation becomes predominant. The upper set of vortices becomes more intense because the potentially unstable stratification is located in this region.

In conclusion, we would like to emphasize an important dynamical feature encountered in the convective regimes when both mechanisms of excitation interact. Close to the thermal explosion border, steady periodic convective regimes are stable with respect to small perturbations but unstable with respect to finite ones. Suppose that a steady regime is obtained for some Fk and Ra_q and the steady regime corresponding to other values of the parameters is to be found. To do that numerically using the parameter-prolongation procedure, one has to choose the steps in Fk and Ra_q small enough. Otherwise, finite perturbations are introduced into the system and the transient process would not lead to a new steady regime. Instead, a thermal explosion would break out. This testifies to the fact that the steady regimes existing close to the thermal explosion border have small ranges of attraction in the phase space of the system. For the case of purely thermogravitational convection, this feature had not been observed.

References

1. Gershuni, G.Z., E.M. Zhukhovitsky and A.K. Kolesnikov Vibrational-convective instability of horizontal fluid layer with internal heat sources, *Trans. Acad. Sci. SSSR (Izvestiya), Mech. Zhidk. Gaza*, 1985, **5**, 3–7.

2. Gershuni, G.Z., E.M. Zhukhovitsky, A.K. Kolesnikov and Yu.S. Yurkov. Vibrational convection in a horizontal layer with internal heat sources, *Int. J. Heat and Mass Transfer*, 1989, **32**(12), 2319–28.
3. Gershuni, G.Z. and E.M. Zhukhovitsky. On vibrational-convective instability of horizontal layer of internally heating fluid, in *Numerical and Experimental Modelling of Hydrodynamic Phenomena in Weightlessness*, Sverdlovsk, 1988, pp. 72–8.
4. Sparrow, E.M., K.J. Goldstein and V.K. Jonsson. Thermal instability in a horizontal fluid layer: effect of boundary conditions and nonlinear temperature profile, *J. Fluid Mech.*, 1964, **18**(4), 513–28.
5. Roberts, P.H. Convection in horizontal layers with internal heat generation, *Theory J. Fluid Mech.*, 1967, **30**(1), 33–47.
6. Gershuni, G.Z. and E.M. Zhukhovitsky. On monotonous and oscillatory instability of two-layer system of immiscible fluids heated from below, *Commun. Acad. Sci. SSSR (Doklady)*, 1982, **265**(2), 302–5.
7. Kozlov, V.G. and S.B. Shatunov. Experimental study of vibrational convection initiation in plane horizontal layer of fluid with internal heat generation, in *Numerical and Experimental Modelling of Hydrodynamic Phenomena in Weightlessness*, Sverdlovsk, 1988, 79–84.
8. Gershuni, G.Z., E.M. Zhukhovitsky and Yu.S. Yurkov. Nonlinear regimes of vibrational thermal convection in horizontal layer with internal heat sources, in *Convective Flows*, Perm, 1989, pp. 45–52.
9. Gershuni, G.Z., E.M. Zhukhovitsky and Yu.S. Yurkov. Finite-amplitude vibrational convection in horizontal fluid layer with internal heat generation, *Modell.in Mechanics*, Novosibirsk, 1990, **4**(21), No. 1, 103–8.
10. Gershuni, G.Z., E.M. Zhukhovitsky and Yu.S. Yurkov. Finite-amplitude convective motions in rectangular cavities with internal heat sources, in *Hydrodynamics*, Perm, 1974, No. 5, pp. 3–23.
11. Frank-Kamenetsky, D.A. *Diffusion and Heat Transfer in Chemical Kinetics*, Nauka, Moscow, 1987, 491 pp.
12. Merzhanov, A.G. and E.A. Stessel. Free convection and thermal explosion in reactive systems, *Astronaut. Acta*, 1973, **18**(3), 191–9.
13. Jones, D.A. The dynamic stability of confined, exothermically reacting fluids, *Int. J. Heat and Mass Transfer*, 1973, **16**(1), 191–9.
14. Yeremin, Ye.A. and A.K. Kolesnikov. On theory of convective stability of horizontal layer of reacting medium, in *Hydrodynamics*, Perm, 1977, No. 10, pp. 76–84.
15. Gershuni, G.Z., E.M. Zhukhovitsky and A.K. Kolesnikov. Convective stability of horizontal layer of reacting medium in high frequency vibrational field, *Phys. of Combustion and Explosion*, 1990, **5**, 91–6.
16. Gershuni, G.Z., E.M. Zhukhovitsky, A.K. Kolesnikov, B.I. Myznikova and Yu.S. Yurkov. On thermovibrational convection in an exothermal liquid in weightlessness, in *Proc. Int. Symp. on Hydromech and Heat/Mass Transfer in Microgravity*, Perm, Moscow, Gordon and Breach, 1992, pp. 63–7.
17. Gershuni, G.Z., E.M. Zhukhovitsky, A.K. Kolesnikov and B.I. Myznikova. Nonlinear regimes of thermovibrational convection in horizontal layer of exothermal liquid, *Phys. of Combustion and Explosion*, 1995, **31**(6), 89–98.
18. Yeremin, Ye.A. Numerical study of finite-amplitude motions and regimes of heat transfer in horizontal layer of reacting fluid, in *Convective Flows*, Perm, 1985, pp. 3–10.

5 Vibrations of Finite Frequencies

All the previous chapters are devoted to various occurrences of the thermovibrational convection. There we focused on the limiting case of high-frequency vibrations and described them on the basis of equations for the averaged fields of velocity, temperature and pressure. The specific feature of this approach is that all the resonance phenomena are excluded from consideration. In the present chapter some problems of thermovibrational convection are studied for vibrations of a finite frequency. Their treatment involves the complete system of Boussinesq equations. For a plane horizontal layer heated from below or from above and exposed to transversal or longitudinal vibrations, the effect of excitation of the parametric resonance is taken into account. The only exception is Section 4, where the convective resonance of a non-parametric origin is discussed.

1 Parametric Excitation of Convective Instability: The Statement of the Problem

1.1 Basic Equations

The present and the next section deal with the problem of convection initiation as a result of mechanical equilibrium stability loss when the acceleration of a mass force is periodically modulated with time at a finite frequency. This may be, for example, a fluid-filled cavity subject to the vertical temperature gradient and executing linear harmonic oscillations in the vertical direction. In the limit of high frequencies, the effect of vibrations with the vertical axis on the stability equilibrium under static gravity is always stabilizing. If the vertical oscillations are of a finite frequency, the effect of parametric instability excitation becomes possible. Generally speaking, an extra instability mechanism—the dynamic one—has been added. This suggests that the effect of vibrations on the convective instability threshold may be both stabilizing or destabilizing. Besides, under definite conditions the appearance of a dynamic instability is conceivable due to the parametric resonance phenomenon [1–4].

Let us write down the equations of convection in the Boussinesq approximation in the system of coordinates attached to the cavity. Vertical oscillations cause the modulation of gravity force acceleration, i.e. the replacement is needed for $g \to g_0(1 + \eta \sin \Omega t)$. Here g_0 is the static gravity acceleration, Ω is the angular frequency and η is the relative amplitude of acceleration modulation: $\eta = b_0 \Omega^2 / g_0$, where b_0 is the vibration amplitude. The pertinent equations are written as

$$\frac{\partial \mathbf{v}}{\partial t} + (\mathbf{v}\nabla)\mathbf{v} = -\frac{\nabla p}{\rho} + \nu \Delta \mathbf{v} + g_0(1 + \eta \sin \Omega t)\beta T \boldsymbol{\gamma}$$

$$\frac{\partial T}{\partial t} + \mathbf{v}\nabla T = \chi \Delta T \tag{5.1}$$

$$\operatorname{div} \mathbf{v} = 0$$

In the state of mechanical equilibrium, $\mathbf{v} = 0$ and $\partial T_0/\partial t = 0$. The equilibrium fields of temperature and pressure satisfy the equations

$$-\frac{\nabla p_0}{\rho} + g_0(1 + \eta \sin \Omega t)\beta T_0 \boldsymbol{\gamma} = 0$$

$$\Delta T_0 = 0 \tag{5.2}$$

From these equations it follows that the temperature gradient in a fluid is vertical and constant: $\nabla T_0 = -A_0 \boldsymbol{\gamma}$. The pressure p_0 is a quadratic function of the vertical coordinate z and is periodically modulated with time. To study the stability of such non-stationary equilibrium, small perturbations are introduced in the ordinary way, and the linearization is carried on about the equilibrium. The following scales are chosen: the linear size L of the cavity for length, $L^2/\sqrt{\nu\chi}$ for time, χ/L for velocity, $A_0 L$ for temperature and $\rho\nu\chi/L^2$ for pressure. Then the dimensionless equations for perturbations take the form

$$\frac{1}{\sqrt{Pr}} \frac{\partial \mathbf{v}}{\partial t} = -\nabla p + \Delta \mathbf{v} + Ra\,(1 + \eta \sin \widetilde{\omega} t) T \boldsymbol{\gamma}$$

$$\sqrt{Pr}\, \frac{\partial T}{\partial t} - (\mathbf{v}\boldsymbol{\gamma}) = \Delta T \tag{5.3}$$

$$\operatorname{div} \mathbf{v} = 0$$

Here $Pr = \nu/\chi$ is the Prandtl number, $Ra = g_0 \beta A L^4/(\nu\chi)$ the Rayleigh number defined through the static gravity acceleration and the equilibrium temperature gradient and $\widetilde{\omega} = \Omega L^2/\sqrt{\nu\chi}$ is the dimensionless frequency. Note that in contrast to the limiting case of high-frequency vibrations, the set of equations now incorporates the frequency and the amplitude as two independent parameters (in the averaged equations there is only one vibrational parameter ε; see Section 1 in Chapter 1).

Let the plane horizontal layer of thickness h be limited by two parallel planes $z = 0$ and $z = 1$; the thickness h of the layer is now taken as the reference scale for length. The origin of the Cartesian coordinate system is placed at the lower plane; the z axis is directed vertically upwards. The horizontal velocity components v_x and v_y as well as the perturbation of pressure are eliminated from the set (5.3) and the perturbations are assumed to be periodic functions of the horizontal coordinates x and y:

$$\begin{aligned} v_z &= v(z,t)\exp[\mathrm{i}(k_1 x + k_2 y)] \\ T &= \theta(z,t)\exp[\mathrm{i}(k_1 x + k_2 y)] \end{aligned} \tag{5.4}$$

For amplitudes $v(z,t)$ and $\theta(z,t)$ the following set is obtained (the prime denotes differentiation with respect to the transversal coordinate z):

$$\begin{gathered} \frac{1}{\sqrt{Pr}}\frac{\partial}{\partial t}\left(v'' - k^2 v\right) = \left(v^{\mathrm{IV}} - 2k^2 v'' + k^4 v\right) - k^2\, Ra\,(1 + \eta \sin \widetilde{\omega} t)\theta \\ \sqrt{Pr}\,\frac{\partial \theta}{\partial t} - v = \left(\theta'' - k^2\theta\right) \\ k^2 = k_1^2 + k_2^2 \end{gathered} \tag{5.5}$$

First, we consider a model case—a layer with free isothermal undeformable boundaries—that admits separation of variables. Indeed, at the boundaries $z = 0$ and $z = 1$ the conditions $v = v'' = \theta = 0$ hold, and the solution corresponding to the first level of the instability spectrum is

$$v(z,t) = a(t)\sin \pi z, \qquad \theta(z,t) = b(t)\sin \pi z \tag{5.6}$$

where the amplitudes $a(t)$ and $b(t)$ are determined by the set of ordinary differential equations of the first order:

$$\begin{gathered} \frac{\pi^2 + k^2}{\sqrt{Pr}}\frac{\mathrm{d}a}{\mathrm{d}t} + \left(\pi^2 + k^2\right)^2 a = k^2\, Ra(1 + \eta \sin \widetilde{\omega} t)b \\ \sqrt{Pr}\,\frac{\mathrm{d}b}{\mathrm{d}t} + \left(\pi^2 + k^2\right)b = a \end{gathered} \tag{5.7}$$

Let us now pass from the amplitudes $a(t)$ and $b(t)$ to two new functions, $g_1(\tau)$ and $g_2(\tau)$, by the replacement

$$a = g_1, \qquad b = l g_2, \qquad t = m\tau \tag{5.8}$$

where

$$m = \frac{1}{\pi^2 + k^2}, \qquad l = \frac{1}{\sqrt{Pr}(\pi^2 + k^2)} \tag{5.9}$$

Then the 'canonical' set may be derived for g_1 and g_2:

$$\begin{aligned} \dot{g}_1 + n g_1 &= R(1 + \eta \sin \omega\tau) g_2 \\ \dot{g}_2 + \frac{1}{n} g_2 &= g_1 \end{aligned} \tag{5.10}$$

Here the dot means differentiation with respect to τ and some new notations are used:

$$n = \sqrt{Pr}, \qquad \omega = \tilde{\omega} m, \qquad R = \frac{Ra}{Ra_c}, \qquad Ra_c = \frac{(\pi^2 + k^2)^3}{k^2} \tag{5.11}$$

Ra_c is the critical value of the Rayleigh number for the layer with free isothermal boundaries under static conditions, i.e. without any modulation.

Now we examine a horizontal layer with rigid isothermal boundaries. The boundary conditions at $z = 0$ and $z = 1$ are $v = v' = \theta = 0$, and so the straightforward separation of variables is not possible. To solve the set (5.5), either the method of finite differences with a one-dimensional grid [5] or the Galerkin-Kantorovitch method [6] may be used. According to the latter, the amplitudes v and θ are presented as series expansions in terms of the basis trial functions $F_i(z)$ and $\Phi_i(z)$:

$$v(z,t) = \sum_i a_i(t) F_i(z), \qquad \theta(z,t) = \sum_i b_i(t) \Phi_i(z) \tag{5.12}$$

with coefficients $a_i(t)$ and $b_i(t)$ depending on time. Substituting (5.12) into (5.5) and requiring the orthogonality of the residuals to the trial functions, the set of the first-order ordinary differential equations is derived for the functions $a_i(t)$ and $b_i(t)$. In the case of a horizontal layer, a satisfactory accuracy is achieved choosing the simplest polynomial approximations for the coordinate-dependent parts:

$$\begin{aligned} v(z,t) &= a(t) F(z), \qquad \theta(z,t) = b(t) \Phi(z) \\ F(z) &= z^2 (1 - z)^2, \qquad \Phi(z) = z(1 - z)\left(1 + z - z^2\right) \end{aligned} \tag{5.13}$$

Here $F(z)$ and $\Phi(z)$ satisfy the appropriate boundary conditions. In deciding on the $\Phi(z)$ approximation, additional conditions are taken into account, $\theta''(0) = \theta''(1) = 0$, provided by the second equation of the set (5.5). The orthogonality requirements of the Galerkin procedure yield the set of equations for $a(t)$ and

$b(t)$ similar to (5.7). It may be converted into the 'canonical' form (5.10) by replacement of variables of the type (5.8). The transformation parameters are

$$m=\left[\frac{31(12+k^2)}{(504+24k^2+k^4)(306+31k^2)}\right]^{1/2}, \qquad l=\frac{11}{62}\frac{m}{\sqrt{Pr}}$$

$$n=\left[\frac{31(504+24k^2+k^4)Pr}{(12+k^2)(306+31k^2)}\right]^{1/2}, \qquad \omega=m\widetilde{\omega} \tag{5.14}$$

$$R=\frac{Ra}{Ra_c}, \qquad Ra_c=\frac{4}{121k^2}\left(504+24k^2+k^4\right)\left(306+31k^2\right)$$

Here Ra_c is the approximate value of the critical Rayleigh number for a layer with rigid boundaries in the absence of modulation. The minimal Ra_c value is achieved at $k_m=3.12$ and is equal to $Ra_{cm}=1719$ (its deviation from the exact value of 1708 is about 0.6%).

Similarly, with the aid of the Galerkin-Kantorovitch technique, the problem of stability in a modulated gravity field is reduced to the 'canonical' set (5.10) for other cavity geometries. As an example, the problem of convective stability is studied in a long vertical cylinder of a circular cross-section in the presence of axial vibrations. As in the static case (see [7]), it is assumed that the main level of the instability spectrum corresponds to the perturbation of the following structure:

$$v_r=v_\varphi=0, \qquad v_z=v(r,\varphi); \qquad T=T(r,\varphi); \qquad p=p(z) \tag{5.15}$$

with $v\sim\cos\varphi$ and $T\sim\cos\varphi$ (the antisymmetrical flow). Here r, φ and z are the cylindrical coordinates. The solution must be finite at $r\to 0$, and at the lateral surface the non-slip condition for the velocity and the Newton heat exchange law are to be satisfied:

$$r=1: \qquad v=0, \qquad \frac{\partial T}{\partial r}=-Bi\,T \tag{5.16}$$

where the radius of the cylinder is chosen as a unit of length and Bi is the Biot number. In accordance with the Galerkin-Kantorovitch approach, the approximate solution may be constructed as

$$v=a(t)J_1(\gamma r)\cos\varphi, \qquad T=b(t)[J_1(\gamma r)-Cr]\cos\varphi \tag{5.17}$$

where J_1 is the Bessel function. The boundary conditions (5.16) indicate that γ is the first root of the equation $J_1(\gamma)=0$, i.e. $\gamma=3.8317$; the coefficient C equals $\gamma J_1'(\gamma)/(Bi+1)$. The set of equations for $a(t)$ and $b(t)$ may be reduced to (5.10) by transformation (5.8) with the parameters

$$m = \frac{n}{\gamma^2 \sqrt{Pr}}, \qquad l = \frac{1}{\gamma^2 n}, \qquad n = \left[\frac{Bi^2 + 6Bi + 5 + 0.5\gamma^2}{(Bi+1)(Bi+3)} Pr \right]^{1/2}$$

$$\omega = \widetilde{\omega} m, \qquad R = \frac{Ra}{Ra_c}, \qquad Ra_c = \gamma^4 \frac{Bi+1}{Bi+3} \tag{5.18}$$

The approximate value Ra_c of the critical Rayleigh number in the absence of vibrations depends on the Biot number. When Bi increases from zero (the adiabatic wall) to infinity (the boundary of high heat conductivity), the critical value Ra_c increases monotonically from 71.85 to 215.6 (for comparison, the exact limiting values are 67.95 and 215.6; see [7]).

Eliminating g_1 from the set (5.10) and denoting $g_2 \equiv f$, the second-order equation with periodic coefficients is obtained:

$$\ddot{f} + 2\varepsilon \dot{f} + (1 - R - r \sin \omega\tau) f = 0 \tag{5.19}$$

Here the coefficient $2\varepsilon = n + 1/n$ plays the role of a friction parameter and the absolute amplitude of modulation $r = R\eta$ is used instead of the relative one η.

It may be shown that the same equation is produced by the mechanical equilibrium stability problem when there is no modulation of the gravity force but the equilibrium temperature gradient is modulated (in the case when the frequency of modulation is small, so that the thickness of the temperature skin layer is much greater than the reference size of the cavity and the equilibrium temperature gradient may be considered as spatially uniform).

1.2 Rectangular Modulation

Thus the problem of description of the evolution of relevant perturbations may be presented as the set of amplitude equations (5.10) or as the equivalent second-order equation (5.19). The problem involves four independent parameters: the reduced Rayleigh number R, the friction parameter ε, the amplitude r and the modulation frequency ω. If the values of these parameters are chosen arbitrarily, the solution is either trivial or damping or unboundedly growing with time. Only a certain relationship between the parameters corresponds to the neutral (periodic) behavior of perturbations, and it is this regime that determines the stability border. To find the stability border in the system with a periodically varying parameter, we apply (see [8]) the method of characteristic exponents.

Let us discuss first the situation where the body force acceleration undergoes a periodic rectangular (step-wise) variation. In this case, in equation (5.19) the replacement $\sin \omega\tau \to \varphi(\tau)$ is to be carried out, where the function $\varphi(\tau)$ is

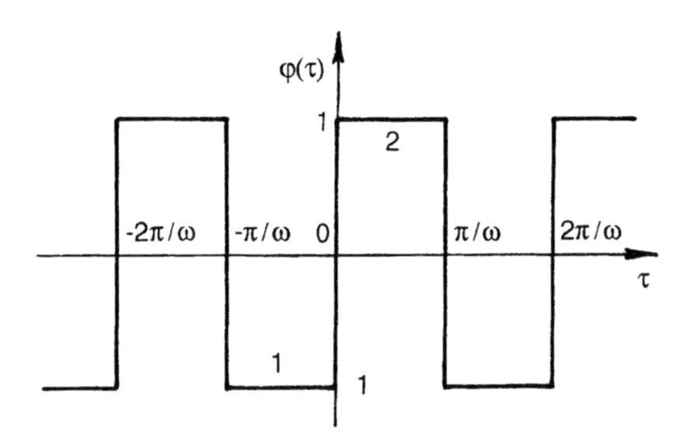

Fig. 129 Graph of the function $\varphi(\tau)$

shown in Fig. 129. During half-period intervals it assumes constant values $\varphi = \pm 1$. Inside the ranges marked by 1 and 2 in Fig. 129, the general solution is of the form:

$$\begin{aligned} f^{(1)} &= \mathrm{e}^{-\varepsilon\tau}(C_1 \sin \alpha\tau + C_2 \cos \alpha\tau) \\ f^{(2)} &= \mathrm{e}^{-\varepsilon\tau}(C_3 \sin \beta\tau + C_4 \cos \beta\tau) \end{aligned} \tag{5.20}$$

where

$$\alpha = \sqrt{1 - R + r - \varepsilon^2}, \qquad \beta = \sqrt{1 - R - r - \varepsilon^2}$$

and C_1 , C_2 , C_3 , C_4 are the constants of integration. To find these constants, the continuity conditions are used at the domain boundary for the solutions and their first derivatives:

$$f^{(2)}(0) = f^{(1)}(0), \qquad \dot{f}^{(2)}(0) = \dot{f}^{(1)}(0) \tag{5.21}$$

as well as the condition of 'normal' solution existence:

$$f^{(2)}\left(\frac{\pi}{\omega}\right) = \rho f^{(1)}\left(-\frac{\pi}{\omega}\right), \quad \dot{f}^{(2)}\left(\frac{\pi}{\omega}\right) = \rho \dot{f}^{(1)}\left(-\frac{\pi}{\omega}\right) \tag{5.22}$$

Relations (5.21) and (5.22) yield a set of linear homogeneous equations for the factor ρ. It may be evaluated from the equation expressing the existence condition of a non-trivial solution of the obtained set:

$$\begin{aligned} &\rho^2 - 2\left(\cos \frac{\pi\alpha}{\omega} \cos \frac{\pi\beta}{\omega} - \frac{\alpha^2 + \beta^2}{2\alpha\beta} \sin \frac{\pi\alpha}{\omega} \sin \frac{\pi\beta}{\omega}\right) \\ &\quad \times \exp\left(-\frac{2\pi\varepsilon}{\omega}\right)\rho + \exp\left(-\frac{4\pi\varepsilon}{\omega}\right) = 0 \end{aligned} \tag{5.23}$$

Periodic solutions emerge at $\rho = \pm 1$. Moreover, from (5.23) one finds the conditions of the existence of periodical solutions:

$$\cos\frac{\pi\alpha}{\omega}\cos\frac{\pi\beta}{\omega} - \frac{\alpha^2+\beta^2}{2\alpha\beta}\sin\frac{\pi\alpha}{\omega}\sin\frac{\pi\beta}{\omega} = \pm\cosh\frac{2\pi\varepsilon}{\omega} \tag{5.24}$$

This expression connects the parameters of the set and determines the equilibrium stability border. The + sign corresponds to the 'integer' periodic solutions ($\rho = 1$) having the period coinciding with that of the modulation. The − sign denotes the 'half-integer' periodic solutions ($\rho = -1$) whose period is twice the period of modulation. Thus the stability border in the case of a rectangular modulation is found exactly.

1.3 Sinusoidal Modulation

Let us turn to the first-order set of equations (5.10). Two particular solutions $(g_1^{(1)}, g_2^{(1)})$ and $(g_1^{(2)}, g_2^{(2)})$ are considered satisfying the initial conditions

$$g_i^{(k)}(0) = \delta_{ik} \tag{5.25}$$

Evidently, these solutions build up a fundamental set. Indeed, the determinant

$$D(\tau) = \begin{vmatrix} g_1^{(1)}(\tau) & g_2^{(1)}(\tau) \\ g_1^{(2)}(\tau) & g_2^{(2)}(\tau) \end{vmatrix}$$

does not vanish. Differentiating $D(\tau)$, using equations (5.10) and the initial condition $D(0) = 1$, one gets

$$D(\tau) = \mathrm{e}^{-2\varepsilon\tau} \neq 0 \qquad \left(2\varepsilon = n + \frac{1}{n}\right) \tag{5.26}$$

Therefore, the general solution of the set (5.10) has the form

$$g_1 = c_1 g_1^{(1)} + c_2 g_1^{(2)}, \qquad g_2 = c_1 g_2^{(1)} + c_2 g_2^{(2)} \tag{5.27}$$

Let us select the 'normal' solutions which are characterized by relations

$$g_1(T) = \rho g_1(0), \qquad g_2(T) = \rho g_2(0) \tag{5.28}$$

where $T = 2\pi/\omega$ is the modulation period. Such a choice leads to the set of linear homogeneous equations for c_1 and c_2. Setting its determinant to zero, we arrive at the equation for the characteristic exponent ρ:

$$\rho^2 - \rho\left[g_1^{(1)}(T) + g_2^{(2)}(T)\right] + \mathrm{e}^{-2\varepsilon T} = 0 \tag{5.29}$$

At $\rho = \pm 1$ the relationship is obtained determining the existence of periodic solutions or, in other words, the stability border:

$$g_1^{(1)}\left(\frac{2\pi}{\omega}\right) + g_2^{(2)}\left(\frac{2\pi}{\omega}\right) = \pm\left[1 + \exp\left(-\frac{4\pi\varepsilon}{\omega}\right)\right] \tag{5.30}$$

The + or – signs mark, respectively, the 'integer' and 'half-integer' solutions. The characteristic relationship (5.30) is the analog of (5.24) derived for the rectangular modulation.

It is not possible to obtain a fundamental set analytically. However, it may be obtained by numerical integration of the set (5.10) using, for example, the Runge–Kutta method if all four parameters are fixed. Practically, one fixes some three of the four parameters and varies the fourth with a small step until the condition (5.30) is attained with the prescribed accuracy. This is a rather efficient procedure to determine the stability border. The approach is equally applicable for constructing periodic solutions for sets of a higher order than (5.10) (see [3]).

2 Parametric Excitation of Convective Instability: Stability Borders

2.1 Rectangular Modulation

In this section some results are reported on the stability border evaluation for rectangular and sine-shaped modulations. First, the rectangular case is analyzed and the stability borders are sought exactly from the characteristic relation (5.24). This formula involves four independent dimensionless parameters: the 'reduced' Rayleigh number R , the friction coefficient ε, the frequency ω and the amplitude

$$r = R\eta = \frac{g_0\,\beta A h^4}{\nu\chi Ra_\mathrm{c}}\frac{b_0\Omega^2}{g_0} = \frac{b_0\Omega^2\beta A h^4}{\nu\chi Ra_\mathrm{c}}$$

i.e. the effective vibrational Rayleigh number defined through the vibration acceleration and frequency ω. The domains of parametric instability excitation are clearly seen in the plane 'amplitude–period of modulation' at fixed R and ε. When R is positive, which means that the cavity is heated from below, three domains should be distinguished: $0 < R < 1$, $1 < R < 1 + \varepsilon^2$ and $R > 1 + \varepsilon^2$. The first of them corresponds to a subcritical heating from below, the second and the third ones to overcritical heating. Figure 130a, b and c exemplifies the stability maps for those three domains.

In the interval $0 < R < 1$ (subcritical heating from below), where vibrations are absent ($r = 0$), one finds a stable equilibrium state. However, as the

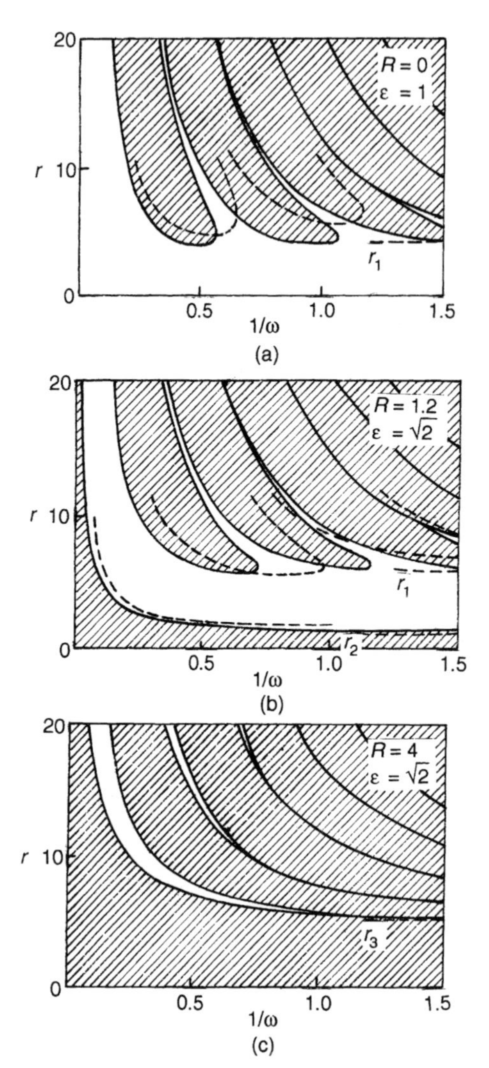

Fig. 130 Stability diagrams in the plane (ω^{-1}, r). Instability regions are shown by hatching; solid lines, rectangular modulation; dashed lines, sinusoidal modulation. (a) $0 < R < 1$, (b) $1 < R < 1 + \varepsilon^2$, (c) $1 + \varepsilon^2 < R$

amplitude r increases, the domains of parametric instability excitation appear in the plane 'amplitude–modulation period' (Fig. 130a). The leftmost domain corresponds to 'half-integer' perturbations. Further on, the 'integer' and 'half-integer' areas alternate. The lower boundary of those domains corresponds to the amplitude $r = r_1 = 3\varepsilon^2 - (R - 1)$ at $1/\omega \to \infty$. If the vibration amplitude exceeds r_1, the zones of parametric excitation occur. The example given in Fig. 130a is for weightlessness ($R = 0$).

If $R > 1$, the equilibrium is unstable under static conditions because the temperature gradient is greater than the critical value. In the presence of modulation, the domains of resonance stabilization emerge in the plane $(1/\omega, r)$. If $1 < R < 1 + \varepsilon^2$, in addition to the resonance excitation areas, the basic instability strip appears adjacent to the axis $r = 0$. The line $r = 0$ itself belongs to this domain. The upper border of the strip at $1/\omega \to \infty$ is determined by the amplitude value $r = r_2 = 2\varepsilon\sqrt{R-1}$. Above this region, the stability area resides. Its width is equal to $r_1 - r_2$. It narrows down and tends to zero as $R \to 1 + \varepsilon^2$. The example of a stability diagram is presented in Fig. 130b. Inside the basic instability strip, the perturbations of the 'integer' type develop whose period of oscillations is equal to the period of vibration.

At $R > 1 + \varepsilon^2$, the structure of the instability areas is given in Fig. 130c. The equilibrium is unstable at practically all values of the amplitude and frequency. Only for $r > r_3 = \varepsilon^2 + (R - 1)$ there exist narrow stripes of resonance parametric stabilization.

Figure 131 displays the threshold amplitudes values r_1, r_2 and r_3 as functions of R. From the formulas for r_1, r_2 and r_3, one determines the critical value of $R(r)$ in the limiting case of very low frequencies $1/\omega \gg 1$. One finds that R increases quadratically in r for $r < 2\varepsilon^2$ and diminishes linearly in r when $r > 2\varepsilon^2$:

$$\begin{aligned} R &= 1 + \frac{r^2}{4\varepsilon^2} \qquad \left(r < 2\varepsilon^2\right) \\ R &= 1 + 3\varepsilon^2 - r \qquad \left(r > 2\varepsilon^2\right) \end{aligned} \tag{5.31}$$

In the limiting case of high frequencies ($\omega^{-1} \ll 1$) and small amplitudes ($r \ll 1$), the critical R value may be obtained by expanding the left and right

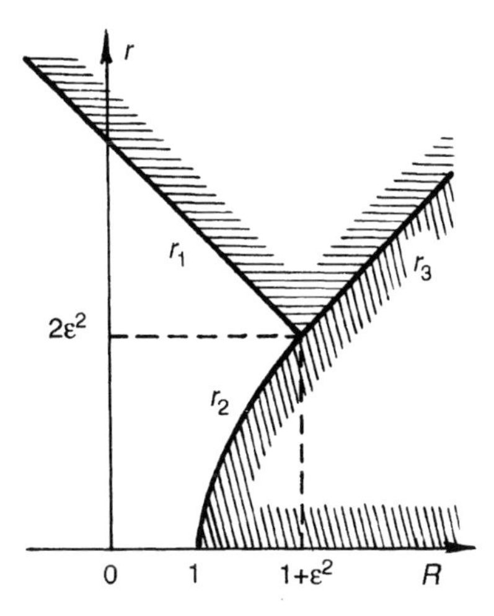

Fig. 131 Threshold values r_1, r_2 and r_3 as functions of R

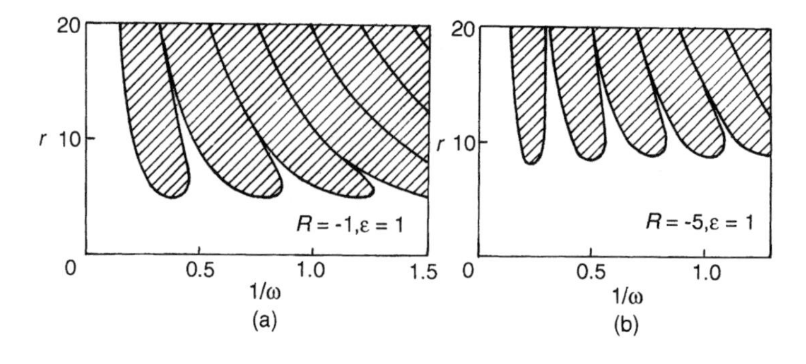

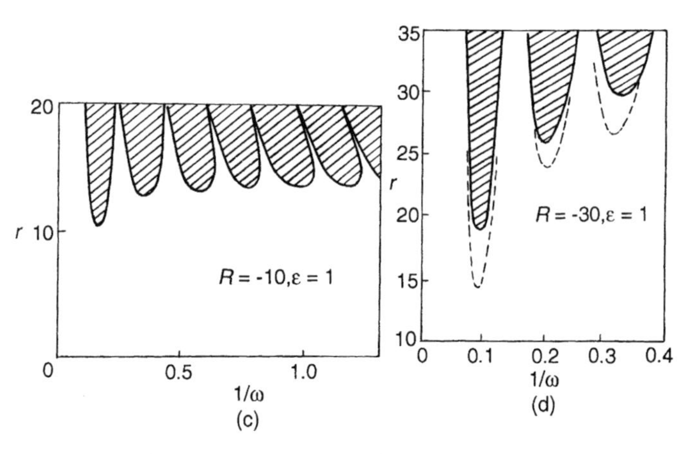

Fig. 132 Maps of stability in the plane (ω^{-1}, r) for the case of heating from above. Instability regions are hatched; solid lines, rectangular modulation; dashed lines, sinusoidal modulation; $\varepsilon = 1$. (a) $R = -1$, (b) $R = -5$, (c) $R = -10$, (d) $R = -30$

parts of (5.24) into power series in $1/\omega$ and r:

$$R = 1 + \frac{r^2}{12\omega^2} \tag{5.32}$$

Now let $R < 0$ where the average corresponds to potentially stable stratification of the layer (Fig. 132a–d, $\varepsilon = 1$). At $R = -1$ (Fig. 132a), the threshold amplitude value is $r_m = 5$, i.e. $r_m > |R|$, and the stratification is potentially unstable for the majority of the period. The situation remains the same at $R = -5$ ($r_m = 8.4$, Fig. 132b) and at $R = -10$ ($r_m = 10.5$, Fig. 132c). The equality $|R| = r_m$ is achieved at $R \approx -11$, and at $|R| > 11$ the layer is stratified potentially stably all throughout the period since $r_m < |R|$ (Fig. 132d, $R = -30$,

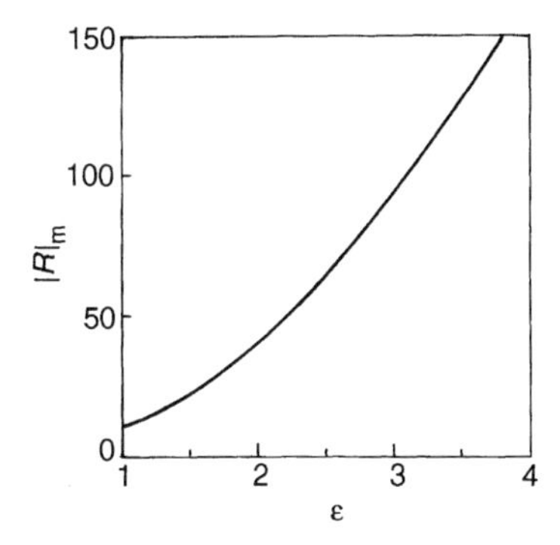

Fig. 133 Threshold $|R|_m$ value as a function of ε

$r_m = 19$). Thus the advent of the parametric excitation strips is the result of the dynamic instability of the system.

The stability diagrams in Fig. 132a to d correspond to the friction parameter $\varepsilon = 1$. Computations performed for ε values varying between 1 and 4 deliver the picture that is much the same. Figure 133 shows the dependence of the boundary value $|R|_m$ on ε. Beginning with $|R|_m$, excitation of the parametric convection takes place for $r_m < |R|$.[†]

With some care one may propose an analogy between the convective system under consideration and Kapitza's pendulum. It is well known [10] that the lowest stable position of the pendulum may be made unstable and, on the contrary, its unstable upper position may be stabilized under appropriate choice of the vibration parameters. In a similar manner, the modulation of gravity force yields the appearance of instability domains at $R < 1$ and resonance stabilization at $R > 1$.

2.2 Sinusoidal Modulation

When the modulation obeys a sinusoidal law, the stability borders are determined by the characteristic relation (5.30). The numerical procedure has been described in the previous section. The borders of stability in the plane $(1/\omega, r)$ are presented in Figs. 130 and 132 together with those for rectangular modulation. It can be seen that the qualitative features, having been discussed in the rectangular case, remain valid for the harmonic modulation as well. At the same time, there are some notable quantitative distinctions in the location of the resonance domains.

The dependence of the critical 'reduced' Rayleigh number R on the dimensionless amplitude η is exemplified in Fig. 134 for fixed $\omega = 1$ and $n = 3.67$.

[†] Note that the possibility to excite instability at large negative R was first predicted in [9] and detailed in [4].

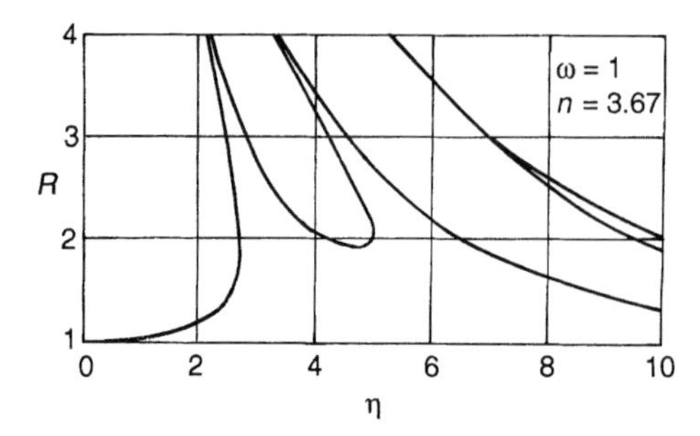

Fig. 134 The 'reduced' critical Rayleigh number R depending on the modulation amplitude η ($\omega = 1$, $n = 3.67$)

Inside the basic instability strip, the critical R value increases with η, i.e. stabilization takes place. When η is large enough ($\eta > 2.7$ for given above ω and n), the instability is related to the parametric resonance excitation.

2.3 Limiting Case of High-Frequency Vibrations

In the present chapter the phenomena are analyzed where the vibration frequency is finite. Passing on to the limit $\omega \to \infty$, the results must coincide with those of Chapter 1. To prove that, we write equation (5.19) in the form

$$\ddot{f} + 2\varepsilon \dot{f} + [1 - R(1 + \eta \sin \omega\tau)] f = 0 \tag{5.33}$$

If there is no modulation, then

$$\ddot{f} + 2\varepsilon \dot{f} + (1 - R) f = 0 \tag{5.34}$$

that yields the stability boundary $R = 1$. In the limiting case of a high-frequency modulation, the averaging technique may be applied directly to (5.33). The solution is sought as a sum of a slow part $f_0(\tau)$ and small fast component $\xi(\tau)$:

$$f(\tau) = f_0(\tau) + \xi(\tau) \tag{5.35}$$

Substituting this expression into (5.33), selecting items incorporating fast parts and retaining only the leading terms, one obtains the following equation:

$$\ddot{\xi} = R\eta f_0 \sin \omega\tau \tag{5.36}$$

Integrating (5.36) twice and bearing in mind that f_0 is the slow function of time, the result is

$$\xi = -\frac{R\eta f_0}{\omega^2} \sin \omega\tau \tag{5.37}$$

Now (5.37) is substituted in (5.35), then in (5.33) and the averaging is carried out with respect to the fast time. It yields

$$\ddot{f}_0 + 2\varepsilon \dot{f}_0 + \left(1 - R - \frac{\eta^2 R^2}{2\omega^2}\right) f_0 = 0 \tag{5.38}$$

From a comparison of (5.38) and (5.34) it is clear that high-frequency vibrations lead to renormalization of the static gravity field. The stability border in the limit of high ω is

$$1 - R = \frac{\eta^2 R^2}{2\omega^2} \tag{5.39}$$

Moreover, one may see that at high frequencies the critical R value is governed by the parameter

$$\left(\frac{\eta}{\omega}\right)^2 \sim b_0^2 \Omega^2$$

In Fig. 135 the stability borders are shown in the plane $(\eta/\omega, R)$. The solid line corresponds to the asymptotic formula (5.39) and the numerical solutions of the general set (5.10) for different ω values are shown by dashed lines. (One has to remember that in the case of finite ω the position of the neutral curves is controlled by both parameters ω and η, not just by their ratio η/ω.) It can be seen that as ω increases, the border of the basic instability domain tends rather rapidly to the asymptote (5.39). In practice, at $\omega > 10$ the high-frequency limit is attained. However, at finite ω there are also the resonance instability areas in addition to the basic one. As ω grows, those areas shift to infinity, and only the basic strip survives.

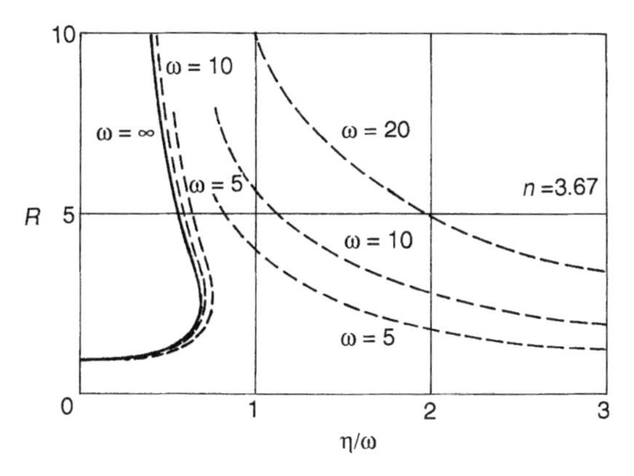

Fig. 135 The critical R value versus η/ω. Solid line, the limiting case of high-frequency vibrations; dashed lines, numerical solutions at finite ω

Let us address now the case of a horizontal layer with rigid isothermal boundaries. In (5.39) the parameters are replaced with $R = Ra/Ra_c$ and $\omega = mh^2\Omega/\sqrt{\nu\chi}$, where Ω is the dimensional frequency of modulation, and Ra_c and m are defined by (5.14). Then

$$Ra = \frac{m^2 Ra_c}{\kappa^2}\left(1 \pm \sqrt{1 - \frac{2\kappa^2}{m^2}}\right) \tag{5.40}$$

Here $\kappa = \omega_0 b_0 \sqrt{\nu\chi}/(g_0 h^2)$ is the dimensionless parameter specifying the effect of high-frequency vibrations on convective stability. Formula (5.40) points out the fact that there is some limiting value $\kappa^* = m/\sqrt{2}$ for which absolute stabilization takes place. Provided the dependences $Ra_c(k)$ and $m(k)$ are known, with the aid of (5.14) the neutral curves $Ra(k)$ may be constructed for different κ. Figure 136 presents a family of neutral curves for several κ. As κ grows, the minimal critical value Ra_m increases (the stabilization) and the critical wavenumber k_m tends to zero (see Chapter 1, Fig. 13). When $\kappa > \kappa^*$, the equilibrium is stable for any value of the vertical temperature gradient and vibration parameters.

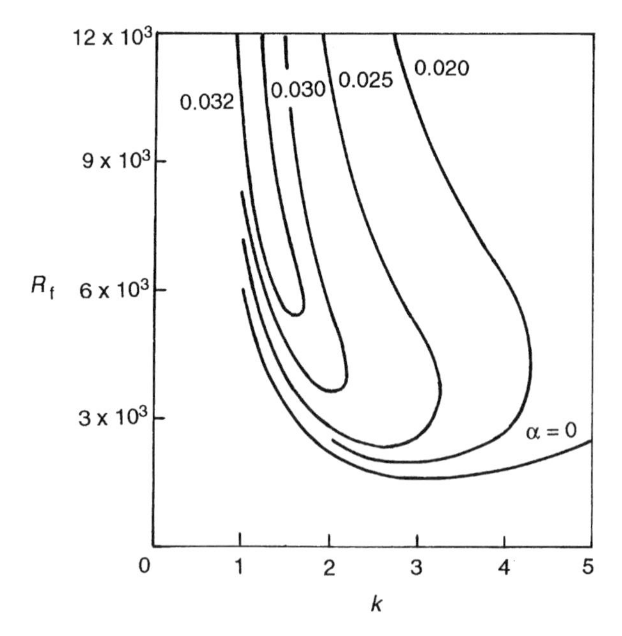

Fig. 136 Neutral curves for a horizontal layer with rigid boundaries for several values of the vibrational parameter κ

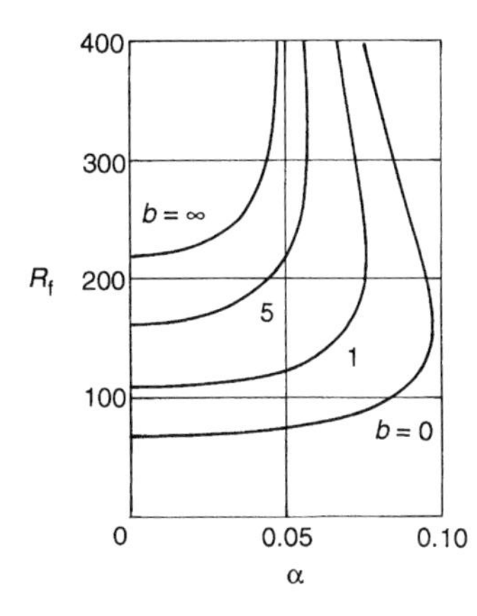

Fig. 137 The critical value of *Ra* as a function of the vibrational parameter κ for a vertical cylinder of a circular cross-section at different values of the Biot number $b = Ka/\lambda$ (K is the heat transfer coefficient, a is the cylinder radius, λ is the heat conductivity coefficient of the fluid)

Formula (5.39) is universal: it allows the stability border to be found for a cavity of an arbitrary shape when the frequency is high (of course, in the framework of a one-function Galerkin approximation). In order to be able to handle a particular situation, the parameters Ra_c and m must be specified. For more details, see [2]. Here we presented the result for the case of an infinitely long vertical cylinder with a circular cross-section. In Fig. 137 the κ dependence of *Ra* is given for different values of the Biot number. As κ increases, the stabilization sets in as for a horizontal layer.

In conclusion of this section we refer to some works [11–17] where the influence of modulation was studied on the Rayleigh–Benard instability. Most of them [11, 13–16] are concerned with the temperature modulation at horizontal boundaries, i.e. with the problem equivalent to that of the present chapter in the limiting case of low frequencies when the boundary layer thickness is greater than the vertical size of the cavity. Paper [12] deals with a vibration-induced modulation of the buoyancy force. This work is close to that of [2], both conceptually (the simplest Galerkin approximations are used) and in the results obtained. In [17] a case of a vibrating infinite vertical cylinder subject to a longitudinal temperature gradient is investigated. The vibration amplitude is taken as the small parameter. The shift of the critical Rayleigh number is proportional to the square of the amplitude with the coefficient depending on the frequency and the Prandtl number.

3 Finite Amplitude Oscillations

The two previous sections dealt with the linear convective instability of a mechanical equilibrium under harmonic modulation of the gravity force. It is obvious that after the equilibrium stability is lost, finite amplitude oscillations develop. To examine such oscillations, an essentially non-linear approach is required. This was done by Burde [18, 19]. The solution of the set of non-linear non-stationary equations describing a free convection was found numerically by the finite difference method for a square cavity with rigid heat-insulated vertical and rigid isothermal horizontal boundaries. The boundary temperature drop corresponded to heating either from below or from above.

A harmonic modulation of the gravity force means that the replacement must be done as $Ra \to Ra(1 + \eta \sin \omega t)$, where Ra is the Rayleigh number defined through the temperature difference and the side h of the cavity, $\eta = \Omega^2 b_0/g$ is the relative amplitude of modulation and $\omega = h^2\Omega/\nu$ is the dimensionless frequency.

The numerical solution is sought of the following boundary problem:

$$\frac{\partial \Delta\psi}{\partial t} + \frac{1}{Pr}\left(\frac{\partial\psi}{\partial y}\frac{\partial\Delta\psi}{\partial x} - \frac{\partial\psi}{\partial x}\frac{\partial\Delta\psi}{\partial y}\right) = \Delta\Delta\psi - Ra\,(1 + \eta\sin\omega t)\frac{\partial T}{\partial x}$$
$$Pr\frac{\partial T}{\partial t} + \left(\frac{\partial\psi}{\partial y}\frac{\partial T}{\partial x} - \frac{\partial\psi}{\partial x}\frac{\partial T}{\partial y}\right) = \Delta T \tag{5.41}$$

$$y = 0: \quad \psi = 0, \quad \frac{\partial\psi}{\partial y} = 0, \quad T = 1$$
$$y = 1: \quad \psi = 0, \quad \frac{\partial\psi}{\partial y} = 0, \quad T = 0 \tag{5.42}$$
$$x = 0, x = 1: \quad \psi = 0, \quad \frac{\partial\psi}{\partial x} = 0, \quad \frac{\partial T}{\partial x} = 0$$

Here ψ is the stream function and x and y are, respectively, the horizontal and the vertical Cartesian coordinates.

An explicit finite difference scheme was used with central differences for all the spatial derivatives and the one-sided right differences for time derivatives. The computations were performed using the equally spaced 16×16 grid, granting the accuracy of the order of 2–3% in the instability border determination. At the initial instant, the perturbation was introduced at some interior grid point and its evolution was monitored. The parameters Ra, Pr and $1/\omega$ were fixed while η varied with a small step. Gradually, the value of η was achieved when the initial perturbation ran its course to the regime of steady convective oscillations. In Fig. 138 the stability borders are shown for $Ra = 3000$ (for a square cavity with heat-insulated lateral boundaries under static gravity, the critical Rayleigh number is $Ra_c = 2600$; see [20] and Section 3 in Chapter 4). Qualitatively, the structure of the instability domains is the same as for a horizontal layer (see the previous section). There exist the basic instability strip

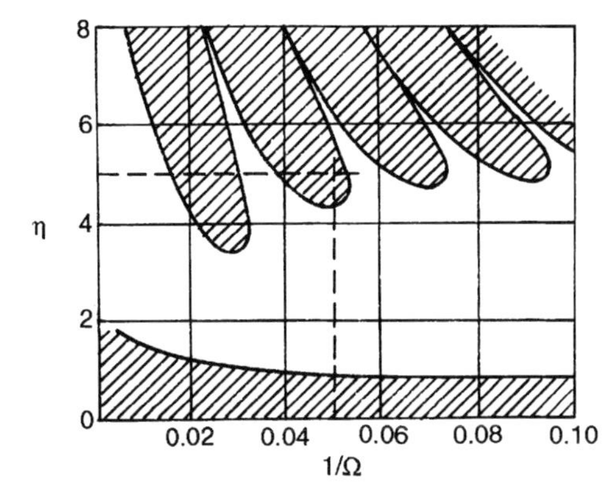

Fig. 138 Stability diagram for a square cavity; $Ra = 3000$, $Pr = 1$

and the areas of parametric resonance excitation. The axis $1/\omega$ belongs to the instability strip since $Ra > Ra_c$.

For the interior points of instability domains, the evolution of the initial perturbation goes through a transition stage to the regime of steady finite amplitude oscillations. Hence, the form of these oscillations may be determined as well as their non-linear characteristics. Following Burde [18, 19], the mean value of the stream function over the grid nodes and the mean value of the stream function over the period are analyzed:

$$\psi_a(t) = \frac{1}{N}\sum_{i,k}\psi_{ik}, \qquad \widetilde{\psi} = \overline{\psi_a(t)}$$

where N is the total number of the grid nodes. Inside of the first integer and all the half-integer domains, the flow is symmetrical, so $\widetilde{\psi} = 0$. Inside the basic strip and the rest of the integer areas, $\widetilde{\psi} \neq 0$.

Another descriptive measure of finite amplitude oscillations is the amplitude $\delta\psi$ which is equal to the half-difference between the maximal and minimal values of $\psi_a(t)$ over the period. In all the instability zones, $\delta\psi \neq 0$.

Figure 139 demonstrates the characteristics of finite amplitude oscillations inside the basic instability strip at $\omega = 20$. When $\eta \to 0$, in the overcritical range the oscillations gradually transform into the motion of constant intensity. Upon that, the amplitude $\delta\psi \to 0$ and the mean intensity $\widetilde{\psi}$ tends to the stationary value corresponding to the case of non-modulated gravity.

In a similar way, one could monitor the non-linear characteristics inside the areas of parametric excitation. In Fig. 140 they are given as functions of $1/\omega$ at $\eta = 5$. In the interior of the first half-integer region, as already noted, $\widetilde{\psi} = 0$. As the domain borders are approached, both characteristics reduce to zero, indicating the soft character of the convection excitation.

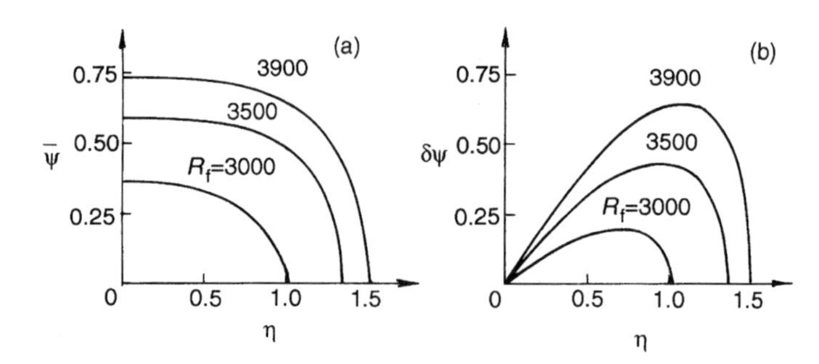

Fig. 139 Characteristics of finite amplitude oscillations in the fundamental instability strip depending on the modulation amplitude ($\Omega = 20$, vertical dashed section in Fig. 138; $Pr = 1$)

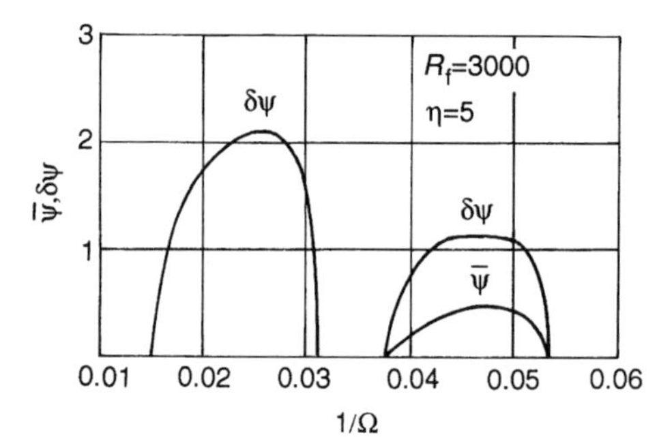

Fig. 140 Characteristics of finite amplitude oscillations in the first (half-integer) and the second (integer) domains ($\eta = 5$, horizontal dashed section in Fig. 138; $Ra = 3000$, $Pr = 1$)

The results displayed in Figs. 138 to 140 correspond to the fixed value of the Prandtl number $Pr = 1$. Burde [19] showed that in the range of small Prandtl numbers there exists an interval of frequencies on the upper border of the half-integer resonance domain, where the convection excitation is of a hard character.

The results presented here are obtained when the temperature modulation is imposed at the horizontal boundaries while the gravity force is constant [21, 22].

4. Resonance Phenomena in a Modulated Gravity Field

In this section the convection is studied in a closed cavity heated from inside in the presence of harmonic vertical vibrations modulating the gravity acceleration.

The problem is formulated as follows. A two-dimensional convective flow is considered in a fluid-filled square cavity $\{0 \le x \le a, 0 \le y \le a\}$, where x is the horizontal coordinate axis and y is the vertical one. The vertical boundaries are maintained at constant different temperatures: $T = 0$ at $x = 0$ and $T = \Theta$ at $x = a$. At the horizontal boundaries $y = 0$ and $y = a$, the temperature varies linearly with $x : T = \Theta x/a$. The cavity performs vertical harmonic oscillations with amplitude b and cyclic frequency Ω. In the proper coordinate system attached to the cavity, the equations describing convection differ from the standard Boussinesq ones by the replacement $g \to g(1 + \eta \cos \Omega t)$, where $\eta = b\Omega^2/g$ is the relative amplitude of the gravity force modulation. The dimensionless equations written in terms of the stream function ψ and temperature T take the form

$$\frac{\partial \Delta\psi}{\partial t} + \left(\frac{\partial\psi}{\partial y}\frac{\partial\Delta\psi}{\partial x} - \frac{\partial\psi}{\partial x}\frac{\partial\Delta\psi}{\partial y}\right) = \Delta\Delta\psi - (Gr + Gr_{\mathrm{v}}\cos\omega t)\frac{\partial T}{\partial x}$$

$$\frac{\partial T}{\partial t} + \left(\frac{\partial\psi}{\partial y}\frac{\partial T}{\partial x} - \frac{\partial\psi}{\partial x}\frac{\partial T}{\partial y}\right) = \frac{1}{Pr}\Delta T \tag{5.43}$$

The scales are chosen as follows: a for length, a^2/ν for time, ν for the stream function, Θ for temperature. Here Gr is the usual (gravitational) Grashof number, Gr_{v} is its vibrational analog, Pr is the Prandtl number, ω is the dimensionless frequency:

$$Gr = \frac{g\beta\Theta a^3}{\nu^2}, \qquad Gr_{\mathrm{v}} = \frac{\beta b\Omega^2\Theta a^3}{\nu^2}, \qquad Pr = \frac{\nu}{\chi}, \qquad \omega = \frac{\Omega a^2}{\nu} \tag{5.44}$$

The boundary conditions are:

$$\begin{aligned} x = 0: &\quad \psi = 0, \quad \frac{\partial\psi}{\partial x} = 0, \quad T = 0 \\ x = 1: &\quad \psi = 0, \quad \frac{\partial\psi}{\partial x} = 0, \quad T = 1 \\ y = 0, y = 1: &\quad \psi = 0, \quad \frac{\partial\psi}{\partial y} = 0, \quad T = x \end{aligned} \tag{5.45}$$

The problem was investigated in [23] to [26]. The approximate solution was produced by means of a finite difference technique using both the explicit and implicit schemes of the alternating direction method. The computations were performed on the grids of 25×25 points for moderate frequencies and 35×35 points for high frequencies ($\omega > 10^3$). The comparison showed that the further grid refinement does not provide any significant improvement in evaluation of the convective oscillation characteristics. For a detailed description of the oscillation structure the time grid step must be small with respect to the period of modulation. In all the calculations the Prandtl number was fixed: $Pr = 1$.

The initial problem was solved numerically. Given some initial perturbation, the evolution, having passed through the transition stage, results in a regular regime of convective oscillations. In this regime, the local and integral properties of the velocity and temperature distributions oscillate near some mean values with the frequency of modulation ω. The oscillations themselves (their structure and intensity depending on the parameters) were under consideration as well as the mean fields. The mean values, in general, differ from those of a static case. This difference is indeed a measure of the vibrational convection contribution.

The mean values of the stream function and dimensionless heat flux are the important instantaneous characteristics of the velocity and temperature fields. The averaged stream function value over the grid points (K is the total number of nodes)

$$\Psi_{\mathrm{m}}(t) = \frac{1}{K}\sum_{i,k} \psi_{ik}$$

indicates the direction and intensity of the instantaneous circulation in the cavity. The dimensionless heat flux is defined as

$$N(t) = -\int_{\Gamma} \frac{\partial T}{\partial n} \mathrm{d}l$$

where Γ denotes the part of the contour where the local heat flux is directed into the cavity. Along with $\Psi_{\mathrm{m}}(t)$ and $N(t)$, their values averaged over the period $\overline{\Psi}_{\mathrm{m}}$ and $\overline{N}$ are evaluated.

Let us begin discussing computation results with the case of weightlessness $Gr = 0$. Figure 141 illustrates the dependence of $\overline{\Psi}_{\mathrm{m}}$ on the frequency for two values of parameter $\varepsilon = 0.5\, Gr_{\mathrm{v}}^2/\omega^2$. We would like to recall that ε is the only vibrational parameter governing convection when the frequency is high, i.e. the averaged approach is applicable. It can be seen that up to $\omega \sim 10^2$ a symmetrical motion of alternating sign takes place with $\overline{\Psi}_{\mathrm{m}} = 0$. In this process, the flow during two sequential half-periods has opposing circulation directions. Such a situation holds up to the critical value ω^* depending on ε. As the frequency exceeds ω^*, the said motion loses its stability and gives way to the flow of a predominant circulation ($\overline{\Psi}_{\mathrm{m}} \neq 0$). Two flow patterns may develop depending on the initial conditions. These two branches differ from one another by the sign of $\overline{\Psi}_{\mathrm{m}}$, but are equivalent from the viewpoint of stability and other properties. In Fig. 141 the branch with $\overline{\Psi}_{\mathrm{m}} > 0$ is presented. Inside the range $\omega \sim 10^2$–10^3, a non-linear resonance takes place. It seems reasonable to ascribe it to the proximity between the period of modulation and the reference time of a fluid element circulation inside the cavity. In the high-frequency limit, all the

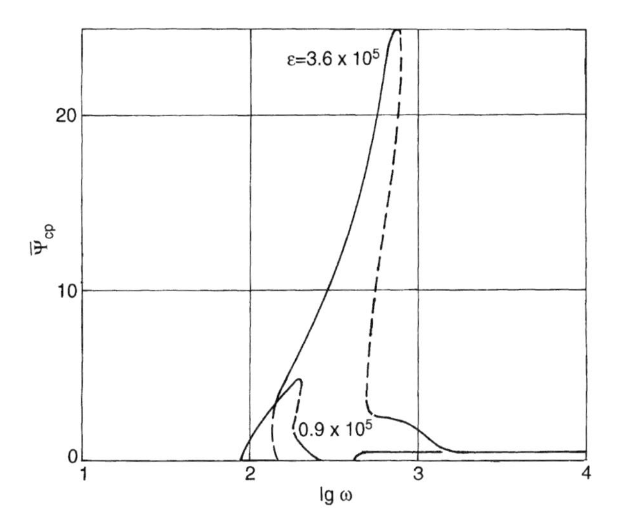

Fig. 141 Frequency-dependent $\overline{\Psi}_{\mathrm{m}}$ for two values of ε (weightlessness, $Gr = 0$)

characteristics including $\overline{\Psi}_{\mathrm{m}}$ cease to depend on ω, i.e. the high-frequency asymptote sets in. In this frequency interval the numerical results agree well both qualitatively and quantitatively with those of Section 1 in Chapter 3, obtained by application of the averaged equations.

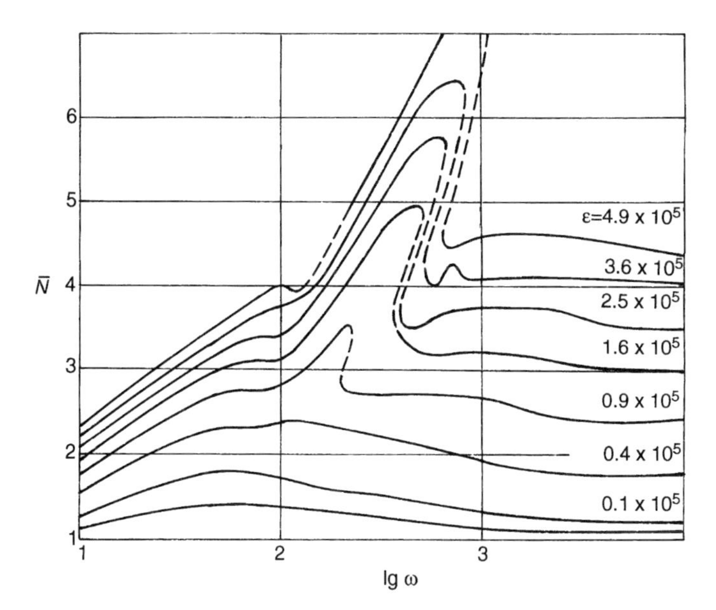

Fig. 142 Frequency-dependent $\overline{N}$ for different values of ε (weightlessness, $Gr = 0$)

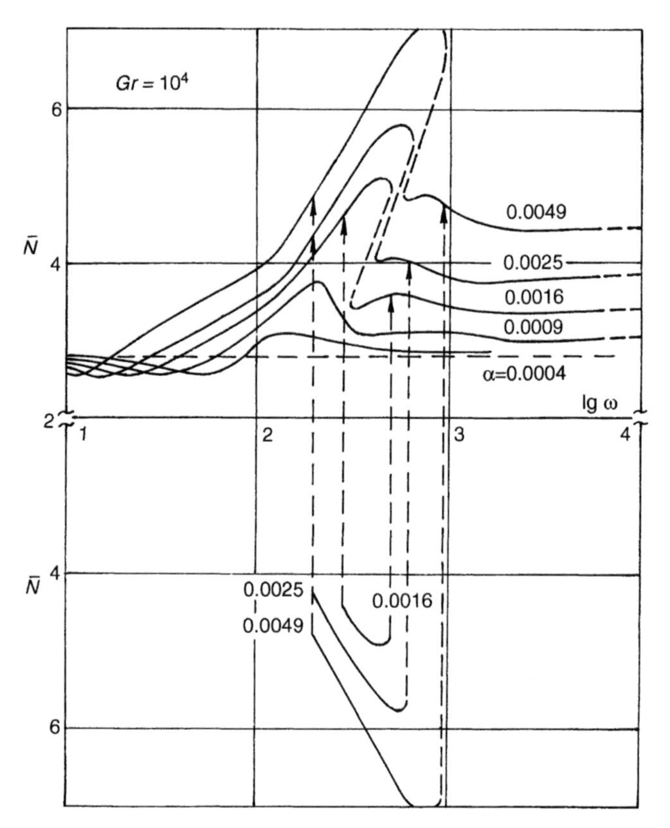

Fig. 143 Frequency-dependent $\overline{N}$ for different values of α; $Gr = 10^4$

The family of curves $\overline{N}(\omega)$ displayed in Fig. 142 for different ε demonstrates clearly the resonance behavior. The unstable parts of curves $\overline{\Psi}_{\mathrm{m}}(\omega)$ and $\overline{N}(\omega)$ are shown in Figs. 141 and 142 by dashed lines.

Let us turn now to the case where the static gravity is present, and it is necessary to deal with the complete first equation of the set (5.43). The next to last term in this equation may be rewritten in the form

$$-Gr\left(1 + \omega\sqrt{2\alpha}\cos\omega t\right)\frac{\partial T}{\partial x}$$

where α is the vibrational parameter:

$$\alpha = \frac{\eta^2}{2\omega^2} = \frac{1}{2}\left(\frac{b\nu\Omega}{ga^2}\right)^2$$

The main distinction from the case of weightlessness is that now there exists circulation in the counter-clockwise direction when the cavity is heated from the right side. Doing so considerably deforms the resonance curves. In Fig. 143 the function $\overline{N}(\omega)$ is given for different α and fixed $Gr = 10^4$. The resonance character is quite visible in the range $\omega \sim 10^2$–10^3. In the resonance domain one may see non-single-valuedness of characteristics, the coexistence of stable and unstable regimes, and hysteresis transitions (shown by arrows). It is interesting to note that in the resonance domain inside a relatively narrow interval of frequencies, the stable regimes of 'improper' circulation are available. There, the vortices rotate clockwise when heated from the right side. Such 'improper' flows may be realized through evolution of an initial perturbation of the appropriate sign. It should be emphasized that the stability range of the 'improper' regimes is frequency bounded: outside it they become unstable and the transition to the stable flow of a regular circulation takes place (arrows). For convenience, the heat flux $\overline{N}$ values corresponding to the 'improper' branches are given in the lower part of the figure, though, of course, $\overline{N} > 0$.

Figure 144 demonstrates the family of curves $\overline{N}(\omega)$ for different Gr values and a fixed value of the vibrational parameter $\alpha = 0.0004$. In Fig. 145 the averaged heat flux $\overline{N}$ is plotted as a function of the vibrational parameter for different frequencies at fixed Gr. As the frequency grows, the resonance features disappear. That is naturally associated with the onset of the high-frequency asymptotes. Finally, in Fig. 146 the family of curves $\overline{\Psi}_m(\alpha)$ is shown

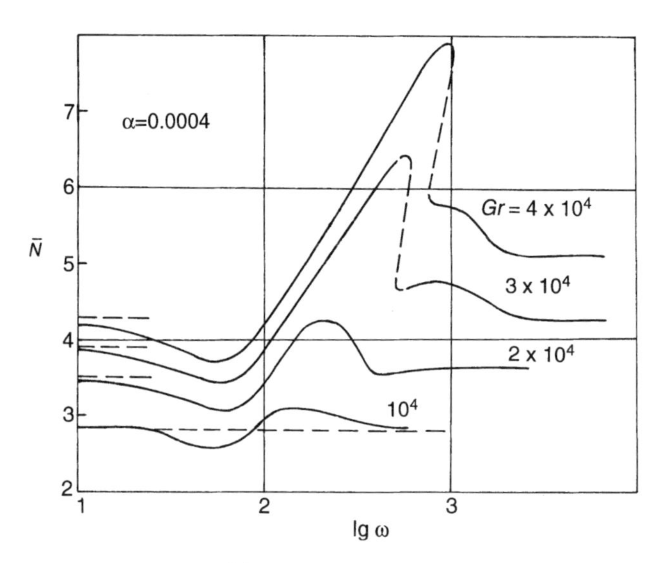

Fig. 144 Frequency-dependent $\overline{N}$ for different values of Gr and a fixed value of the vibrational parameter $\alpha = 0.0004$

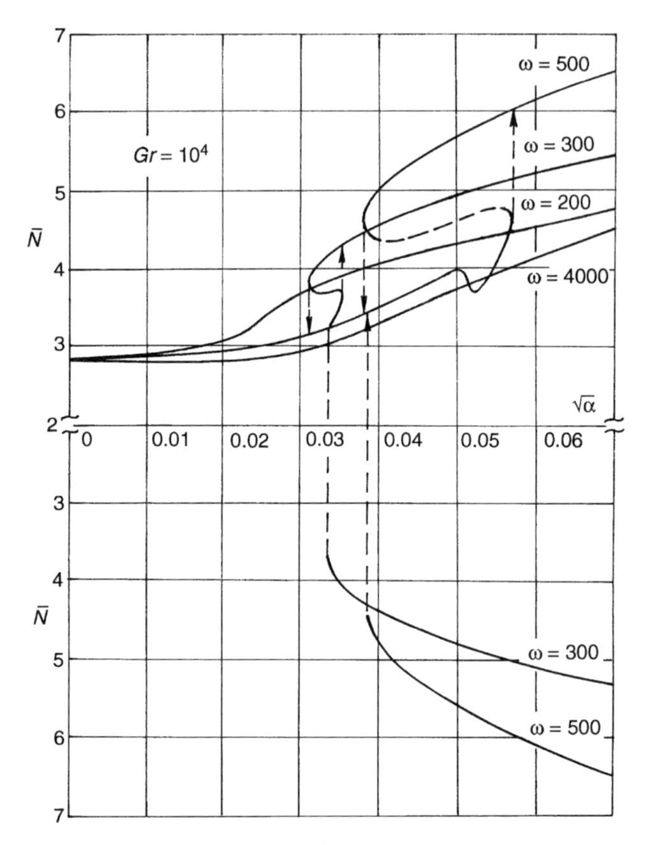

Fig. 145 $\overline{N}$ versus vibrational parameter for different frequencies; $Gr = 10^4$

for different ω. It is also seen that the complicated pattern associated with the non-linear resonance dies out as the frequency goes higher.

Resonance phenomena in a variable acceleration field were described later in [27] and [28].

5 Vibrational Convective Instability at Finite Frequencies

From Chapter 1 and on, the stability problems of mechanical equilibrium and flows were considered with the aid of a set of averaged equations. The results obtained correspond to the limiting case of high frequencies. This approach to the stability problems eliminates all the effects of parametric excitation which are essentially important in the range of low and finite frequencies. The present

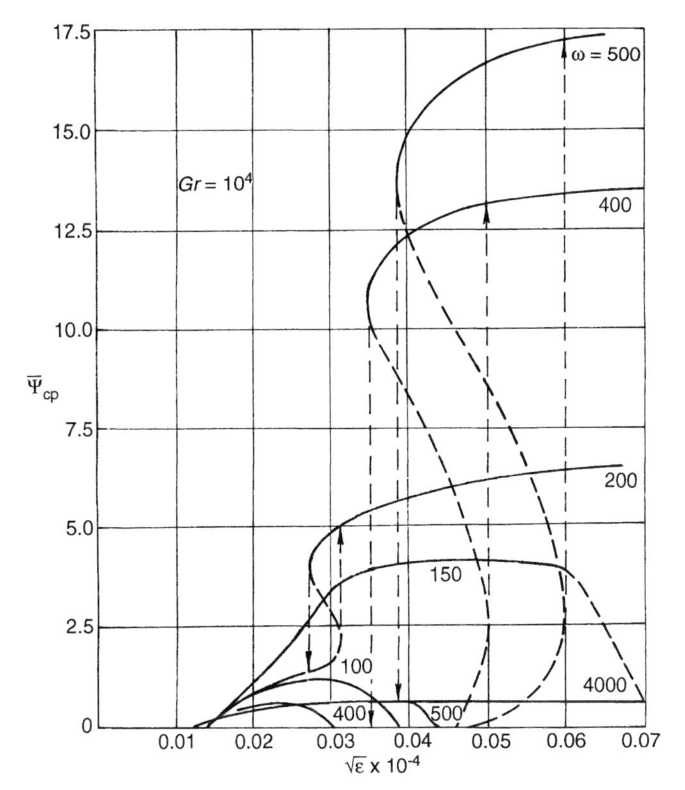

Fig. 146 $\overline{\Psi}_{\mathrm{m}}$ versus vibrational parameter for different frequencies; $Gr = 10^4$

section deals with the problem of the vibrational convective instability in its complete statement, i.e. without recourse to the averaging technique. Naturally, when reduced to the high-frequency limit, the results coincide with those of Chapter 1.

The standard problem of convection initiation is studied for a plane infinite horizontal layer subject to a static gravity field and longitudinal vibrations in the presence of a transversal temperature difference (the high-frequency limit of the problem is discussed in Section 5 in Chapter 1). A plane horizontal layer is bounded by rigid parallel planes $z = \pm h$, i.e. h is the half-thickness of the layer. The planes are maintained at constant difference temperatures $T = \mp \Theta$. The layer as a whole, including its boundaries, oscillates linearly and harmonically in the direction of the longitudinal x axis with the angular frequency Ω and displacement amplitude b. To write down the equations of convection in the Boussinesq approximation in the proper (oscillating) coordinate system, it is necessary as usual to replace the gravity acceleration $\mathbf{g}$ by $\mathbf{g} + b\Omega^2 \sin \Omega t \cdot \mathbf{n}$, where **n** is the unit vector along the vibration axis. The dimensionless variables are introduced with the units h for length, h^2/ν for time, χ/h for velocity, Θ for temperature and $\rho\nu\chi/h^2$ for pressure. Then the dimensionless equations acquire the form

$$\frac{\partial \mathbf{v}}{\partial t} + \frac{1}{Pr}(\mathbf{v}\nabla)\mathbf{v} = -\nabla p + \Delta\mathbf{v} + Ra\, T\gamma + \overline{R}_v T \sin\omega t \cdot \mathbf{n}$$

$$Pr\frac{\partial T}{\partial t} + \mathbf{v}\nabla T = \Delta T \tag{5.46}$$

$$\text{div } \mathbf{v} = 0$$

Here Ra is the conventional Rayleigh number and $\overline{R}_v$ is its vibrational analog:

$$Ra = \frac{g\beta\Theta h^3}{\nu\chi}, \qquad \overline{R}_v = \frac{b\Omega^2\beta\Theta h^3}{\nu\chi}$$

Other notations are standard. The set (5.46) has the basic solution describing an oscillatory plane-parallel flow. This solution is of the structure $T = T_0(z)$, $\mathbf{v} = \mathbf{v}_0\{v_0(z,t), 0, 0\}$, which satisfies the boundary conditions

$$z = \pm 1: \qquad T_0 = \mp 1, \qquad v_0 = 0 \tag{5.47}$$

and the condition of closeness of the oscillatory flow:

$$\int_{-1}^{1} v_0 \, \mathrm{d}z = 0 \tag{5.48}$$

From the boundary problem (5.46) to (5.48) it follows that

$$T_0 = -z, \qquad v_0 = \frac{\overline{R}_v}{\omega} V_0 \tag{5.49}$$

It is denoted here:

$$V_0 = (z + Z_1)\cos\omega t + Z_2 \sin\omega t$$

$$Z_1 = \frac{\cosh\alpha \cdot \cos\beta - \cosh\beta\cos\alpha}{\cos 2\kappa - \cosh 2\kappa}, \qquad Z_2 = \frac{\sinh\alpha \cdot \sin\beta - \sinh\beta\sin\alpha}{\cos 2\kappa - \cosh 2\kappa}$$

$$\alpha = \kappa(1+z), \qquad \beta = \kappa(1-z), \qquad \kappa = \sqrt{\frac{\omega}{2}}$$

When $\omega \to \infty$, this solution describes the 'quasi-equilibrium', i.e. the state with the zero mean velocity but a non-vanishing oscillatory component.

To study the stability of the basic state (5.49), small perturbations are introduced. In the limiting case of high-frequency vibrations (see Section 3 in Chapter 1), it has been found that two-dimensional perturbations are the most unstable under weightlessness. It can be shown that the same property is valid as well for the case considered here of finite vibration frequencies. Therefore, let the basic state (5.49) be perturbed by small two-dimensional perturbations $T'(x,z,t)$, $p'(x,z,t)$, $\mathbf{v}'(v_x', 0, v_z')$, where $v_x' = v_x'(x,z,t)$ and $v_z' = v_z'(x,z,t)$. Substituting the perturbed fields into the set (5.46) and linearizing about the basic state (5.49), the set of equations for perturbations is derived. We write it down in terms of the stream function ψ' and temperature T' of perturbations of the 'normal' type:

$$\begin{aligned} \psi'(x,z,t) &= \varphi(z,t)\exp(-\mathrm{i}kx) \\ T'(x,z,t) &= \theta(z,t)\exp(-\mathrm{i}kx) \end{aligned} \tag{5.50}$$

Here $\varphi(z,t)$ and $\theta(z,t)$ are the amplitudes depending on the transversal coordinate z and time t, and k is the wavenumber. After substitution of (5.50) into the set of equations for perturbations, the boundary problem is obtained for the amplitudes:

$$\begin{aligned} &\frac{\partial}{\partial t}D\varphi - \frac{\mathrm{i}k\overline{R}_\mathrm{v}}{\omega\, Pr}V_0 D\varphi + \frac{\mathrm{i}k\overline{R}_\mathrm{v}}{Pr}F_0\varphi = D^2\varphi + \overline{R}_\mathrm{v}\sin\omega t\cdot\theta' + \mathrm{i}k\, Ra\,\theta \\ &Pr\frac{\partial\theta}{\partial t} - \mathrm{i}k\varphi - \frac{\mathrm{i}k\overline{R}_\mathrm{v}}{\omega}V_0\theta = D\theta \end{aligned} \tag{5.51}$$

Here the prime denotes differentiation with respect to the transversal coordinate z, D is the operator $D = \mathrm{d}^2/\mathrm{d}z^2 - k^2$, and V_0 and F_0 are the coefficients varying periodically with time:

$$V_0 = (z + Z_1)\cos\omega t + Z_2\sin\omega t$$

$$F_0 = -Z_1\sin\omega t + Z_2\cos\omega t$$

The boundary conditions for the amplitudes at the rigid isothermal planes are

$$z = \pm 1: \qquad \varphi = \varphi' = 0, \qquad \theta = 0 \tag{5.52}$$

The amplitude boundary problem (5.51) and (5.52) allows the behavior of the 'normal' perturbations to be monitored. The stability border may be found as well. According to the Floquet theory, the critical values of the parameters are determined from the existence condition of a periodic solution of the amplitude problem. To construct those solutions, a numerical approach was developed in [29, 30] based on the Galerkin–Kantorovitch method. The spatial basis is formed from the eigenfunctions of the amplitude problem for a quiescent layer using a certain type of normalization. Therefore, the basis for the stream function approximation is chosen as the solution of the following eigenvalue problem [31]:

$$\Delta^2\varphi_i = -\mu_i\Delta\varphi_i$$
$$\varphi_i(\pm 1) = \varphi_i'(\pm 1) = 0$$

The trial functions for the temperature amplitude coincide with the eigenfunctions of the problem:

$$\Delta\theta_j = -\nu_j\theta_j, \qquad \theta_j(\pm 1) = 0$$

The solution is built as a superposition of the spatial trial functions with time-dependent coefficients. It comprises up to 20 trial functions φ_i and θ_j. The orthogonalization procedure of the Galerkin method yields a set of ordinary differential equations for the time-dependent coefficients. The set was integrated by the Runge–Kutta method. If the parameters of the system are chosen arbitrarily, then the initial conditions evolve to the solution describing either damping or growing oscillations. In the first situation one refers to stability, in the second to instability. The combination of parameters that provides the periodic solution determines the stability border. The search for the periodic solution may be facilitated by determining a multiplicator being a function of all the parameters. Interpolating the multiplicator values to ± 1, the set of conditions may be found corresponding to the periodic solution or, what is the same, to the stability border.

We begin the discussion from the case of complete weightlessness $Ra = 0$. Only one mechanism of convective instability excitation remains, the thermovibrational one, and $\overline{R}_v$ plays the role of the regime parameter. Under high frequencies the instability, by its physical nature, is connected with stratification in the vibrational field. When in the solution of the complete amplitude problem (5.51) and (5.52) a limiting transition $\omega \to \infty$ is performed, the attained asymptote coincides with the result of the average approach. In this range, instead of $\overline{R}_v$, it is convenient to use the Rayleigh number Ra_v entering the averaged equations. While doing that, one has to take into account the relationships

$$Ra_v = \frac{(\beta b\Omega\Theta h)^2}{2\nu\chi}, \qquad \overline{R}_v = \frac{\beta b\Omega^2\Theta h^3}{\nu\chi}, \qquad Ra_v = \frac{\overline{R}_v^2}{2\omega^2 Pr}$$

The neutral curves in the plane (k, Ra_v) have minima at $k = k_m$ (cellular instability). The minimal critical value of Ra_v is Ra_{vm}. Figure 147 demonstrates the critical parameters Ra_{vm} and k_m as the functions of the vibration frequency ω at $Pr = 1$. When $\omega \to \infty$, the parameters tend to the asymptotic values already known from the averaged method: $Ra_{vm} = 2129/16 = 133.1$; $k_m =$

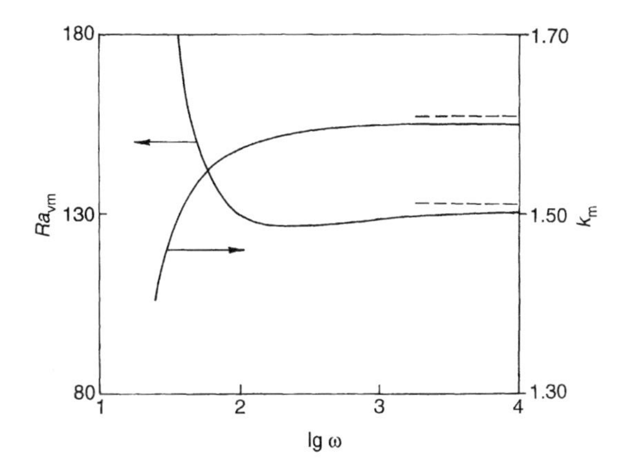

Fig. 147 Instability parameters Ra_{vm} and k_{m} in the high-frequency range; $Pr = 1$

$3.23/2 = 1.61$. (One should recall that in this section the length and temperature scales are taken as the half of the layer thickness h and the half-difference of the temperature Θ, respectively.) One sees that Ra_{vm} achieves its asymptote non-monotonically; the dependence $Ra_{\mathrm{vm}}(\omega)$ has the minimum at $\omega \approx 200$.

In the opposite limiting case of low frequencies, the basic flow has the form

$$v_0 = \overline{R}_{\mathrm{v}} \frac{z^3 - z}{6} \sin \omega t + O(\omega), \qquad T_0 = -x \tag{5.53}$$

Thus a slowly oscillating plane-parallel flow occurs with the cubic velocity profile that involves two counter-flows with the inflection point in the middle of the layer instead of the stream with a developed Stokes boundary layers. In the static case, the instability of such a flow has a non-viscous origin; it is studied in detail in [32]. If the velocity profile is modulated slowly, the asymptotic Wentzel–Kramers–Brillouin method [33] may be applied.

Let the set of amplitude equations (5.51) be written in the matrix form at $Ra = 0$:

$$\frac{\partial}{\partial t}\mathbf{A}\mathbf{u} = \mathbf{N}\mathbf{u} \tag{5.54}$$

where the explicit form of vector $\mathbf{u}$, matrices $\mathbf{A}$ and $\mathbf{N}$ is

$$\mathbf{u} = \begin{pmatrix} \varphi \\ \theta \end{pmatrix}, \qquad \mathbf{A} = \begin{pmatrix} D & 0 \\ 0 & Pr \end{pmatrix},$$

$$\mathbf{N} = \begin{pmatrix} D^2 + \dfrac{\mathrm{i}k\overline{R}_{\mathrm{v}}}{\omega Pr} V_0 D - \dfrac{\mathrm{i}k\overline{R}_{\mathrm{v}}}{Pr} F_0 & \overline{R}_{\mathrm{v}} \sin \omega t \dfrac{\partial}{\partial z} \\ \mathrm{i}k & D + \dfrac{\mathrm{i}k\overline{R}_{\mathrm{v}}}{\omega} V_0 \end{pmatrix}$$

If to introduce a variable $\tau = \omega t$, equation (5.54) assumes the form

$$\omega \frac{\partial}{\partial \tau} \mathbf{A}\mathbf{u} = \mathbf{N}\mathbf{u} \tag{5.55}$$

where $\mathbf{N}$ is now the matrix of a 2π-periodic operator with respect to τ. The boundary conditions are

$$z = \pm 1: \qquad \varphi = \varphi' = 0, \qquad \theta = 0 \tag{5.56}$$

At small ω it is possible to expand N and the critical number $\overline{R}_{\mathrm{v}}$ in powers of ω:

$$\begin{aligned} \overline{R}_{\mathrm{v}} &= R_0 + \omega R_1 + \omega^2 R_2 + \cdots \\ N &= N_0 + \omega N_1 + \omega^2 N_2 + \cdots \end{aligned} \tag{5.57}$$

The solution of (5.55) is sought as follows:

$$\begin{aligned} u &= \exp\left[\frac{1}{\omega}\int_0^{\tau} \lambda(\tau')\, \mathrm{d}\tau'\right]\left(w_0 + \omega w_1 + \omega^2 w_2 + \cdots\right) \\ \lambda &= \lambda_0 + \omega\lambda_1 + \omega^2\lambda_2 + \cdots \end{aligned} \tag{5.58}$$

Here λ—a 2π-periodic function of τ—is a characteristic decrement of perturbations of the quasi-static ('frozen') flow with the cubic velocity profile (5.53). After substitution of (5.57) and (5.58) in (5.55), the equations of successive approximations are derived. In the zeroth order with respect to ω the equation is written as

$$\lambda_0\, \mathbf{A} w_0 = \mathbf{N}_0 w_0 \tag{5.59}$$

Upon adding the appropriate boundary conditions, the eigenvalue problem is obtained for $\lambda_0(\tau)$. According to the Floquet theory, the stability boundary in the zeroth order may be found from the integral relation

$$\int_0^{2\pi} \lambda_{0r}(R_0 \sin \omega\tau)\, \mathrm{d}\tau = 0 \tag{5.60}$$

where λ_{0r} is the real part of λ_0. To solve (5.59), the Galerkin method was applied using the same, as described earlier, spatial basis involving up to 20 trial functions. As a result, at $Pr = 1$ the critical value $\overline{R}_{\mathrm{vm}}$ minimized over k is obtained at the minimal wavenumber k_{m}:

$$\left(\overline{R}_{\mathrm{vm}}\right)_{\omega \to 0} = 764, \qquad k_{\mathrm{m}} = 1.33$$

The behavior of the critical Rayleigh number over the whole range of frequencies is illustrated by Fig. 148, where the function $\overline{R}_{\mathrm{vm}}(\omega^{-1})$ is presented.

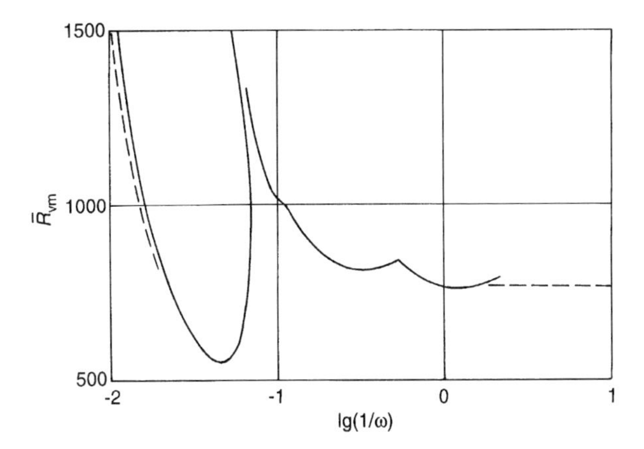

Fig. 148 Stability border $\overline{Ra}_{vm}$ as a function of the period in the range of intermediate frequencies; $Pr = 1$

The Prandtl number is fixed at $Pr = 1$; the instability region is above the curve. The asymptotes of high and low frequencies are shown by dashed lines. When the frequency is high, the crisis is determined by the value $Ra_{vm} = \overline{R}_v^2/(2\omega^2 Pr) = 133.1$. Thence, it follows that $\overline{R}_{vm} = 16.32(1/\omega)^{-1}$. This hyperbola is drawn in Fig. 148 as a dashed curve. The low-frequency asymptote $\overline{R}_{vm} = 764$ is also given in this figure by the dashed horizontal line. In the intermediate interval of frequencies, the zones of parametric instability excitation are visible. The period of neutral perturbation oscillation within these regions coincides with the period of vibration. Nevertheless, it is worth while to note that the perturbations behave differently with time in the neighboring instability zones. As the computations show, the real and imaginary parts of perturbations may have different periods of oscillations inside the neighboring domains—they either coincide or differ by a factor of two.

Let us turn now to the situation where the static gravity field exists. The analysis of perturbation evolution should be based on the complete set of the amplitude equations (5.51) when $Ra \neq 0$.

If there is no vibration ($\overline{R}_v = 0$), it is well known [20] that the set (5.51) describes the decay of perturbations when heated from above ($Ra < 0$), which may be monotonic or oscillatory. When the cavity is heated from below ($Ra > 0$), the principle of monotonicity of perturbations is valid, and there is some critical value of the Rayleigh number determining the border for monotonic instability. For example, for a horizontal layer heated from below, the critical Rayleigh number is $Ra_m = 106.7$. Now the problem is to investigate the effect of longitudinal vibrations of finite frequency and amplitude on the convective instability. The frequency and the amplitude are obviously

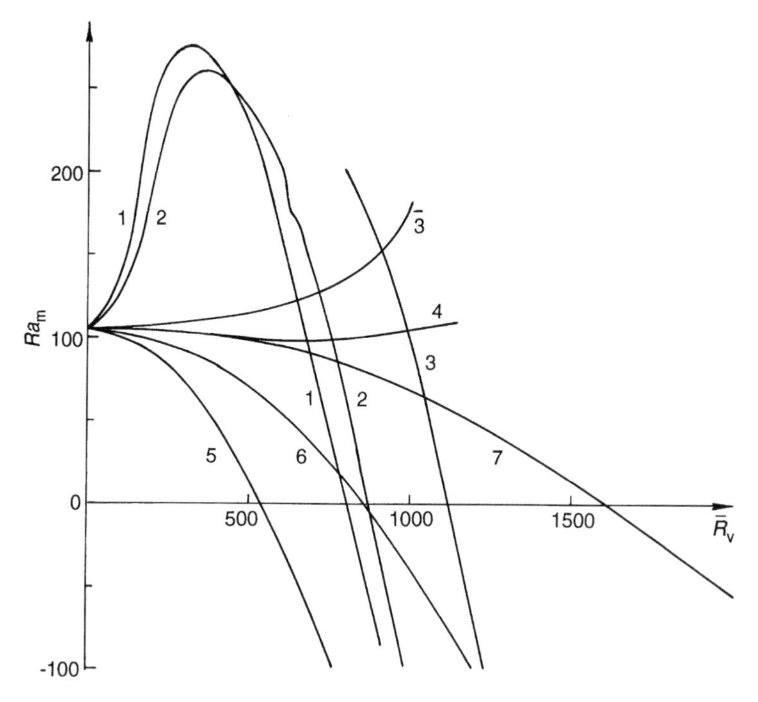

Fig. 149 Amplitude-dependent critical Rayleigh number at several values of frequency; $Pr = 1$. 1 to 7: $\omega = \pi$, 2π, 4π, 4.1π, 8π, 16π, 32π

independent parameters, whereas in the high-frequency limit the product $b\Omega$ appears as the single parameter.

Figure 149 displays the dependence of the critical Rayleigh number minimized with respect to the wavenumber on amplitude and frequency of vibrations. The function $Ra_{\mathrm{m}}(\overline{R}_{\mathrm{v}})$ is plotted for several values of frequency (curves 1 to 7 correspond to $\omega = \pi$, 2π, 4π, 4.1π, 8π, 16π and 32π). At small $\overline{R}_{\mathrm{v}}$ this dependence is parabolic:

$$Ra_{\mathrm{m}} = Ra_0 + d\overline{R}_{\mathrm{v}}^2 \tag{5.61}$$

where d is determined by the rest parameters. The Prandtl number is fixed at $Pr = 1$. It turns out that $d > 0$ when $\omega < \omega^*$ and $d < 0$ when $\omega > \omega^*$, where $\omega^* \approx 4.1\pi$. In the range of finite $\overline{R}_{\mathrm{v}}$ the law (5.61) is violated: if $\omega < \omega^*$ the curves have a maxima at some values of $\overline{R}_{\mathrm{v}}$, and if the frequency is relatively high, the curves go to the range of negative Ra corresponding to heating from above.

Finally, the stability borders are represented in Fig. 150 in the plane (Ra, Ra_{v}) for several values of frequency. The stability lines are practically straight, as in the case $\omega \to \infty$ (see Section 5 in Chapter 1). They continue on to the range

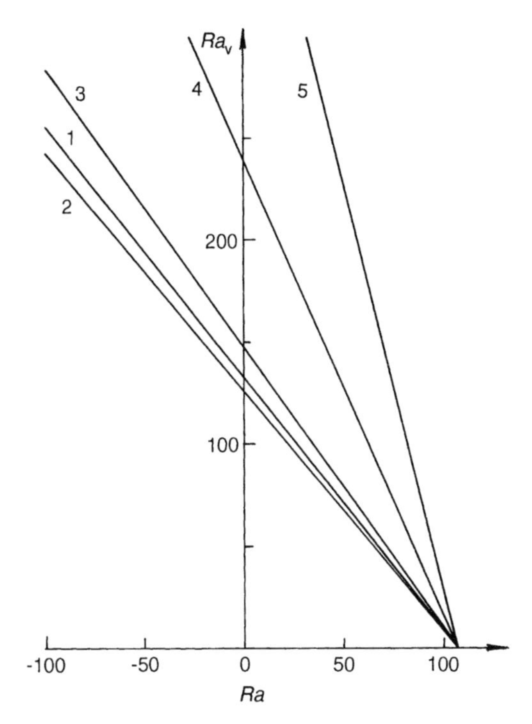

Fig. 150 Stability boundaries in the plane (Ra, Ra_v) at several values of frequency; $Pr = 1$. 1, $\omega \to \infty$; 2, $\omega = 32\pi$; 3, $\omega = 16\pi$; 4, $\omega = 8\pi$; 5, $\omega = 6\pi$

$Ra < 0$. This means that in a cavity heated from above, the specific vibrational instability mechanism acts at finite frequencies as well as in the limit $\omega \to \infty$.

It can be seen that in contrast to the case $\omega \to \infty$, where the vibrational mechanism always operates in a destabilizing manner, at finite frequencies (due to parametric excitation) both destabilization and stabilization of equilibrium are possible depending on the amplitude and the frequency of modulation.

In conclusion, let us refer to papers [27], [28], [34] to [40] devoted to the 'g-jitter' effect in thermal convection. In most of these papers, the problem is treated by means of direct modeling on the basis of non-stationary non-linear equations. This scope of problem attracts particular interest in connection with investigations of hydrodynamics and heat transfer under reduced gravity and in weightlessness, especially in crystal growth processes.

References

1. Gershuni, G.Z. and E.M. Zhukhovitsky. On parametric excitation of convective instability, *Prikl. Math. Mech.*, 1963, **27**(5), 779–83.

2. Gershuni, G.Z., E.M. Zhukhovitsky and Yu.S. Yurkov. On convective stability in the presence of periodically varying parameter, *Prikl. Math. Mech.*, 1970, **34**(3), 470–80.
3. Gershuni, G.Z., E.M. Zhukhovitsky and Yu.S. Yurkov. On numerical determination of convective stability boundaries in system with periodically varying parameter, in *Hydrodynamics*, Perm, 1971, pp. 29–37.
4. Gershuni, G.Z., A.P. Durymanova and E.M. Zhukhovitsky. On parametric excitation of convective instability when heated from above, in *Convective Flows*, Perm, 1985, pp. 14–8.
5. Gershuni, G.Z., E.M. Zhukhovitsky and Yu.S. Yurkov. On parametric excitation of convective instability close to fluid surface, in *Modern Problems of Thermal Gravitational Convection*, Minsk, 1974, pp. 19–25.
6. Kantorovitch, L.V. and V.I. Krylov. *Approximate Methods of Higher Analysis*, Gostechtheorizdat, Moscow, 1949, 695 pp.
7. Ostroumov, G.A. *Free Convection under the Inner Problem Conditions*, Gostechtheorizdat, Moscow, 1952, 284 pp.
8. Coddington, E.A. and N. Levinson. *Theory of Ordinary Differential Equations*, IIL, Moscow, 1958, 475 pp.
9. Markman, G.S. and A.L. Urintsev. On finite-amplitude convection initiation in vibrating fluid layer heated from above, *Trans. Acad. Sci. SSSR (Izvestiya), Mech. Zhidk. Gasa*, 1978, **1**, 27–35.
10. Landau, L.D. and E.M. Lifshits. *Mechanics*, Nauka, Moscow, 1988, 208 pp.
11. Venezian, G. Effect of modulation on the onset of thermal convection, *J. Fluid Mech.*, 1969, **35**, Pt 2.
12. Gresho, P.M. and R.L. Sani. The effect of gravity modulation on the stability of heated fluid layer, *J. Fluid Mech.*, 1970, **40**, Pt 4, 783–806.
13. Rosenblat, S. and D.M. Herbert. Low-frequency modulation of thermal convection instability, *J. Fluid Mech.*, 1970, **43**, Pt 2, 385–98.
14. Rosenblat, S. and G.A. Tanaka. Modulation of thermal convection instability, *Phys. Fluids*, 1971, **14**(7), 1319–22.
15. Yih, C.-S. and C.-H. Li. Instability of unsteady flows or configuration, Pt 2, Convective instability, *J. Fluid Mech.*, 1972, **54**, Pt 1, 143–52.
16. Roppo, M.N., S.H. Davis and S. Rosenblat. Benard convection with time-periodic heating, *Phys. Fluids*, 1984, **27**, 796–803.
17. Wadih, M. and B. Roux. Natural convection in a long vertical cylinder under gravity modulation, *J. Fluid Mech.*, 1988, **193**, 391–415.
18. Burde, G.I. Numerical study of convection appearing in modulated field of external forces, *Trans. Acad. Sci. SSSR (Izvestiya), Mech. Zhidk. Gasa*, 1970, **2**, 196–201.
19. Burde, G.I. On finite-amplitude convection onset in modulated gravity force, *Trans. Acad. Sci. SSSR (Izvestiya), Mech. Zhidk. Gasa*, 1972, **4**, 124–34.
20. Gershuni, G.Z. and E.M. Zhukhovitsky. *Convective Stability of Incompressible Fluid*, Nauka, Moscow, 1972, 392 pp.
21. Burde, G.I. Numerical study of convection onset under temperature modulation at horizontal boundaries, *Trans. Acad. Sci. SSSR (Izvestiya), Mech. Zhidk. Gasa*, 1971, **1**, 144–50.
22. Burde, G.I. Numerical study of convection in the conditions of periodic modulation of external force, in *Hydrodynamics*, Perm, 1971, No. **3**, 75–96.
23. Gershuni, G.Z., E.M. Zhukhovitsky and Yu.S. Yurkov. Convection in closed cavity executing vertical oscillations, in *Convective Flows and Hydrodynamic Stability*, Sverdlovsk, 1979, pp. 55–60.

24. Gershuni, G.Z., E.M. Zhukhovitsky and Yu.S. Yurkov. Numerical study of free convection in a closed cavity executing vertical oscillations, in *Numerical Methods in Viscous Fluid Dynamics*, Novosibirsk, 1979, pp. 85–96.
25. Gershuni, G.Z., E.M. Zhukhovitsky and Yu.S. Yurkov. Convective oscillations in a closed cavity in modulated gravity field, in *Convective Flows*, Perm, 1979, pp. 73–80.
26. Yurkov, Yu.S. Vibrational thermal convection in a square cavity in weightlessness (finite frequencies), in *Convective Flows*, Perm, 1981, pp. 98–103.
27. Avduyevsky, V.S., A.V. Korol'kov, V.S. Kuptsova and V.V. Savitchev. Investigation of thermal gravitational convection in the field of variable weak accelerations, *J. Appl. Mech. Techn. Phys.*, 1987, **1**, 54–9.
28. Feonychev, A.I. and S.V. Ermakov. Parametric resonance in nonisothermal fluid subjected to harmonic mass force, *Adv. Space Res.*, 1991, **11**(7), 177–80.
29. Gershuni, G.Z., I.O. Keller and B.L. Smorodin. On vibrational convective instability of plane fluid layer; finite frequencies *Commun. AN SSSR (Doklady)*. 1996, **348**, pp. 194–6.
30. Gershuni, G.Z., I.O. Keller and B.L. Smorodin. On vibrational convective instability of plane horizontal layer at finite frequencies of vibration. *Trans. Acad. Sci. SSSR (Izvestiya). Mech. Zhidk. Gasa*, 1997, N5, pp. 44–51.
31. Petrov, G.I. Application of Galerkin method to the viscous flow stability problem, *Prikl. Math. Mech.*, 1940, **4**(3), 3–12.
32. Gershuni, G.Z., E.M. Zhukhovitsky and A.A. Nepomnyashchy. *Stability of Convective Flows*, Nauka, Moscow, 1989, 319 pp.
33. Landau, L.D. and E.M. Lifshitz. *Quantum Mechanics*, Nauka, Moscow, 1989, 723 pp.
34. Spradley, L.W., S.V. Bourgeois and F.N. Lin. Space processing convection evaluation: G-jitter convection in confined fluids in low gravity, AIAA paper, 1975, pp. 675–95.
35. Kamotani, Y., A. Prasad and S. Ostrach. Thermal convection in an enclosure due to vibration aboard spacecraft, *AIAA J.*, 1981, **19**(4), 511–16.
36. Biringen, S. and G. Danabasoglu. Computation of convective flow with gravity modulation in rectangular cavities, *J. Thermophysics*, 1990, **4**(3), 357–65.
37. Biringen, S. and L.J. Peltier. Numerical simulation of 3-D Benard convection with gravitational modulation, *Phys. Fluids*, 1990, **A2**(5), 754–64.
38. Tsau, F., S. Elghobashi and W.A. Sirignano. Effects of G-jitter on a thermal buoyant flow. AIAA paper, 1990, No. 0653, pp. 1–25.
39. Dolgikh, G.A., A.I. Feonychev, N.R. Storozhev and E.V. Zharikov. Effect of vibration of crystal during Bridgman growth under microgravity conditions, *Mater. Sci. Forum*, 1991, **77**, 43–50.
40. Alexander, I.D. and Y.Q. Zhang. Sensitivity of a non-isothermal liquid bridge to residual acceleration, *Micrograv. Sci. Technol.*, 1991, **IV**(2), 128-9.

6 Thermovibrational Convection in the General Case of Arbitrary Vibrations

As shown in Chapter 1, the 'vibrational force' emerges in the equations of motion when non-uniformities of both the density and pulsation velocity field simultaneously exist. In the preceding chapters situations were considered when a vessel completely filled with a fluid undergoes periodic translational motion, i.e. does not change its orientation. This is the most important case of vibrations for which there exists a non-inertial reference frame that does not rotate and with respect to which the domain boundary is at rest. In what follows we shall call this type of motion 'uniform vibrations'.

Under uniform vibrations, a uniform fluid would have been quiescent in the reference frame of the vessel, and in the laboratory frame the pulsation velocity field would have been uniform. Therefore, under uniform vibrations, the non-uniformity of the pulsation velocity field results from the density non-uniformity. For the non-uniform density induced by non-isothermality, the 'vibrational force' turns out to be of the second order with respect to the parameter $\beta\Theta$.

An entirely different situation takes place for 'non-uniform vibrations', i.e. those under which different parts of the fluid boundary move according to different laws. Examples are non-translational vibrations of a vessel completely filled with a fluid, oscillations of a solid body immersed in a fluid, vibrations of a partially filled vessel or vibrations of a set of immiscible fluids with a deformable interface boundary. A common feature for those systems is that there does not exist any framework retaining its orientation within which all the fluid boundaries are at rest. Generally speaking, under such conditions, even in a uniform fluid, the pulsation velocity field would have been non-uniform.

The coupling of the non-uniformities of the pulsation velocity field with the non-uniform vorticity could generate an average flow. If the frequency of the vibrations is sufficiently high and their amplitude is not too large, the pulsation vorticity should concentrate near the boundaries of the domain occupied by the fluid. This phenomenon has been discovered experimentally by Faraday [1] for the flow near a rigid wall. The corresponding theoretical investigation was done by Rayleigh [2, 3] for acoustic waves in a channel and afterwards by Schlichting [4, 5] for a flow near an oscillating cylinder. Later on, the problem was studied extensively in a number of works. The review on the average flows generated by interaction of an oscillating flow with a rigid surface may be found in [6].

Generation of the average vorticity also occurs near a free surface along which surface waves propagate. The theory of this phenomenon was developed by Longuet-Higgins [7, 8]. The said effects are inherent to the case of a non-uniformly heated fluid subject to non-uniform vibrations. Due to that, they must be taken into account when considering the thermovibrational convection. Moreover, one may expect that in the case of non-uniform vibrations the 'vibrational force' would be of the first order with respect to the non-isothermality parameter. This makes the non-uniform vibrations qualitatively different from the uniform ones and requires that the averaged equations of the thermovibrational convection should be derived anew [9–11]. This latter objective is the main goal of this chapter.

The equations of thermovibrational convection, originally obtained in [12] and described in Chapter 1, are based on the Boussinesq approximation. This means that the temperature dependence of density is taken into account only in the body forces in the reference frame of the vessel. If the vibrations are accompanied by deformations of the domain occupied by the fluid, the Boussinesq approximation cannot be formulated at all since there is no proper reference frame and consequently the effect of vibrations cannot be reduced to any body forces. Moreover, even when the proper reference frame does exist, as, for example, happens in the case of swing vibrations, the validity of the Boussinesq approximation for non-uniform vibrations is doubtful. The point is that the fluid accelerations and, which matters even more, their non-uniformities are not small.

The encountered situation resembles that in the theory of thermal buoyancy convection in a fluid with a deformable free surface. As shown in [13] to [17], the Boussinesq approximation becomes incorrect when allowance is made for the deformations of the fluid free surface. The formal use of the Boussinesq approach with an objective to take into account the free surface deformations might yield physically wrong conclusions.

Bearing this in mind, we begin from the analysis of a simple problem from which one may easily see the inconsistency of the Boussinesq approximation.

1 Inconsistency of the Boussinesq Approximation in the Case of a Fluid with a Free Surface

Let us consider the motion of a fluid in a horizontal layer with a free upper boundary and a rigid lower one. The fluid moves in the horizontal plane

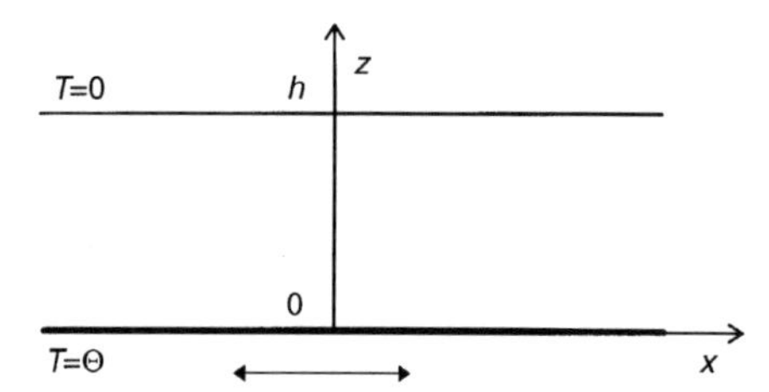

Fig. 151 Geometry of the problem and the coordinate system

according to a harmonic law with the frequency Ω and amplitude b (Fig. 151). The layer boundaries are maintained at constant temperatures Θ (the lower) and 0 (the upper). We shall seek the solution in which the free surface remains plane, the velocity is parallel to the vibration direction and the temperature depends only on the transversal coordinate. In the reference frame set on the lower boundary, the behavior of the fluid is described by the following equations:

$$\rho \frac{\partial u}{\partial t} = -\frac{\partial p}{\partial x} + \eta \frac{\partial^2 u}{\partial z^2} - b\rho\Omega^2 \cos \Omega t \tag{6.1}$$

$$-\frac{\partial p}{\partial z} - \rho g = 0 \tag{6.2}$$

$$\frac{\partial T}{\partial t} = \chi \frac{\partial^2 T}{\partial z^2} \tag{6.3}$$

$$\frac{\partial u}{\partial x} = 0 \tag{6.4}$$

with the boundary conditions:

$$\text{at} \quad z = 0: \qquad u = 0, \qquad T = \Theta \tag{6.5}$$

$$\text{at} \quad z = h: \qquad p = 0, \qquad T = 0, \qquad \frac{\partial u}{\partial z} = 0 \tag{6.6}$$

Here $\rho = \rho(T)$ is the temperature-dependent density of the fluid, h is the layer thickness, the axis x is directed along the vibration axis and axis z is vertical.

Apparently, for arbitrary initial conditions the linear steady-state temperature distribution $T = \Theta(h - z)/h$ settles after a transient process. The second equation and the boundary conditions on the free surface imply that $\partial p/\partial x = 0$.

For a steady-state motion we may set:

$$u = U(z)\,\mathrm{e}^{\mathrm{i}t} + U^*(z)\,\mathrm{e}^{-\mathrm{i}t} \tag{6.7}$$

This results in the problem for $U(z)$:

$$\mathrm{i}\Omega\rho U = \eta\frac{\mathrm{d}^2 U}{\mathrm{d}z^2} - \frac{1}{2}\rho b\Omega^2 \tag{6.8}$$

$$\text{at } z = 0: \qquad U = 0 \tag{6.9}$$

$$\text{at } z = h: \qquad \frac{\mathrm{d}U}{\mathrm{d}z} = 0 \tag{6.10}$$

The solution may be presented in the following form:

$$U = \frac{\mathrm{i}}{2} b\Omega[1 + V(\xi)] \tag{6.11}$$

where $\xi = z/h$ is the dimensionless vertical coordinate and V obeys the equation

$$\frac{\mathrm{i}\Omega}{\eta} h^2 \rho V = \frac{\mathrm{d}^2 V}{\mathrm{d}\xi^2} \tag{6.12}$$

with the boundary conditions:

$$V(0) = -1, \quad \frac{\mathrm{d}V(1)}{\mathrm{d}\xi} = 0 \tag{6.13}$$

The explicit solution of equation (6.12) cannot be obtained for an arbitrary temperature dependence of density. Due to this, we restrict ourselves to the case of weakly non-isothermal conditions. Then the equation of state is written as

$$\rho = \rho_0\big[1 - \beta(T - \Theta)\big] = \rho_0(1 + \beta\Theta\xi) \tag{6.14}$$

where ρ_0 is the density corresponding to the temperature at the lower boundary and β is the thermal expansion coefficient.

We assume that $\beta\theta \ll 1$ and expand the solution in a series with respect to $\beta\Theta$:

$$V = V_0 + \beta\Theta V_1 + \cdots \tag{6.15}$$

Since the first term here describes the isothermal flow, the effect of density non-uniformities is rendered solely by V_1. Substituting expansion (6.15) into equation (6.12), we obtain in the zeroth order:

$$V_0 = -\frac{\cosh\alpha(\xi - 1)}{\cosh\alpha}, \quad \alpha = (1 + \mathrm{i})\sqrt{\frac{\Omega\rho h^2}{2\eta}} \tag{6.16}$$

In the next order the non-isothermal effects emerge:

$$V_1 = \frac{\xi \cosh \alpha(\xi - 1) - \alpha\xi^2 \sinh \alpha(\xi - 1)}{4 \cosh \alpha} - \frac{(1 - \alpha^2) \sinh \alpha\xi}{4\alpha \cosh^2 \alpha} \tag{6.17}$$

If we had used the Boussinesq approximation, i.e. in (6.8) taken into account the temperature dependence of density only in the body forces, we would have apparently obtained the same form of V_0 but V_1 would have had another form, namely:

$$V_{1b} = \xi - \frac{\sinh \alpha\xi}{\alpha \cosh \alpha} \tag{6.18}$$

The difference between the velocities V_1 and V_{1b} is determined by the ratio of the vibration penetration depth $\delta = (\nu/\Omega)^{1/2}$ to the layer thickness h. For the limiting case $\delta \gg h$, the expressions for V_1 and V_{1b} coincide. In the opposite limit the expressions (6.17) and (6.18) differ. The Boussinesq approximation gives $V_{1b} \to \xi$, which, taking into account the intertial properties of the fluid, yields $V_1 \to 0$. Dependencies of the real parts of V_1 and V_{1b} on ξ are presented in Figs. 152 and 153 for $h/\delta = 10$. It can be seen that the velocity profiles essentially differ in their structure as well as in the characteristic values.

Thus, even in a simple situation like the afore-considered one, the Boussinesq approximation correctly describes a weakly non-isothermal fluid only at low vibration frequencies, where $\delta \gg h$ and the inertial properties of a fluid are only weakly pronounced.

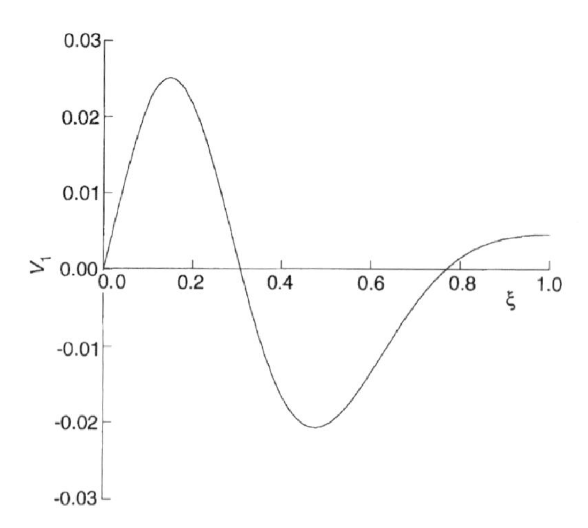

Fig. 152 Distribution of the convective part of the velocity across the layer. The exact solution

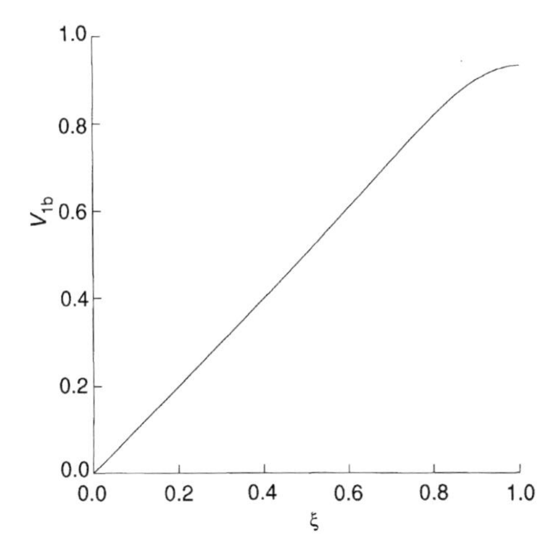

Fig. 153 Distribution of the convective part of the velocity across the layer. The Boussinesq approximation

Note that in a number of studies (see, for example, [18]) where vibrations of a layer with a free boundary are considered, the condition of the normal stress balance ($p = 0$) is replaced by the condition of zero flux of a fluid through the layer cross-section. Such a condition would have been reasonable if the gravity force were so great that the following inequality would hold:

$$\frac{b\Omega^2 L}{gh} \ll 1 \tag{6.19}$$

where L is the layer length.

Indeed, at vibrations of a bounded layer the fluid accumulates near one boundary during one half of the period and near the other boundary during the other half. The excess of the surface level over the average mean position ζ may be estimated from comparison of the pulsation variation of pressure along the layer $\rho b\Omega^2 L$ and the hydrostatic pressure $\rho g\zeta$. Assuming that ζ is much less than h, one obtains the estimation (6.19).

If, as well, $h \ll L$, the finite layer may be replaced by an infinite one. However, the restriction from above on the frequency, arising in practically important situations, makes the allowance for the thermovibrational convection futile. Actually, as known (see Chapter 1), in the Boussinesq approximation the intensity of the thermal vibrational convection is determined by the parameter $Gr_v = (b\Omega\beta\Theta h/\nu)^2/2$, the ratio of which to the ordinary Grashof number $Gr = g\beta\Theta h^3/\nu^2$ may be written in the form:

$$\frac{Gr_v}{Gr} = \frac{b\Omega^2 L}{2gh}\,\beta\Theta\,\frac{b}{L}$$

If the condition (6.19) holds, the value of Gr_v/Gr is the product of small parameters and consequently the role of the thermovibrational convection is negligible.

2 Pulsation Fields of Velocity and Temperature in a Non-uniform Fluid

Let us consider the behavior of a non-uniform fluid subject to vibrations of an arbitrary kind. Suppose that the amplitude of vibrations b is small in comparison with the reference spatial scale of the problem L:

$$b \ll L \tag{6.20}$$

In Chapter 1 when deriving the equations of vibrational convection, we restricted the vibration amplitude by the condition

$$b \ll \frac{L}{\beta\Theta}$$

Although both conditions are intended to ensure the smallness of the non-linear terms in the equations for pulsations, they are different. The difference is due to the fact that for the case of uniform vibrations the pulsation velocity is small in the reference frame to the same extent as the density non-uniformities. Because of that, the corresponding non-linearity begins to really matter at greater values of the vibrational velocity than in the case of vibrations of an arbitrary kind.

As in Chapter 1, we assume that the vibration frequency Ω is so high that the thickness of the viscous skin layer $\delta = \sqrt{\nu/\Omega}$ is small:

$$\Omega \gg \frac{\nu}{L^2} \tag{6.21}$$

At the same time, we assume the reference spatial scale of the problem to be small in comparison with the sound wavelength $\lambda = 2\pi c/\Omega$ (c is the sound velocity):

$$\Omega \ll \frac{2\pi c}{L} \tag{6.22}$$

so that the isothermal compressibility effects may be neglected.

Although the given estimations are double-bounded, the range of admissible values for frequency and size is large enough. For example, at $\Omega \sim 10^3 \mathrm{s}^{-1}$ in water δ does not exceed a hundred micrometers whereas λ equals several meters.

The behavior of a non-uniform fluid subject to vibrations is described by the set of equations

$$\rho\left(\frac{\partial \mathbf{v}}{\partial t}+\mathbf{v}\cdot\nabla\mathbf{v}\right)=-\nabla p+\eta\Delta\mathbf{v}+\left(\zeta+\frac{\eta}{3}\right)\nabla\,\mathrm{div}\,\mathbf{v}+\rho\mathbf{g} \tag{6.23}$$

$$\frac{\partial T}{\partial t}+\mathbf{v}\cdot\nabla T=\chi\Delta T \tag{6.24}$$

$$\frac{\partial \rho}{\partial t}+\mathrm{div}\,(\rho\mathbf{v})=0 \tag{6.25}$$

where ζ is the second viscosity, the density ρ is the prescribed function of temperature and the rest of the material parameters of the fluid are assumed to be constant.

Equations (6.23) to (6.25) are written with respect to an inertial (laboratory) reference frame. The effect of vibrations must be taken into account through boundary conditions. Provided the vibration amplitude is small, they may be imposed on the surface corresponding to the boundary averaged position. The boundary conditions are discussed in more detail in Chapter 7.

At high-frequency vibrations one may assume that the time dependences of all the dynamic variables are determined by just two time scales, viz. oscillatory ($\sim 1/\Omega$) and dissipative ($\sim L^2/\nu$), where $\nu=\eta/\rho^*$ is defined through some reference density ρ^*. In accordance with the idea of the multiscale method [19] we replace the operator of time differentiation by the sum of two operators:

$$\Omega\frac{\partial}{\partial \tau}+\frac{\partial}{\partial t}$$

where $\tau=\Omega t$ is the 'fast' time.

Let us present all the hydrodynamic fields as the sums of pulsation and averaged components where averaging is performed with respect to the 'fast' time τ:

$$\begin{aligned}\mathbf{v}&=\overline{\mathbf{v}}+\widetilde{\mathbf{v}}\\ p&=\overline{p}+\widetilde{p}\\ T&=\overline{T}+\widetilde{T}\\ \rho&=\overline{\rho}+\widetilde{\rho}\end{aligned} \tag{6.26}$$

and write down the pulsation parts of equations (6.23) to (6.25):

$$\Omega\overline{\rho}\frac{\partial \widetilde{\mathbf{v}}}{\partial \tau}+\widetilde{\rho}\frac{\partial \overline{\mathbf{v}}}{\partial t}+\{\rho(\mathbf{v}\nabla)\mathbf{v}\}=-\nabla\widetilde{p}+\eta\Delta\widetilde{\mathbf{v}}+(\zeta+\tfrac{1}{3}\eta)\nabla\ \mathrm{div}\ \widetilde{\mathbf{v}}+\widetilde{\rho}\mathbf{g} \tag{6.27}$$

$$\Omega \frac{\partial \widetilde{T}}{\partial \tau} + \overline{\mathbf{v}} \nabla \widetilde{T} + \widetilde{\mathbf{v}} \nabla \overline{T} + \{\widetilde{\mathbf{v}} \nabla \widetilde{T}\} = \chi \Delta \widetilde{T} \tag{6.28}$$

$$\Omega \frac{\partial \widetilde{\rho}}{\partial \tau} + \operatorname{div}(\overline{\rho}\, \widetilde{\mathbf{v}}) + \operatorname{div}(\widetilde{\rho}\, \overline{\mathbf{v}}) + \operatorname{div}\{\widetilde{\rho}\, \widetilde{\mathbf{v}}\} = 0 \tag{6.29}$$

where the braces denote extraction of the pulsation part, i.e. $\{F(t)\} = F(t) - \overline{F}(t)$. In these equations nothing is neglected except for the density dependence on the pressure.

Let us employ inequalities (6.20) and (6.21) to simplify the obtained equations. While doing estimations, we shall assume the pulsation velocity to be of the order of $b\Omega$, and the spatial derivatives of that of the inverse reference size. The latter means that so far we refrain from considering phenomena inside the boundary layers and also do not allow for turbulent pulsations. In addition we suppose that the velocities of the average flow are not too great, so that the following inequality holds:

$$\overline{v} \ll \Omega L \tag{6.30}$$

As an example, let us do the estimations for equation (6.29). One finds that its four terms, from left to right, have the following orders of magnitude: $\Omega\widetilde{\rho}$, $b\Omega\overline{\rho}/L$, $\overline{v}\widetilde{\rho}/L$ and $b\Omega\widetilde{\rho}/L$. It is clear that due to inequality (6.30), the third term is small compared to the first, and the last term is small because of inequality (6.20). Thus, comparison of the remaining terms shows that the density pulsations are at least a/L times smaller than the reference density. The remark 'at least' is due to the fact that, as will be clarified in what follows, the divergence of the pulsation velocity is small. Given that, in the estimation for $\widetilde{\rho}$ one should replace $\overline{\rho}$ by the density increment caused by non-isothermality.

Therefore, we conclude that the density pulsations are small. Similarly, with the aid of equation (6.28) one ensures the smallness of the temperature pulsations.

Finally, let us consider the equation of motion (6.27). In view of the obtained estimation for $\widetilde{\rho}$, one may neglect the second term. The third term may be omitted due to inequality (6.20), and inequality (6.21) allows the viscous contributions to be neglected. We also suppose that in equation (6.27) the last term may be omitted, which is equivalent to the inequality

$$\Omega^2 \gg \frac{g\beta\Theta}{L}. \tag{6.31}$$

Condition (6.31) holds in all practically important cases and could be violated only for very small cavities under enhanced gravity. The term with the pulsation pressure gradient should be retained because, first, we have no *a priori* estimation for pressure pulsations and, second, we expect that the pulsation pressure adjusts itself to the changes of the acceleration field in the fluid.

Thus, retaining only leading terms in the equation of motion (6.27) one obtains

$$\Omega \overline{\rho} \frac{\partial \widetilde{\mathbf{v}}}{\partial \tau} = -\nabla \widetilde{p} \tag{6.32}$$

However, this equation in itself is not sufficient to determine the pulsation velocity field. To obtain the closing equation, let us make use of the fact that the density depends solely on temperature. Combining equations (6.24) and (6.25) we get

$$\operatorname{div} \mathbf{v} = \beta \chi \Delta T \tag{6.33}$$

As a consequence of the assumptions adopted in the present section, from equation (6.33) it follows that the leading part of the pulsation velocity obeys the incompressibility condition

$$\operatorname{div} \widetilde{\mathbf{v}} = 0 \tag{6.34}$$

Equations (6.32) and (6.34) make a complete set to determine the pulsation velocity field. It can be seen that by their form they coincide with the dynamic equations for an inviscid fluid.

To equations (6.32) and (6.34) one has to add the boundary conditions. It has been already mentioned that by assuming the smallness of the vibration amplitude, the boundary conditions may be shifted from the true profile of the boundary (at a given moment of time) to its average position. Hence one has to require that at each point of the rigid parts of the boundary the normal component of the pulsation velocity should coincide with that of the boundary velocity:

$$\mathbf{n} \cdot \widetilde{\mathbf{v}}|_S = \mathbf{n} \cdot \mathbf{w} \tag{6.35}$$

where $\mathbf{n}$ is the unit vector normal to the average boundary position and $\mathbf{w}$ is the velocity of the boundary points.

We do not have to impose any conditions on the tangential components of the pulsation velocity due to reduction of the order of the corresponding pulsation equations. The boundary conditions for the full velocity should be satisfied in the skin layer near the rigid boundary. The boundary conditions on a free surface will be discussed in Section 1 in Chapter 8.

Thus, to evaluate the leading part of the pulsation velocity one needs to solve a linear non-stationary boundary-value problem. We remark that the coordinate-dependent function in the right-hand side of (6.35) which determines the law of vibrations cannot be entirely arbitrary. Indeed, integrating equation (6.34) over the volume occupied by the fluid, applying the Gauss theorem and using (6.35), we arrive at the condition for the oscillations of the boundary:

$$\oint_S \mathbf{w} \cdot d\mathbf{S} = 0$$

Its meaning is evident: vibrations must not be accompanied by variations of the volume occupied by the fluid.

It is noteworthy that, unlike the general problem for an inviscid fluid, on the solution of the problem under study an additional condition is imposed, namely the average must equal zero. Let us show that this requirement implies the potentiality of the vector $(\overline{\rho}\,\widetilde{\mathbf{v}})$. For that, we apply the curl operation to equation (6.32):

$$\frac{\partial}{\partial \tau} \operatorname{curl} (\overline{\rho}\,\widetilde{\mathbf{v}}) = 0 \tag{6.36}$$

Here we have taken into account the fact that $\overline{\rho}$ does not depend on τ.

Equation (6.36) leads to

$$\operatorname{curl} (\overline{\rho}\,\widetilde{\mathbf{v}}) = \mathbf{A} \tag{6.37}$$

where $\mathbf{A}$ is a coordinate function that does not depend on the 'fast' time. Averaging (6.37) over the 'fast' time τ and accounting for the condition of the zero average for the pulsation velocity, we find that $\mathbf{A} = 0$, and consequently

$$\operatorname{curl} (\overline{\rho}\,\widetilde{\mathbf{v}}) = 0 \tag{6.38}$$

Thus the field of the pulsation density of momentum is irrotational. This means potentiality, at least in the case of simply connected cavities.

2.1 Vibrations with a Uniform Phase

The simplest situation occurs when all the cavity boundaries are rigid and the time law of motion is the same for all points of the boundary. Let us discuss this case in more detail.

Assume that the normal velocity of any point of the boundary may be presented as

$$\mathbf{n} \cdot \mathbf{w} = f(\tau) Q(\mathbf{r}) \tag{6.39}$$

where f is a function with the zero average depending only on the 'fast' time and Q is a coordinate function of a boundary point determining the oscillation amplitude of the considered part of the boundary. Then one may seek the solution of the problem in the form

$$\widetilde{\mathbf{v}} = f(\tau) \mathbf{V}(\mathbf{r}), \qquad \widetilde{p} = \frac{\partial f}{\partial \tau} \Phi(\mathbf{r}) \tag{6.40}$$

Therefore, the time behavior of the pulsation velocity is the same in all points of the cavity. We shall refer to such a motion as the 'vibrations with a uniform phase'.

The following boundary-value problem is obtained for the stationary fields **V** and Φ:

$$\Omega \overline{\rho} \mathbf{V} = -\nabla \Phi \tag{6.41}$$

$$\text{div } \mathbf{V} = 0 \tag{6.42}$$

$$\mathbf{V} \cdot \mathbf{n}|_S = Q \tag{6.43}$$

It can be seen that the pulsation pressure amplitude Φ plays the role of a potential for $\overline{\rho}\mathbf{V}$. Moreover, this does not hold only for simply connected domains.

Eliminating the field **V** from (6.41) to (6.43) we arrive at the problem for a potential in the form

$$\text{div}\left(\frac{1}{\overline{\rho}} \nabla \Phi\right) = 0 \tag{6.44}$$

$$\left.\frac{\partial \Phi}{\partial n}\right|_S = -\Omega \overline{\rho} Q \tag{6.45}$$

If the density non-uniformity is related only to the fluid non-isothermality, then in the adopted approximation one has to treat $\overline{\rho}$ as a constant. This leads to further simplification of the problem and instead of equation (6.44) yields the Laplace equation

$$\Delta \Phi = 0 \tag{6.46}$$

If different parts of the boundary move according to different time laws (e.g. at some points the law is $\cos \tau$ and at other points it is $\sin \tau$), the separation of variables in (6.40) is not possible any longer. In these cases we may apply a superposition principle and reduce the non-stationary problem to several stationary ones. Naturally, in this case one may not speak of uniformity of the pulsation phase. Circularly polarized oscillations of a solid body immersed in a fluid yield a typical example of vibrations with a non-uniform phase. This case will be discussed in more detail in Section 2 in Chapter 9.

2.2 Second Approximation for the Pulsation Field

The above-formulated problem concerns the leading part of the isothermal pulsation field in a fluid. However, in a number of cases this approximation turns out to be insufficient. Let us, for example, consider a closed cavity completely filled with a fluid that undergoes translational vibrations. This means that at any moment of time all the points of the boundary have the same velocities (in both modulus and direction). It happens, for example, in the cases of linearly and circularly polarized vibrations. Thence the vector **w** in the boundary condition (6.35) depends only upon time and the problem (6.32), (6.34) and (6.35) has an obvious solution (for constant $\overline{\rho}$):

$$\widetilde{\mathbf{v}} = \mathbf{w} \tag{6.47}$$

In other words, the fluid moves as a rigid body together with the cavity. Clearly, such a pulsation field cannot produce any non-trivial effects. This means that for such vibrations one needs to consider a higher approximation in order to find corrections to the pulsation velocity caused by non-isothermality. This very important particular case has been studied in the preceding chapters. The relation between this case and the general one will be discussed in Chapter 8.

Another case where one needs to calculate the corrections to the pulsation velocity is the case of vibrations with a non-uniform phase. In this situation the non-linear coupling between pulsations and the average flow might yield significant effects [20, 21], vanishing, however, if the vibration phase is uniform.

To find the corrections to the pulsation velocity $\widetilde{\mathbf{v}}'$, we retain in equation (6.27) the terms containing the velocity of the average flow and neglect, as earlier, the dissipative terms and the pulsations of body forces. Thus we obtain for $\widetilde{\mathbf{v}}'$ the following equations:

$$\Omega\overline{\rho}\frac{\partial\widetilde{\mathbf{v}}'}{\partial\tau} + \overline{\rho}(\overline{\mathbf{v}}\nabla)\widetilde{\mathbf{v}} + \overline{\rho}(\widetilde{\mathbf{v}}\nabla)\overline{\mathbf{v}} = -\nabla\widetilde{p}' \tag{6.48}$$

$$\operatorname{div}\widetilde{\mathbf{v}}' = 0 \tag{6.49}$$

To find the explicit form of the corrections, it is appropriate to introduce the additional vector field **q** by the relations

$$\Omega\frac{\partial\mathbf{q}}{\partial\tau} = \widetilde{\mathbf{v}}, \qquad \overline{\mathbf{q}} = 0 \tag{6.50}$$

Obviously, in a unique manner equation (6.50) defines **q** through the leading part of the pulsation field, and

$$\operatorname{div}\mathbf{q} = 0, \qquad \operatorname{curl}\mathbf{q} = 0 \tag{6.51}$$

Note that by the order of magnitude vector **q** coincides with the displacement of the fluid particles participating in a pulsation motion.

Substituting (6.50) into (6.48) and integrating over the 'fast' time, we arrive at the following expression for $\widetilde{\mathbf{v}}'$:

$$\widetilde{\mathbf{v}}' = -(\mathbf{q}\nabla)\overline{\mathbf{v}} - (\overline{\mathbf{v}}\nabla)\mathbf{q} + \nabla\varphi' \tag{6.52}$$

Here we have introduced the gradient term of the potential φ'. For its evaluation one should use equation (6.49):

$$\Delta\varphi' = \text{div}\ \{(\mathbf{q}\nabla)\overline{\mathbf{v}} + (\overline{\mathbf{v}}\nabla)\mathbf{q}\} \tag{6.53}$$

Equation (6.53), together with the boundary condition of impermeability, determines φ'. Actually, however, it is not necessary to solve the problem for φ', since in all the situations where one might need $\widetilde{\mathbf{v}}'$ the gradient terms do not contribute. Note that with the aid of (6.51) one may rewrite expression (6.52) for $\widetilde{\mathbf{v}}'$ as

$$\widetilde{\mathbf{v}}' = \mathbf{q} \times \boldsymbol{\omega} + \cdots \tag{6.54}$$

where the dots stand for the gradient terms and the notation $\boldsymbol{\omega}$ is introduced for the vorticity of the average flow ($\boldsymbol{\omega} = \text{curl}\ \overline{\mathbf{v}}$). From (6.54) it is clear that the discussed correction to the pulsation velocity results from the interaction between pulsations and vorticity of the average flow. If U is the reference value of the average velocity, then the reference value for $\widetilde{\mathbf{v}}'$ is aU/L. The resulting contribution of $\widetilde{\mathbf{v}}'$ to the average effect may be significant due to interaction between $\widetilde{\mathbf{v}}'$ and the leading part of the pulsation velocity. The latter may greatly exceed the average velocity. Calculations of the resulting coupling of pulsations with the average vorticity will be performed in the following section simultaneously with the calculations of interaction between the velocity pulsations and density non-uniformities.

3 Averaged Equations of Motion for an Isothermally Incompressible Fluid

Consider averaging the equation of motion. Averaging equation (6.43) with respect to the 'fast' time yields

$$\overline{\rho}\frac{\partial\overline{\mathbf{v}}}{\partial t} + \Omega\overline{\widetilde{\rho}\frac{\partial\widetilde{\mathbf{v}}}{\partial\tau}} + \overline{\rho}(\overline{\mathbf{v}}\nabla)\overline{\mathbf{v}} + \overline{\widetilde{\rho}(\widetilde{\mathbf{v}}\nabla)}\,\overline{\mathbf{v}} + \overline{\widetilde{\rho}(\overline{\mathbf{v}}\nabla)\widetilde{\mathbf{v}}} + \overline{\rho}\,\overline{(\widetilde{\mathbf{v}}\nabla)\widetilde{\mathbf{v}}} + \overline{\widetilde{\rho}(\widetilde{\mathbf{v}}\nabla)\widetilde{\mathbf{v}}}$$
$$= -\nabla\overline{p} + \eta\Delta\overline{\mathbf{v}} + (\zeta + \tfrac{1}{3}\eta)\cdot\nabla\ \text{div}\ \overline{\mathbf{v}} + \overline{\rho}\mathbf{g} \tag{6.55}$$

Before estimating individual terms in (6.55), let us improve the estimation for the density pulsations made in the preceding section. Retaining only the leading terms in (6.29) we obtain

$$\Omega\frac{\partial\widetilde{\rho}}{\partial\tau} + \text{div}\ (\overline{\rho}\widetilde{\mathbf{v}}) = 0 \tag{6.56}$$

This equation uniquely determines $\widetilde{\rho}$, but an explicit expression for it is not necessary.

Of all the terms of equation (6.55), let us select the leading ones which contain the pulsation velocity:

$$\Omega\overline{\widetilde{\rho}\,\frac{\partial\widetilde{\mathbf{v}}}{\partial\tau}}+\overline{\rho}\,\overline{(\widetilde{\mathbf{v}}\nabla)\widetilde{\mathbf{v}}} \tag{6.57}$$

Shifting the time differentiation to the density pulsations under the sign of averaging and using (6.56), one may rewrite (6.57) as

$$\overline{\widetilde{\mathbf{v}}(\widetilde{\mathbf{v}}\nabla)\overline{\rho}}+\overline{\rho}\,\overline{(\widetilde{\mathbf{v}}\nabla)\widetilde{\mathbf{v}}} \tag{6.58}$$

Finally, using formula (6.38) and applying the standard vector analysis formulas, for the contribution of pulsations to the equation of motion we obtain

$$\frac{1}{2}\overline{\widetilde{\mathbf{v}}^2}\,\nabla\overline{\rho}+\nabla\left(\rho\frac{\overline{\mathbf{v}^2}}{2}\right) \tag{6.59}$$

3.1 Pulsational Transport of Vorticity

In the presented calculations we have taken into account only the leading part of the pulsation velocity. In the general case of a variable density, the corrections to the pulsation velocity will yield contributions negligibly small relative to the terms included in (6.59). If non-uniformities of the average velocity are small or absent, the corrections to the pulsation velocity, which have been found in the preceding section, should be taken into account. The corresponding calculation leads to the following result [22]. Together with the terms taken into account in (6.59) in the left-hand side of the equation of motion one more term emerges:

$$\rho\mathbf{S}\times\operatorname{curl}\overline{\mathbf{v}} \tag{6.60}$$

where the following notation is introduced:

$$\mathbf{S}=\overline{(\mathbf{q}\nabla)\widetilde{\mathbf{v}}} \tag{6.61}$$

Therefore, we obtain the averaged equation of motion in the following form:

$$\overline{\rho}\left[\frac{\partial\overline{\mathbf{v}}}{\partial t}+(\overline{\mathbf{v}}\nabla)\overline{\mathbf{v}}-\mathbf{S}\times\operatorname{curl}\overline{\mathbf{v}}\right]+\tfrac{1}{2}\overline{\widetilde{\mathbf{v}}^2}\nabla\overline{\rho}=-\nabla\overline{p}+\eta\Delta\overline{\mathbf{v}}+(\zeta+\tfrac{1}{3}\eta)\nabla\operatorname{div}\overline{\mathbf{v}}+\overline{\rho}\mathbf{g} \tag{6.62}$$

In equation (6.62) the gradient terms are included in the average pressure.

In the case of a weak non-isothermality when the Boussinesq parameter is small, $\beta\theta \ll 1$, equation (6.33) turns into the incompressibility equation:

$$\operatorname{div}\,\overline{\mathbf{v}}=0 \tag{6.63}$$

and in (6.62) ρ may be considered as constant everywhere except for the term with the gradient $\overline{p}$ (where density increments are multiplied by a large value $\overline{\widetilde{\mathbf{v}}^2} \sim b^2\Omega^2$) and in the body forces. This results in the following equation for the average motion:

$$\overline{\rho}\left[\frac{\partial \overline{\mathbf{v}}}{\partial t} + (\overline{\mathbf{v}}\nabla)\overline{\mathbf{v}} - \mathbf{S} \times \text{ curl } \overline{\mathbf{v}}\right] - \tfrac{1}{2}\beta\overline{\widetilde{\mathbf{v}}^2}\,\nabla T = -\nabla\overline{p} + \eta\Delta\overline{\rho} - \overline{\rho}\mathbf{g}\beta T \quad (6.64)$$

Thus, the averaged equation of a fluid motion under high-frequency vibrations differs from the ordinary Navier–Stokes equation by two additional terms. The first one is related to the non-uniformity of the average flow vorticity. Vector **S** included in this term vanishes in the case of pulsations with a uniform phase. Indeed, in this case the pulsation velocity field and the field **q** may be presented as the product of time- and coordinate-dependent functions:

$$\begin{aligned} \mathbf{q} &= F(\tau)\mathbf{f}_1(\mathbf{r}) \\ \mathbf{v} &= \Omega\frac{\mathrm{d}F}{\mathrm{d}\tau}\mathbf{f}_1(\mathbf{r}) \end{aligned} \quad (6.65)$$

Substitution of (6.65) into the definition of vector **S** (6.61) yields:

$$\mathbf{S} = \overline{F\frac{\mathrm{d}F}{\mathrm{d}\tau}}(\mathbf{f}_1\nabla)\mathbf{f}_1 = 0 \quad (6.66)$$

If the vibration phase is non-uniform, vector **S**, in general, is non-zero and is proportional to the gradient of the phase.

Vector **S** has a simple physical meaning. When the fluid performs a pulsation motion, we may consider the average motion velocity in two different aspects. The first is the average velocity in a fixed point (Euler's viewpoint). In this case the operation of averaging incorporates velocities of various fluid elements which pass the fixed point at some moment in time. Therefore defined average velocity forms the vector field **v** considered in this chapter. However, another viewpoint (that of Lagrange) is possible as well when we follow a fixed fluid element and find the average value of its velocity: the field $\overline{\mathbf{v}}_{\mathrm{L}}$.

It is natural that the vector fields $\overline{\mathbf{v}}$ and $\overline{\mathbf{v}}_{\mathrm{L}}$ differ. Moreover, it can be shown [23] that the difference $(\overline{\mathbf{v}}_{\mathrm{L}} - \overline{\mathbf{v}})$ exactly coincides with the quantity **S** introduced by us.

Let us write down the equation of motion of a particle carried along by the fluid. Let $\mathbf{r}(t)$ denote the particle coordinates at the time moment t and $\mathbf{r}_0$ its coordinates at the moment t_0. Then

$$\frac{\mathrm{d}\mathbf{r}}{\mathrm{d}t} = \mathbf{v}(\mathbf{r},\, t) \quad (6.67)$$

Equation (6.67) is a non-linear differential equation with respect to $\mathbf{r}$. We seek its solution by means of successive approximations, bearing in mind that we are interested in the solution during one period of vibrations and the particle displacement during this time is small compared to the reference scales of the field $\mathbf{v}$ variation. In the first approximation, instead of a current vector $\mathbf{r}$ one may substitute in the right-hand side of (6.67) its value at the initial time moment. Then for $\mathbf{r}^{(1)}$ we have

$$\mathbf{r}^{(1)}(t) = \overline{\mathbf{v}}(\mathbf{r}_{\mathrm{i}})\,|_{t=t_0} + \mathbf{q}(\mathbf{r}_0, t) - \mathbf{q}(\mathbf{r}_0, t_0) \tag{6.68}$$

Since $\mathbf{f}$ is a periodic function, at the moment $t = t_0 + T$, $T = 2\pi/\Omega$ only the contribution of the average flow to $\mathbf{r}^{(1)}$ remains. Thus, to determine the contribution of pulsations, one should perform one more step of the successive approximation method.

Let us assume $\overline{\mathbf{v}} = 0$ for simplicity. Then to evaluate the correction of the second approximation we obtain the equation

$$\frac{\mathrm{d}\mathbf{r}^{(2)}}{\mathrm{d}t} = \mathbf{r}^{(1)}(t) \cdot \nabla\,\widetilde{\mathbf{v}}(\mathbf{r}_0, t) \tag{6.69}$$

Integrating (6.69) over time from t_0 to $(t_0 + T)$ and taking into account the definition of the pulsational transport, for the vector $\mathbf{S}$ one gets the equation

$$\mathbf{r}^{(2)}(t_0 + T) = T \cdot \mathbf{S} \tag{6.70}$$

Thus, the average displacement of the particle per unit of time, i.e. the average velocity of the particle motion, coincides with $\mathbf{S}$. In the presented calculation a strict periodicity of the pulsation field was assumed. The obtained result clarifies the meaning of the vector $\mathbf{S}$. It should be noted, however, that the definition of $\mathbf{S}$ introduced earlier by formula (6.61) holds for non-periodic pulsations as well. We would like to emphasize that $\mathbf{S}$ could be non-zero even if the Euler average velocity equals zero. In this case $\mathbf{S}$ coincides with the Lagrange velocity and, in particular, gives the average velocity of motion for the particles suspended in a fluid. Therefore, the particles used for flow visualization can demonstrate the average motion even when the average fluid velocity is equal to zero at every point.

There is a simple situation for which the effect of the average transport for the case where the average velocity equals zero had been discovered long ago. Let us consider a wave of a small amplitude b_ω, propagating along the surface of an infinite inviscid fluid. We choose the x axis along the direction of the wave propagation and the z axis vertically upwards. The deviations of the fluid surface from a flat horizontal one are described by the formula

$$z = \zeta(x,\ t) = b_\omega \sin(\Omega t - kx) \tag{6.71}$$

Considering the fluid motion as irrotational, we may present the velocity potential in the form

$$\varphi = A\cos(\Omega t - kx)\mathrm{e}^{-kz} \tag{6.72}$$

The potential (6.72) satisfies the Laplace equation. To determine the constant b the kinematic boundary condition should be used:

$$\frac{\partial \zeta}{\partial t} = \frac{\partial \varphi}{\partial z} \qquad \text{at } z = 0 \tag{6.73}$$

Hence

$$A = \frac{\Omega}{k} b_\omega \tag{6.74}$$

Evaluation of vector **S** for the found velocity distribution yields the following expression:

$$\mathbf{S} = b_\omega^2 \Omega \mathbf{k} \mathrm{e}^{2kz} \tag{6.75}$$

Note that this is precisely one of those cases when the average fluid velocity equals zero at every point.

Thus, a wave propagating along a fluid surface entrains the suspended particles. This is the phenomenon discovered by Stokes in 1847 [24].

Clearly, not only particles but any non-uniformities, e.g. those of temperature or of an admixture concentration, would be transported in the same way. This is valid as well for the average vorticity. The appearance of the term containing the pulsational transport vector **S** in the averaged motion equation reflects this fact.

3.2 'Vibrational Force'

Let us discuss now the second additional term in the averaged equation of motion (6.62) containing the density non-uniformities. First, we remark that the vibrational force $\mathbf{f}_\mathrm{v} = \frac{1}{2}\overline{\tilde{\mathbf{v}}^2}\nabla\bar{\rho}$ may affect the average flow only when both the density and kinetic energy of pulsations are non-uniform. Indeed, in the opposite case curl $\mathbf{f}_\mathrm{v}$ will be equal to zero and the vibrational force will yield only renormalization of the average pressure. Let us write down the vibrational force in a form that explicitly singles out the non-uniformities of the kinetic energy of pulsations. Omitting potential terms, one finds that

$$\mathbf{f}_\mathrm{v} = -\tfrac{1}{2}\bar{\rho}\nabla\overline{\tilde{\mathbf{v}}^2} \tag{6.76}$$

One may see that the vibrational force is directed opposite to the kinetic energy gradient. Formula (6.76) proves the vibrational force to be a body force.

Moreover, the squared pulsation velocity might be interpreted as the potential of an effective gravitational field. In some cases this analogy allows the behavior of a non-uniform system under vibrations to be predicted; namely, the system tends to a state in which the energy of the effective gravitational field, i.e. the kinetic energy of pulsations, is minimal.

In those cases where the density gradient is parallel to the vibration direction, the pulsation energy turns to zero. Then the tendency to establish the states in which constant density surfaces are orthogonal to the vibration direction is observed. For the systems in which the density non-uniformity is related to the existence of interface, this tendency was found in [25] to [27].

Another interesting analogy is worth mentioning. As may be seen from the structure of the vibrational force, the effect of vibrations on a non-uniform fluid proves to be similar to the electric field influence on a liquid dielectric with a non-uniform dielectric permeability. Indeed, as known [28], the electric field acts on a dielectric with a volumic force:

$$\mathbf{F} = -\frac{1}{8\pi} E^2 \nabla \varepsilon \tag{6.77}$$

where ε is the dielectric permeability and $\mathbf{E}$ is the electric field strength inside a medium satisfying the equations

$$\text{curl } \mathbf{E} = 0, \qquad \text{div } (\varepsilon \mathbf{E}) = 0 \tag{6.78}$$

Upon substitution,

$$\mathbf{E} = 4\pi\rho\mathbf{V}, \qquad \varepsilon = \frac{1}{4\pi\rho}$$

equations (6.78) are transformed into the equations for the pulsation field and the ponderomotive force (6.77) is transformed into the 'vibrational' force in equation (6.62), though inversion of the sign.

For the case of weakly non-isothermal conditions and a nearly constant electric field such an analogy was noted in [29] for the problem of small perturbations of equilibrium. For the cases when the density non-uniformity is related to the existence of an interface, the electric-vibrational analogy was investigated in [30].

4 Averaged Equation for the Heat Transfer

Let us treat averaging of the heat transfer equation. The general equation of heat transfer can be written in the form

$$\frac{\partial T}{\partial t} + \mathbf{v} \cdot \nabla T = \chi \Delta T \tag{6.79}$$

We decompose the temperature (as has already been mentioned in Section 2) into the average and pulsation parts:

$$T = \overline{T} + \widetilde{T} \tag{6.80}$$

Applying the operation of averaging over 'fast' time to equation (6.79), we obtain

$$\frac{\partial \overline{T}}{\partial t} + \overline{\mathbf{v}} \cdot \nabla \overline{T} + \overline{\widetilde{\mathbf{v}} \cdot \nabla \widetilde{T}} = \chi \Delta \overline{T} \tag{6.81}$$

Subtracting this equation from (6.79), we arrive at the following equation for pulsations:

$$\Omega \frac{\partial \widetilde{T}}{\partial \tau} + \overline{\mathbf{v}} \cdot \nabla \widetilde{T} + \widetilde{\mathbf{v}} \cdot \nabla \overline{T} + \widetilde{\mathbf{v}} \cdot \nabla \widetilde{T} - \overline{\widetilde{\mathbf{v}} \cdot \nabla \widetilde{T}} = \chi \Delta \widetilde{T} \tag{6.82}$$

Let us estimate the terms entering equation (6.82). Denoting the reference scales for temperature pulsations, temperature difference and average velocity as θ, Θ and U, respectively, for areas outside the boundary layer one gets

$$\Omega\theta, \qquad U\theta/L, \qquad b\Omega\theta/L, \qquad b\Omega\theta/L, \qquad b\Omega\theta/L, \qquad \chi\theta/L^2 \tag{6.83}$$

From (6.83) it follows that the second term is small in comparison with the first one due to the restriction on the average velocity (6.30) introduced earlier. The fourth and the fifth terms are small as compared to the first due to the smallness of the vibrational amplitude. Finally, we assume that both the temperature skin-layer thickness $\delta_T = \sqrt{\chi/\Omega}$ and the dynamical skin-layer thickness are much smaller than the reference size. Then, the dissipative term in (6.82) is also small relative to the first term on the left-hand side. Thus, in the leading order only two terms survive in (6.82):

$$\Omega \frac{\partial \widetilde{T}}{\partial \tau} + \widetilde{\mathbf{v}} \cdot \nabla \overline{T} = 0 \tag{6.84}$$

Using the auxiliary field $\mathbf{q}$ introduced in Section 3,

$$\widetilde{\mathbf{v}} = \Omega \frac{\partial \mathbf{q}}{\partial \tau}$$

we obtain from (6.84) the following explicit expression for the temperature pulsations:

$$\widetilde{T} = -\mathbf{q} \cdot \nabla \overline{T} \tag{6.85}$$

The physical meaning of this result is simple: the major source of fast change of temperature in a fixed point is the arrival of the fluid element possessing a different temperature.

We are then ready to evaluate the pulsation contribution to the averaged equation of heat transfer (6.81). Note, first of all, that the corresponding term may be rewritten in a divergent form:

$$\overline{\widetilde{\mathbf{v}} \cdot \nabla \widetilde{T}} = \operatorname{div} \left(\overline{\widetilde{\mathbf{v}} \widetilde{T}} \right) = \operatorname{div} \mathbf{E} \tag{6.86}$$

where we introduced the notation $\mathbf{E} = \overline{\widetilde{\mathbf{v}} \widetilde{T}}$ for the vector of pulsational heat transport. Using expression (6.85), obtained earlier for the temperature pulsations, we can rewrite the definition of the vector $\mathbf{E}$:

$$E_i = M_{ij} \nabla_i \overline{T} \tag{6.87}$$

where tensor M_{ij} is determined by the formula

$$M_{ij} = -\overline{\widetilde{v}_i \, q_j} \tag{6.88}$$

Thus, the pulsational heat transport proves to be proportional to the gradient of the average temperature.

It is interesting to note that exactly the same hypothesis is usually accepted in semiempirical models of turbulence. In this case, tensor M_{ij} is called a turbulent thermal diffusivity, and for it further hypotheses are used. In the simplest case of an isotropic turbulence it is assumed that tensor M_{ij} is proportional to the unit tensor $M_{ij} = \chi_T \, \delta_{ij}$, where χ_T is the turbulent coefficient of thermal diffusivity. Adoption of such a hypothesis means simply a renormalization of the thermal diffusivity coefficient. The situation is quite different in the case of a pulsation field generated by vibrations. Tensor M_{ij} is defined explicitly by formula (6.38) and is not proportional to a unit tensor. Moreover, this tensor is antisymmetric. Indeed, writing down the averaging operation in the integral form, we have

$$M_{ij} = -\frac{1}{\Delta t} \int_t^{t+\Delta t} \widetilde{v}_i \, f_j \, \mathrm{d}t \tag{6.89}$$

where Δt is the interval of averaging, large in comparison with the vibration period, but small relative to the reference thermal times. Recalling the relation between $\widetilde{\mathbf{v}}$ and $\mathbf{f}$ and integrating (6.89) by parts, one obtains

$$M_{ij} = \overline{q_i \, \widetilde{v}_j} = -M_{ji} \tag{6.90}$$

Due to antisymmetry of the tensor M_{ij}, the divergence of the heat flux vector does not contain second spatial derivatives of the average temperature. Indeed,

$$\operatorname{div} \mathbf{E} = \nabla_i (M_{ij} \nabla_j T) = (\nabla_i M_{ij}) \nabla_j T + M_{ij} \nabla_i \nabla_j T \tag{6.91}$$

The last term in (6.91) vanishes because of the antisymmetry of M_{ij}. Finally, let us calculate the divergence of tensor M_{ij} entering (6.91). Using definition (6.88) of M_{ij}, we have

$$\nabla_i M_{ij} = \nabla_i \overline{(q_i \widetilde{v}_j)} = \overline{q_i \nabla_i \widetilde{v}} = S_j \tag{6.92}$$

where the solenoidality of the field $\mathbf{q}$ is taken into account.

Finally, we come to the following averaged heat transfer equation:

$$\frac{\partial \overline{T}}{\partial t} + (\overline{\mathbf{v}} + \mathbf{S}) \cdot \nabla \overline{T} = \chi \Delta \overline{T} \tag{6.93}$$

As in the case of the pulsational transport of vorticity, the pulsation contribution to the average transport of heat is determined by the vector **S**. This should have been expected taking into account the physical meaning of vector **S** as the velocity vector of the average Lagrange transport.

In conclusion, we remark that all the results obtained in this section equally apply to the case of admixture diffusion processes.

5 The Complete Set of Equations of Thermovibrational Convection

In the preceding sections of this chapter the averaged equations of motion and heat transfer as well as equations for the pulsation velocity field have been derived. In this section we summarize the obtained results and formulate the general equations of thermovibrational convection and their simplified variants for some particular cases.

5.1 General Equations for Thermovibrational Convection

It follows from Sections 2 to 4 that in the general case of arbitrary vibrations the complete set of equations of thermovibrational convection has the following form:

$$\rho \frac{\partial \mathbf{v}}{\partial t} + \rho \mathbf{v} \cdot \nabla \mathbf{v} - \rho \mathbf{\Omega} \times \mathbf{S} + \tfrac{1}{2} \overline{\widetilde{\mathbf{v}}^2} \nabla \rho = -\nabla p + \eta \Delta \mathbf{v} + (\zeta + \tfrac{1}{3}\eta) \nabla \operatorname{div} \mathbf{v} + \rho \mathbf{g} \tag{6.94}$$

$$\frac{\partial T}{\partial t} + (\mathbf{v} + \mathbf{S}) \cdot \nabla T = \chi \Delta T \tag{6.95}$$

$$\operatorname{div} \mathbf{v} = \beta \chi \Delta T \tag{6.96}$$

$$\operatorname{curl} \rho \widetilde{\mathbf{v}} = 0 \tag{6.97}$$

$$\operatorname{div} \widetilde{\mathbf{v}} = 0 \tag{6.98}$$

where a bar, denoting averaging, is omitted over $\mathbf{v}$, p, T, ρ, $\mathbf{\Omega} = \text{curl}\ \mathbf{v}$ and the pulsational transport vector $\mathbf{S}$ is defined by the relations

$$\mathbf{S} = \overline{\mathbf{q} \cdot \nabla \widetilde{\mathbf{v}}}, \qquad \widetilde{\mathbf{v}} = \frac{\partial \mathbf{q}}{\partial t}, \qquad \overline{\mathbf{q}} = 0 \tag{6.99}$$

We emphasize that the averaged equation of motion (6.94) differs from the conventional Navier–Stokes equation by the presence of two terms responsible for different mechanisms of pulsation action on the average flow. The first of these terms involving the pulsational transport vector $\mathbf{S}$ describes the non-linear interaction of pulsations with the average vorticity. This effect cannot be responsible for the average flow generation but it may modify it significantly. The second term containing the kinetic energy of pulsations, unlike the first one, describes the volumic generation of the average flow due to dissipation of the pulsation energy at the density non-uniformities. The dimensionless parameters determining the role of the said effects are the pulsational Reynolds number $Re_{\mathrm{p}} = b^2\Omega/L$ and the vibrational Galilei number $Ga_{\mathrm{v}} = b^2\Omega^2 L^2 \delta\rho/(\eta\nu)$, where $\delta\rho$ is the reference density difference.

The heat transfer equation changes its form as well. It incorporates the term describing the effect of pulsational transport of the temperature non-uniformities. We recall that the phase non-uniformity of the pulsation velocity field is the necessary condition of existence of the pulsational transport effect.

The continuity equation (6.96) is written taking into account the fact that the density depends solely upon temperature. Of course, it can also be presented in a more familiar form:

$$\frac{\partial \rho}{\partial t} + (\mathbf{v} + \mathbf{S}) \cdot \nabla \rho + \rho\ \text{div}\ \mathbf{v} = 0 \tag{6.100}$$

Here, as well as in the heat transfer equation, the pulsational transport effect is taken into account. Generally, it should be noted that vector $\rho(\mathbf{v} + \mathbf{S})$ plays the role of the mass flux density—a circumstance that should be kept in mind while formulating the closure, impermeability, etc., conditions.

Finally, the equations for the pulsation field have the same form as those of an inviscid incompressible fluid with a coordinate-dependent density. The solution of these equations depends on the 'fast' time, as imposed by the vibration law, and is to be taken into account through boundary conditions. Due to the presence of the average density in these equations, their solution depends, generally, on the 'slow' time as well and it should be obtained together with the solution of the averaged equations.

5.2 Thermovibrational Convection for Weakly Non-isothermal Conditions and Vibrations of a General Type

While writing down the set of equations (6.94) to (6.98) it has been assumed that ρ is only temperature dependent. Since in practically important situations the density variations due to heating are small, it is natural to adopt an approximation similar to the Boussinesq one. Specifically, we set $\beta\theta \ll 1$, where θ is the reference temperature difference, and retain the terms caused by the density variations only in those terms of equations (6.94) to (6.98) that contain a large factor, i.e. in the vibrational force and in the body force (gravity). From a formal point of view, this means that we consider the limiting transition where simultaneously $\beta\theta \to 0$ and the dimensionless parameters $\rho_0 g L^3/\eta^2$ (Galilei number) and $b^2\Omega^2 L^2 \rho_0^2/\eta^2$ (squared dimensionless velocity of vibrations) tend to infinity. In the remaining terms we may set the density equal to its value ρ_0 at the reference temperature. In this limit the set of equations describing the thermovibrational convection simplifies to

$$\frac{\partial \mathbf{v}}{\partial t} + \mathbf{v}\cdot\nabla\mathbf{v} - \boldsymbol{\Omega}\times\mathbf{S} - \frac{1}{2}\beta\overline{\widetilde{\mathbf{v}}^2}\nabla T = -\frac{1}{\rho_0}\nabla p + \nu\Delta\mathbf{v} - \beta T\mathbf{g} \tag{6.101}$$

$$\frac{\partial T}{\partial t} + (\mathbf{v} + \mathbf{S})\cdot\nabla T = \chi\Delta T \tag{6.102}$$

$$\operatorname{div}\,\mathbf{v} = 0 \tag{6.103}$$

$$\operatorname{curl}\,\widetilde{\mathbf{v}} = 0 \tag{6.104}$$

$$\operatorname{div}\,\widetilde{\mathbf{v}} = 0 \tag{6.105}$$

where $\nu = \eta/\rho_0$.

The obtained set of equations retains the main features noted in the discussion of the general equations, which are the effects of pulsational transport and interaction of pulsations with the density non-uniformities. The occurring distinctions stem from simplifications done in the continuity equation and, as a consequence, in the dissipative terms in the equation of motion. The latter now have the same form as in the conventional equations for the thermal buoyancy convection in the Boussinesq approximation. Since the density variations are now due solely to the temperature variations, the vibrational Grashof number $Gr_v = b^2\Omega^2 L^2 \beta\theta/\nu^2$ will be the governing parameter instead of the vibrational Galilei number.

It can be seen that, in the limiting case of weak non-isothermality, the equations for the pulsation field decouple from the equations for the average components and may be solved independently if it is consistent with the boundary conditions discussed in the next chapter.

5.3 Thermovibrational Convection for Weakly Non-isothermal Conditions and Vibrations with Uniform Phase

Let us consider one more possible simplification of the problem related to the case of monochromatic vibrations with uniform phase. In this case the pulsation velocity field admits the separation of variables. Let us write it down in the form

$$\widetilde{\mathbf{v}} = b\Omega \mathbf{V} \sin \Omega t \tag{6.106}$$

where $\mathbf{V}$ is the dimensionless amplitude of the pulsation velocity which depends on coordinates.

The equations for $\mathbf{V}$ formally coincide with (6.104) and (6.105), and the average of the square of pulsation velocity may be written as

$$\overline{\widetilde{\mathbf{v}}^2} = \tfrac{1}{2} b^2 \Omega^2 V^2 \tag{6.107}$$

The pulsational transport vector $\mathbf{S}$ vanishes for the vibrations with a uniform phase.

Finally, we arrive at the following set of equations:

$$\frac{\partial \mathbf{v}}{\partial t} + \mathbf{v} \cdot \nabla \mathbf{v} - \frac{1}{4} b^2 \Omega^2 \beta \mathbf{V}^2 \nabla T = -\frac{1}{\rho_0} \nabla p + \nu \Delta \mathbf{v} - \beta T \mathbf{g} \tag{6.108}$$

$$\frac{\partial T}{\partial t} + \mathbf{v} \cdot \nabla T = \chi \Delta T \tag{6.109}$$

$$\operatorname{div} \mathbf{v} = 0 \tag{6.110}$$

$$\operatorname{curl} \mathbf{V} = 0 \tag{6.111}$$

$$\operatorname{div} \mathbf{V} = 0 \tag{6.112}$$

Unlike the preceding cases, equations (6.108) to (6.112) do not contain the 'fast' time.

5.4 Thermovibrational Convection for Weakly Non-isothermal Conditions in the Particular Cases when the Pulsation Velocity Field is Uniform

In all the above-discussed cases, the pulsational field with which the density non-uniformities interact, does not depend on thermal conditions. For weakly non-isothermal conditions, this results in a thermovibrational effect of the first order with respect to the Boussinesq parameter $\beta\theta$. However, there are certain situations where an isothermal pulsation field proves to be uniform and vanishes in the reference frame of the vessel. In such cases it is necessary to find corrections to the pulsation field caused by non-isothermality. For this purpose let us return to the general equations of the pulsation field in a non-uniform fluid (6.97) and (6.98). Rewriting (6.97) in the form

$$\nabla \rho \times \widetilde{\mathbf{v}} + \rho \operatorname{curl} \widetilde{\mathbf{v}} = 0 \tag{6.113}$$

and taking into account the temperature dependence of density, we obtain

$$-\beta\nabla T \times \widetilde{\mathbf{v}} + \mathrm{curl}\ \widetilde{\mathbf{v}} = 0 \tag{6.114}$$

The first term in (6.114) is proportional to the small parameter $\beta\theta$ and therefore the pulsation velocity entering this term may be written only in the leading order. If a vessel filled with a fluid oscillates in the direction $\mathbf{n}$, one may write the leading part $\widetilde{\mathbf{v}}^{(0)}$ of the pulsation velocity as

$$\widetilde{\mathbf{v}}^{(0)} = f(\tau)\mathbf{n} \tag{6.115}$$

where $f(\tau)$ is a periodic function of the 'fast' time. Denoting the pulsation velocity correction as $\widetilde{\mathbf{v}}^{(1)}$ and substituting (6.115) into (6.114), we come to the following equation for $\widetilde{\mathbf{v}}^{(1)}$:

$$\mathrm{curl}\ \widetilde{\mathbf{v}}^{(1)} = f(\tau)\beta\nabla T \times \mathbf{n} \tag{6.116}$$

Apparently, the solution of this equation may be written in the form

$$\widetilde{\mathbf{v}}^{(1)} = f(\tau)\mathbf{W} \tag{6.117}$$

where $\mathbf{W}$ satisfies the equality

$$\mathrm{curl}\ \mathbf{W} = \beta\nabla T \times \mathbf{n} \tag{6.118}$$

It follows from (6.118) that the field $\mathbf{W}$ proves to be solenoidal:

$$\mathrm{div}\ \mathbf{W} = 0 \tag{6.119}$$

As mentioned above, the pulsational transport vector $\mathbf{S}$ turns into zero for vibrations with a uniform phase. The particular vibrations considered in this subsection are all of this type. Substituting (6.117) into (6.114) and, as earlier, replacing ρ by ρ_0 everywhere except for the terms with the vibrational and gravity forces, we arrive at the equations

$$\frac{\partial \mathbf{v}}{\partial t} + \mathbf{v}\cdot\nabla\mathbf{v} - \beta\overline{f^2}(\mathbf{n}\cdot\mathbf{W})\nabla T = -\frac{1}{\rho_0}\nabla p + \nu\Delta\mathbf{v} - \beta T\mathbf{g} \tag{6.120}$$

$$\frac{\partial T}{\partial t} + \mathbf{v}\cdot\nabla T = \chi\Delta T \tag{6.121}$$

$$\mathrm{div}\ \mathbf{v} = 0 \tag{6.122}$$

$$\mathrm{curl}\ \mathbf{W} = \beta\nabla T \times \mathbf{n} \tag{6.123}$$

$$\mathrm{div}\ \mathbf{W} = 0 \tag{6.124}$$

While writing down the vibrational force, we have omitted the term quadratic with respect to $\widetilde{\mathbf{v}}^{(1)}$ and included the gradient term into the pressure part.

The obtained set of equations almost coincides with the equations of thermovibrational convection described in Chapter 1, differing from the latter only in the form of the vibrational force. Let us compare the set (6.120) to (6.124) with the equations of thermovibrational convection for the case of weak non-isothermal conditions and vibrations of a general kind (6.101) to (6.105). Both sets of equations are derived from the general equations (6.94) to (6.98) in the approximation of weak non-isothermality. However, they are related to different limiting cases.

While deriving equations (6.101) to (6.105) the non-uniformities of the pulsation velocity due to non-isothermality were assumed to be much smaller than those caused by non-thermal factors, viz. deformable boundaries, complex types of vibrations, etc. In the case of equations (6.120) to (6.124) the opposite assumption was used, i.e. the situations were discussed where non-isothermality was considered as the only reason for the pulsation field non-uniformities. Of course, it is not difficult to imagine a situation where both mentioned factors play equally important roles. Such cases require special analysis which could be performed on the basis of the general equations (6.94) to (6.98). An example of such a situation will be considered in Chapter 8.

Let us discuss the most important distinctions of equations (6.101) to (6.105) and (6.120) to (6.124). First, in the case of vibrations of a general kind the part of the pulsation velocity needed for calculation of the average effects is determined irrespective of the average flow, whereas in the case of uniform vibrations the equations for pulsation and average components must be solved compatibly. Second, the proportionality of the field $\mathbf{W}$ to the non-isothermality parameter $\beta\theta$ leads to the quadratic dependence of the vibrational force on $\beta\theta$ in the case of uniform vibrations. For the vibrations of a general kind this depend-ence is linear. This means a change in the intrinsic symmetry of equations which, in turn, may change the symmetry of the solution.

Consider, for example, a situation where a geometrical symmetry exists with respect to inversion of the horizontal coordinate. The vibrations are assumed to be horizontal and a lateral heating is imposed. In this case the set of equations (6.120) to (6.124) admits the solutions of the antisymmetrical type, since the horizontal component of the velocity changes its sign when the horizontal coordinate is inverted. Simultaneously, the vibration force in (6.101) will generate a motion of the opposite symmetry. However, the non-linear terms do not allow such symmetry. Due to that, the resulting solutions will be symmetrical for a creeping flow and will not possess any definite symmetry for non-linear regimes. Finally, we would like to note that the general case of vibrations requires that the pulsational transport effect (absent for uniform vibrations) must be taken into account if the vibration phase is non-uniform.

References

1. Faraday, M. On a peculiar class acoustical figures and on certain forms assumed by a group of particles upon vibrating elastic surface, *Phil. Trans. Roy. Soc. London*, 1831, **121**, 209–318.
2. Lord Rayleigh. On the circulation of air observed in Kundt's tubes, and on some allied acoustical problems, *Phil. Trans. Roy. Soc. London*, 1883, **A175**, 1–21.
3. Lord Rayleigh. *Theory of Sound*, 2nd edn, Dover Publications, New York, 1945, Vols. 1 and 2.
4. Schlichting H. Berechnung ebener periodischer Grenzschichtstromungen, *Z. Phys.*, 1932, **33**, 327.
5. Schlichting, H. *Boundary Layer Theory*, Pergamon, 1955.
6. Lighthill, M.J. Acoustic streaming, *J. Sound. Vib.*, 1978, **61**, 391–418.
7. Longuet-Higgins, M.S. Mass transport in water waves, *Phil. Trans. Roy. Soc. London*, 1953, **A245**, 535–81.
8. Longuet-Higgins, M.S. Mass transport in the boundary layer at a free oscillatory surface, *J. Fluid Mech.*, 1960, **8**, 293–306.
9. Lyubimov, D.V. Thermovibrational flows in a fluid with a free surface, *Microgravity Quart.*, 1994, **4**(2), 107–12.
10. Lyubimov, D.V. A new approach in the vibrational convection theory, *C. R. Acad. Sci. Paris*, 1995, **320**, Ser. IIb, 271–6.
11. Lyubimov, D.V. Convective flows under the influence of high frequency vibrations, *Eur. J. Mech., B/Fluids*, 1995, **14**(4), 439–58.
12. Zenkovskaya, S.M. and I.B. Simomenko. On the influence of high frequency vibrations on the onset of convection, *Izv. AN SSSR, Mekhanika Zhidkosti i Gaza*, 1966, **5**, 51–5.
13. Drazin, P.G. and W.N. Reid. *Hydrodynamic Stability*, Cambridge University Press, Cambridge, 1981.
14. Nepomnyashii, A.A. On the long wave convective instability in a horizontal layer with deformable boundary, in *Convective Flows*, Perm, 1983, pp. 25–31.
15. Rasenat, S., F.H. Busse and I. Rehberg. A theoretical and experimental study of double-layer convection, *J. Fluid Mech.*, 1989, **199**, 519–40.
16. Andreev, V.K., O.P. Kaptsov, V.V. Pukhnachev and A.A. Rodionov. *Application of Theoretical Group Methods in Hydrodynamics*, Nauka, Novosibirsk, 1994, pp. 216.
17. Lobov, N.I., D.V. Lyubimov and T.P. Lyubimova. Convective instability of a system of horizontal layers of immiscible fluids with deformable interfaces, *Fluid Dynamics*, 1996, **31**(2), 186–92.
18. Briskman, V.A. Vibration-thermocapillary convection and stability, in *Proc. of the First Int. Symp. on Hydromechanics and Heat/Mass Transfer in Microgravity*, Perm, Moscow, 1991, Gordon and Breach, London, 1992, pp. 111–19.
19. Nayfeh, A.H. *Introduction to Perturbation Techniques*, John Wiley, Chichester, 1981.
20. Dore, B.D. Double boundary layer in standing surface waves, *Pure Appl. Geophys.*, 1976, **114**, 629–37.
21. Iskandarany, M., P.L.-F. Lin. Mass transport in two-dimensional water waves, *J. Fluid Mech.*, 1991, **231**, 395–415.
22. Lyubimov, D.V. Vibrational flows in non-uniform systems. General approach. The role in heat/mass transfer processes in microgravity and terrestrial conditions, in *IX Eur. Symposium on Gravity Dependent Phenomena in Physical Sciences*, Abstracts, Berlin, 1995, p. 143.

23. Batchelor, G.K. *An Introduction to Fluid Dynamics*, Cambridge University Press, 1970.
24. Stokes, G.G. On the theory of oscillatory waves, *Trans. Camb. Phil. Soc.*, 1847, **8**, 441–55.
25. Lyubimov, D.V., M.V. Savvina and A.A. Cherepanov. On quasi-equilibrium shape of a free fluid surface in modulated gravity field, in *Problems of Hydrodynamics and Heat/Mass Transfer with Free Boundaries*, Novosibirsk, 1987, 97–105.
26. Lyubimov, D.V. and T.P. Lyubimova. One method for numerical modelling in the problems with deformable fluid interfaces, *Modelling in Mechanics*, 1990, **4**(21), 1, 136–40.
27. Lyubimov, D.V., T.P. Lyubimova and A.Yu. Lapin. Numerical investigation on quasi-equilibrium shapes of interface of rotating fluids in axial vibrational field, in *Nonlinear Problems of Dynamic of Viscous Fluid*, Perm, 1990, pp. 43–52.
28. Landau, L.D. and E.M. Lifshitz. *Electrodynamics of Continuous Media*, Pergamon Press, Oxford, 1960.
29. Gershuni, G.Z., E.M. Zhukhovitsky and A.A. Nepomnyaschii. *Stability of Convective Flows*, Nauka, Moscow, 1989.
30. Lyubimov, D.V. and A.A. Cherepanov. *Izv. AN SSSR, Mekhanika Zhidkosti i Gaza*, 1986, **6**, 8–13.

7 The Problem of Boundary Conditions

For a description of average flows generated by vibrations, the correct formulation of boundary conditions is of special importance. A simple statement like 'since on the rigid wall the velocity vanishes at any moment of time, the same should hold for the average velocity' is absolutely inappropriate, however evident it might seem. The point is that, as was noted in the preceding chapter when deriving the averaged equations, this conclusion is correct only for the interior points of the fluid, located in the bulk outside the skin layer thickness. Therefore the boundary conditions are to be set, strictly speaking, not on the boundary itself but on the external border of the skin layer where, of course, the velocity values differ from those on the actual boundary. The fact that we set these conditions at the coordinate values corresponding to the true boundary (or its average position if it oscillates) using the smallness of the skin layer thickness does not change the essence. Thus, to obtain the effective boundary conditions one needs to study the flow in a dynamic skin layer near a rigid or free surface. Consideration of problems of this kind constitutes the subject of the present chapter.

1 Boundary Conditions at a Rigid Wall

The pulsating flow near a rigid surface was considered first in 1883 by Rayleigh [1]. He investigated generation of the average flow in the near-wall layer in a channel with a standing acoustic wave. In this early study a surprising specific phenomenon was discovered. Although the effect of a non-zero average velocity appearing on the external border of the skin layer is related to the viscous effect, nevertheless this velocity itself does not depend on viscosity. Actually, if one recalls the usual no-slip boundary condition, it is also ascribed by its origin to the viscosity of a fluid, but does not contain it. Ultimately, the velocity is not determined by the length scale on which the pulsation energy losses occur but by those losses themselves.

Later Schlichting made a significant contribution to the investigation of this problem. He considered the average flows that arose near a cylinder oscillating in an incompressible fluid [2]. The structure of the skin layer was explicitly analyzed and the formula was derived for the effective average velocity on the external border of the skin layer. It is noteworthy that this formula holds not only for an oscillating cylinder but for other cases as well of one-dimensional pulsating flow with a uniform phase near a rigid surface. Schlichting's study has revealed that the compressibility of a medium has no impact on the considered phenomenon. The only circumstance that matters is that a flow should contain a pulsating component. However, until now the problems of the near-wall generation of average flows are often discussed in handbooks in the sections devoted to physical acoustics. Subsequently, in papers by a number of authors (see, for example, [3] and [4]) the Schlichting formula was generalized for the case of either non-uniform phase of the pulsation field or for a non-one-dimensional flow. Here we present the derivation of the generalized formula for the average effective velocity, simultaneously taking into account the effects of phase non-uniformity and non-one-dimensionality of a flow.

1.1 Equations for the Average and Pulsation Fields near a Rigid Wall

Consider a flow near a rigid surface containing the pulsation component, so that the conditions adopted in this book of smallness of the skin layer thickness $\delta = \sqrt{\nu/\Omega}$ and vibration amplitude b in comparison with the reference length scales of the hydrodynamic fields hold. The ratio b/δ is taken to be a free parameter. Assuming also δ to be small as compared to the curvature radius of the rigid surface, we restrict ourselves to consideration of a flat surface.

Let us choose a Cartesian set of coordinates so that the z axis is directed along the surface normal vector which lies in the xy plane. It is convenient to single out explicitly the normal component of the velocity. Thus, we shall write the velocity vector in the form $(\mathbf{v}, w)$, where $\mathbf{v}$ is the vector of tangential components and w is the z component of the velocity. Projecting the Navier–Stokes equation on to the z axis and xy plane we have

$$\frac{\partial w}{\partial t} + (\mathbf{v} \cdot \nabla) w + w\frac{\partial w}{\partial z} = -\frac{\partial \Phi}{\partial z} + \nu \Delta w + \nu \frac{\partial^2 w}{\partial z^2} + F \tag{7.1}$$

$$\frac{\partial \mathbf{v}}{\partial t} + (\mathbf{v} \cdot \nabla) \mathbf{v} + w\frac{\partial \mathbf{v}}{\partial z} = -\nabla \Phi + \nu \Delta \mathbf{v} + \nu \frac{\partial^2 \mathbf{v}}{\partial z^2} + \mathbf{f} \tag{7.2}$$

Here $\Phi = p/\rho$ and $(\mathbf{f},\ F)$ is the body force vector.

The continuity equation may be written as

$$\text{div } \mathbf{v} + \frac{\partial w}{\partial z} = 0 \tag{7.3}$$

In the set (7.1) and (7.2) the operators Δ and ∇ act in the xy plane. The considered reference frame is unique for the given part of the rigid surface so that F and $\mathbf{f}$ incorporate the inertia forces.

Let us decompose all the fields into the average and pulsation parts:

$$w = \overline{w} + \widetilde{w}$$

$$\mathbf{v} = \overline{\mathbf{v}} + \widetilde{\mathbf{v}}$$

$$\Phi = \overline{\Phi} + \widetilde{\Phi}$$

$$F = \overline{F} + \widetilde{f}$$

$$\mathbf{f} = \overline{\mathbf{f}} + \widetilde{\mathbf{f}}$$

Averaging (7.1) to (7.3) over the 'fast' time $\tau = \Omega t$ we obtain

$$\frac{\partial \overline{w}}{\partial t} + (\overline{\mathbf{v}} \cdot \nabla)\overline{w} + \overline{(\widetilde{\mathbf{v}}\nabla)\widetilde{w}} + \overline{w}\frac{\overline{\partial w}}{\partial z} + \overline{\widetilde{w}\frac{\partial \widetilde{w}}{\partial z}} = -\frac{\partial \overline{\Phi}}{\partial z} + \nu\Delta\overline{w} + \nu\frac{\partial^2 \overline{w}}{\partial z^2} + \overline{F} \tag{7.4}$$

$$\frac{\partial \overline{\mathbf{v}}}{\partial t} + (\overline{\mathbf{v}} \cdot \nabla)\overline{\mathbf{v}} + \overline{(\widetilde{\mathbf{v}}\nabla)\widetilde{\mathbf{v}}} + \overline{w}\frac{\partial \overline{\mathbf{v}}}{\partial z} + \overline{\widetilde{w}\frac{\partial \widetilde{\mathbf{v}}}{\partial z}} = -\nabla\overline{\Phi} + \nu\Delta\overline{\mathbf{v}} + \nu\frac{\partial^2 \overline{\mathbf{v}}}{\partial z^2} + \overline{\mathbf{f}} \tag{7.5}$$

$$\text{div } \mathbf{v} + \frac{\partial \overline{w}}{\partial z} = 0 \tag{7.6}$$

Subtraction of the averaged equations (7.4) to (7.6) from the complete ones (7.1) to (7.3) yields the equations for the pulsation components:

$$\begin{aligned} &\Omega\frac{\partial \widetilde{w}}{\partial \tau} + (\overline{\mathbf{v}} \cdot \nabla)\widetilde{w} + (\mathbf{v} \cdot \nabla)\overline{w} + \overline{w}\frac{\partial \widetilde{w}}{\partial z} + \widetilde{w}\frac{\partial \overline{w}}{\partial z} + \left\{(\widetilde{\mathbf{v}} \cdot \nabla)\widetilde{w} + \widetilde{w}\frac{\partial \widetilde{w}}{\partial z}\right\} \\ &\quad = -\frac{\partial \widetilde{\Phi}}{\partial z} + \nu\Delta\widetilde{w} + \nu\frac{\partial^2 \widetilde{w}}{\partial z^2} + \widetilde{F} \end{aligned} \tag{7.7}$$

$$\begin{aligned} &\Omega\frac{\partial \widetilde{\mathbf{v}}}{\partial \tau} + (\overline{\mathbf{v}} \cdot \nabla)\widetilde{\mathbf{v}} + (\widetilde{\mathbf{v}} \cdot \nabla)\overline{\mathbf{v}} + \overline{w}\frac{\partial \widetilde{\mathbf{v}}}{\partial z} + \widetilde{w}\frac{\partial \overline{\mathbf{v}}}{\partial z} + \left\{(\widetilde{\mathbf{v}} \cdot \nabla)\widetilde{\mathbf{v}} + \widetilde{w}\frac{\partial \widetilde{\mathbf{v}}}{\partial z}\right\} \\ &\quad = -\nabla\widetilde{\Phi} + \nu\Delta\widetilde{\mathbf{v}} + \nu\frac{\partial^2 \widetilde{\mathbf{v}}}{\partial z^2} + \widetilde{\mathbf{f}} \end{aligned} \tag{7.8}$$

$$\text{div } \widetilde{\mathbf{v}} + \frac{\partial \widetilde{w}}{\partial z} = 0 \tag{7.9}$$

Here, as in Section 2 in Chapter 6, the braces denote extraction of the pulsation part.

Since the problem possesses two reference length scales, to construct a uniformly valid approximation it is necessary to use a method similar to the multiscale technique. Here we take it in the form of the method of matched asymptotic expansions.

Recall that according to the idea of this method [5] one may seek the solution in the form of a series with respect to a small parameter separately in the skin layer region and in the external zone. Moreover, inside the skin layer, the stretched coordinates are introduced so that the stretching transformation itself involves a small parameter. A uniformly valid approximation will emerge if the mentioned expansions have overlapping validity ranges so that matching conditions may be imposed. The ratio of the skin layer thickness to the external size $\varepsilon = \delta/L$ may work as a natural small parameter of the problem.

1.2 Intrinsic Expansion for Pulsations

Let us write equations (7.7) to (7.9) in the dimensionless form, choosing as scales the external size L for the xy coordinates, the skin layer thickness δ for the z coordinate, the vibration velocity $b\Omega$ for pulsation velocities, ν/L for the average velocities, $b\Omega^2$ for the body forces and $\rho L b\Omega^2$ for the pulsation pressure. Thus we get

$$\frac{\partial \widetilde{w}}{\partial \tau} + \varepsilon^2(\overline{\mathbf{v}}\nabla)\widetilde{w} + \varepsilon^2(\widetilde{\mathbf{v}}\nabla)\overline{w} + \varepsilon\overline{w}\frac{\partial \widetilde{w}}{\partial \xi} + \varepsilon\widetilde{w}\frac{\partial \overline{w}}{\partial \xi} + \left\{S\varepsilon(\widetilde{\mathbf{v}}\nabla)\widetilde{w} + S\widetilde{w}\frac{\partial \widetilde{w}}{\partial \xi}\right\}$$

$$= -\frac{1}{\varepsilon}\frac{\partial \widetilde{\Phi}}{\partial \xi} + \varepsilon^2\Delta\widetilde{w} + \frac{\partial^2 \widetilde{w}}{\partial \xi^2} + \widetilde{F} \tag{7.10}$$

$$\frac{\partial \widetilde{\mathbf{v}}}{\partial \tau} + \varepsilon^2(\overline{\mathbf{v}}\nabla)\widetilde{\mathbf{v}} + \varepsilon^2(\widetilde{\mathbf{v}}\nabla)\overline{\mathbf{v}} + \varepsilon\overline{w}\frac{\partial \widetilde{\mathbf{v}}}{\partial \xi} + \varepsilon\widetilde{w}\frac{\partial \overline{\mathbf{v}}}{\partial \xi} + \left\{S\varepsilon(\widetilde{\mathbf{v}}\nabla)\widetilde{\mathbf{v}} + S\widetilde{w}\frac{\partial \widetilde{\mathbf{v}}}{\partial \xi}\right\}$$

$$= -\nabla\widetilde{\Phi} + \varepsilon^2\Delta\widetilde{\mathbf{v}} + \frac{\partial^2 \widetilde{\mathbf{v}}}{\partial \xi^2} + \widetilde{\mathbf{f}} \tag{7.11}$$

$$\frac{\partial \widetilde{w}}{\partial \xi} + \varepsilon \operatorname{div} \widetilde{\mathbf{v}} = 0 \tag{7.12}$$

Here the notations $S \equiv \sqrt{Re_{\mathrm{p}}} = b/\delta$ and ξ for the stretched vertical coordinate are introduced.

We seek the solution in the form of a series with respect to the small parameter ε:

$$\begin{aligned}\widetilde{\mathbf{v}} &= \widetilde{\mathbf{v}}_0 + \varepsilon\widetilde{\mathbf{v}}_1 + \cdots \\ \widetilde{w} &= \widetilde{w}_0 + \varepsilon\widetilde{w}_1 + \cdots \\ \widetilde{\Phi} &= \widetilde{\Phi}_0 + \varepsilon\widetilde{\Phi}_1 + \cdots\end{aligned} \tag{7.13}$$

We also assume that the average flow velocity and body forces are expanded into similar series:

$$\begin{aligned}
\overline{\mathbf{v}} &= \overline{\mathbf{v}}_0 + \varepsilon \overline{\mathbf{v}}_1 + \cdots \\
\overline{w} &= \overline{w}_0 + \varepsilon \overline{w}_1 + \cdots \\
\widetilde{F} &= \widetilde{F}_0 + \varepsilon \widetilde{F}_1 + \cdots \\
\widetilde{\mathbf{f}} &= \widetilde{\mathbf{f}}_0 + \varepsilon \widetilde{\mathbf{f}}_1 + \cdots
\end{aligned} \tag{7.14}$$

In the leading order from (7.10) we get

$$\frac{\partial \widetilde{\Phi}_0}{\partial \xi} = 0 \tag{7.15}$$

i.e. as it is customary in the boundary layer theory, the pressure does not vary along the transverse direction.

In the zeroth order from (7.12) we obtain

$$\frac{\partial \widetilde{w}_0}{\partial \xi} = 0 \tag{7.16}$$

Taking into account the boundary condition of impermeability ($\widetilde{w} = 0$ at $\xi = 0$) we conclude that in the leading order the pulsation velocity component normal to the boundary is absent: $\widetilde{w}_0 = 0$.

In the next order we have, from the same equation,

$$\frac{\partial \widetilde{w}_1}{\partial \xi} = -\operatorname{div} \widetilde{\mathbf{v}}_0 \tag{7.17}$$

Equation (7.11) yields, in the leading (zeroth) order,

$$\frac{\partial \widetilde{\mathbf{v}}_0}{\partial \tau} = -\nabla \widetilde{\Phi}_0 + \frac{\partial^2 \widetilde{\mathbf{v}}_0}{\partial \xi^2} + \widetilde{\mathbf{f}}_0 \tag{7.18}$$

Focusing on monochromatic vibrations, we present the time dependence of the variable fields in (7.17) and (7.18) in the form

$$\begin{aligned}
\widetilde{\mathbf{v}}_0 &= \mathbf{V} \mathrm{e}^{\mathrm{i}\tau} + \mathbf{V}^* \mathrm{e}^{-\mathrm{i}\tau} \\
\widetilde{\Phi}_0 &= \Pi \mathrm{e}^{\mathrm{i}\tau} + \Pi^* \mathrm{e}^{-\mathrm{i}\tau} \\
\widetilde{w}_0 &= W \mathrm{e}^{\mathrm{i}\tau} + W^* \mathrm{e}^{-\mathrm{i}\tau} \\
\widetilde{\mathbf{f}}_0 &= \mathbf{A} \mathrm{e}^{\mathrm{i}\tau} + \mathbf{A}^* \mathrm{e}^{-\mathrm{i}\tau}
\end{aligned} \tag{7.19}$$

Then equations (7.17) and (7.18) can be written as

$$\frac{\partial W}{\partial \xi} = -\operatorname{div} \mathbf{V} \tag{7.20}$$

$$\mathrm{i}\mathbf{V} = -\nabla \Pi + \frac{\partial^2 \mathbf{V}}{\partial \xi^2} + \mathbf{A} \tag{7.21}$$

To make available the explicit solution of equations (7.20) and (7.21) one needs to find the dependence of ξ on the body forces. It should be remembered that the body forces include the gravity and inertia forces which, in their turn, may be divided into translational, centrifugal, Coriolis and those forces containing the angular acceleration. All these forces do not depend on viscosity and consequently do not contain the parameter δ. Therefore, for these forces, the expansion with respect to ε in fact coincides with the one with respect to z. (Recall that $z = \delta\xi$.) This means that $\widetilde{\mathbf{f}}_0$ does not depend on ξ, which allows the general solution of equation (7.21) to be written in the form

$$\mathbf{V} = \mathbf{U}(1 - \mathrm{e}^{-\alpha\xi}) \tag{7.22}$$

where we have taken (7.15) into account and have used the no-slip boundary condition. Here $\mathbf{U}$ is an arbitrary vector independent of ξ but may perhaps depend on coordinates in the rigid boundary plane, $\alpha \equiv (1+i)/\sqrt{2}$.

The obtained results for $\widetilde{\mathbf{v}}_0$ and $\widetilde{w}_0$ determine the one-term intrinsic expansion of the pulsation velocity. If we were interested only in the latter, we might have stopped here. However, bearing in mind that derivation of the average velocity would later be necessary, one needs to calculate the normal component of the pulsation velocity to the next order. Substituting (7.22) in (7.20), integrating over ξ and taking into account the boundary conditions we obtain

$$\mathbf{W} = \left(\frac{1}{\alpha} - \xi - \frac{1}{\alpha}\mathrm{e}^{-\alpha\xi}\right) \operatorname{div} \mathbf{U} \tag{7.23}$$

Thus, we have the pulsation velocity distribution in the boundary layer in the first non-vanishing order:

$$\widetilde{\mathbf{v}} = \mathbf{U}(1 - \mathrm{e}^{-\alpha\xi})\mathrm{e}^{\mathrm{i}\tau} + \text{c.c.} + \cdots \tag{7.24}$$

$$\widetilde{w} = \varepsilon\left(\frac{1}{\alpha} - \xi - \frac{1}{\alpha}\mathrm{e}^{-\alpha\xi}\right)\mathrm{e}^{\mathrm{i}\tau} \operatorname{div} \mathbf{U} + \text{c.c.} + \cdots \tag{7.25}$$

Here c.c. denotes the complex conjugate terms and the dots denote those of the higher orders of smallness.

The distributions (7.24) and (7.25) demonstrate the coordinate dependence inherent in the boundary layers. It is qualitatively described by Fig. 154, where

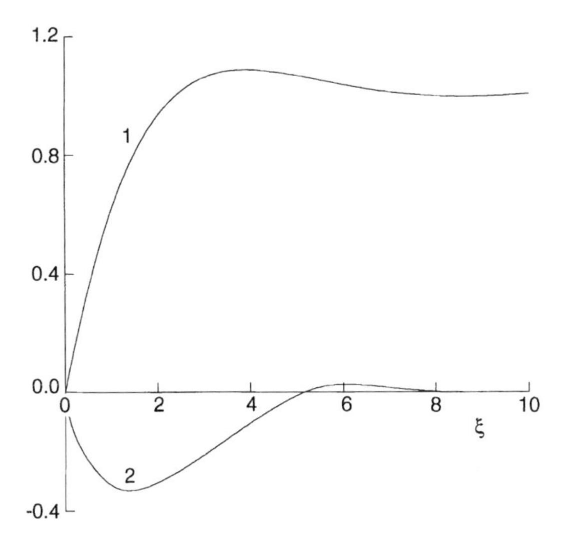

Fig. 154 Distribution of the pulsation velocity in the boundary layer for two different moments of time

velocity distributions obtained for two different time moments are presented. In formula (7.25) we note the presence of the secular term proportional to ξ. The appearance of such a term in the zeroth order would point to the inconsistency of the expansion. However, in accordance with the principle of minimal singularity [6] its appearance in the first order simply means that the validity range of the intrinsic expansion is limited by the layer boundaries. The obtained solution should be matched to the external expansion.

1.3 External Expansion for Pulsations

In fact, we shall not construct the external expansion for the pulsation velocity but restrict ourselves to the zeroth order in it.

To obtain this expansion for the vertical coordinate the external scale should be used. In terms of dimensionless variables, this means that, first, one makes the replacement $\varepsilon\xi = z$ in equations (7.10) to (7.12) and, second, collects the terms of the same order in ε. In the leading (zeroth) order we obtain

$$\frac{\partial \widetilde{\mathbf{v}}_0}{\partial \tau} = -\nabla \widetilde{\Phi}_0 + \mathbf{f}_0 \tag{7.26}$$

$$\frac{\partial \widetilde{w}_0}{\partial \tau} = -\frac{\partial \widetilde{\Phi}_0}{\partial z} + \widetilde{F}_0 \tag{7.27}$$

$$\text{div } \widetilde{\mathbf{v}}_0 + \frac{\partial \widetilde{w}_0}{\partial z} = 0 \tag{7.28}$$

As might have been expected, these equations coincide with those for the pulsation velocity field obtained in Section 2 in Chapter 6 except for notations.

Matching the external and intrinsic expansions we first find that

$$\widetilde{w}_0 = 0 \qquad \text{at } z = 0 \tag{7.29}$$

which justifies the conditions for the pulsation field applied in the preceding sections. Second, we find that, for $z = 0$,

$$\widetilde{\mathbf{v}}_0 = \mathbf{U}\mathrm{e}^{\mathrm{i}\tau} + \mathbf{U}^*\mathrm{e}^{-\mathrm{i}\tau} \tag{7.30}$$

which imparts a meaning to the arbitrary vector $\mathbf{U}$ which has appeared in the intrinsic expansion.

1.4 Intrinsic Expansion for the Average Velocity

Consider the construction of the matched expansions for the average velocity. Rewriting equations (7.4) to (7.6) in terms of dimensionless coordinates and using the intrinsic coordinate ζ we have

$$\begin{aligned}\varepsilon^2\frac{\partial\overline{w}}{\partial t} + \varepsilon^2(\overline{\mathbf{v}}\nabla)\overline{w} + Re_{\mathrm{p}}\overline{(\widetilde{\mathbf{v}}\nabla)\widetilde{w}} + \varepsilon\overline{w}\frac{\partial\overline{w}}{\partial\xi} + \frac{Re_{\mathrm{p}}}{\varepsilon}\overline{\widetilde{w}\frac{\partial\widetilde{w}}{\partial\xi}} \\ = -\frac{1}{\varepsilon}\frac{\partial\overline{\Phi}}{\partial\xi} + \varepsilon^2\Delta w + \frac{\partial^2\overline{w}}{\partial\xi^2} + \overline{F}\end{aligned} \tag{7.31}$$

$$\begin{aligned}\varepsilon^2\frac{\partial\mathbf{v}}{\partial t} + \varepsilon^2(\overline{\mathbf{v}}\nabla)\overline{\mathbf{v}} + Re_{\mathrm{p}}\overline{(\widetilde{\mathbf{v}}\nabla)\widetilde{\mathbf{v}}} + \varepsilon\overline{w}\frac{\partial\overline{\mathbf{v}}}{\partial\xi} + \frac{Re_{\mathrm{p}}}{\varepsilon}\overline{\widetilde{w}\frac{\partial\widetilde{\mathbf{v}}}{\partial\xi}} \\ = -\nabla\overline{\Phi} + \varepsilon^2\Delta\overline{\mathbf{v}} + \frac{\partial^2\overline{\mathbf{v}}}{\partial\xi^2} + \overline{\mathbf{f}}\end{aligned} \tag{7.32}$$

$$\frac{\partial\overline{w}}{\partial\xi} + \varepsilon \operatorname{div} \overline{\mathbf{v}} = 0 \tag{7.33}$$

Here the reference time L^2/ν is adopted as a time scale.

From (7.31) in the main order of expansion with respect to ε it follows that

$$\frac{\partial\overline{\Phi}_0}{\partial\xi} = 0 \tag{7.34}$$

and equation (7.33) results in

$$\frac{\partial\overline{w}_0}{\partial\xi} = 0 \tag{7.35}$$

The latter relation and the boundary conditions on the rigid wall yield

$$\overline{w}_0 = 0 \tag{7.36}$$

In the zeroth order in ε, from equation (7.32) we obtain:

$$\frac{\partial^2 \overline{\mathbf{v}}_0}{\partial \xi^2} = \nabla \overline{\Phi}_0 - \overline{\mathbf{f}}_0 + Re_\mathrm{p}\left[\overline{\widetilde{w}_1 \frac{\partial \widetilde{\mathbf{v}}_0}{\partial \xi}} + \overline{(\widetilde{\mathbf{v}}_0 \nabla)\widetilde{\mathbf{v}}_0}\right] \tag{7.37}$$

The relation for the pulsation velocity (7.16) has already been used here.

Averaging the right-hand side of (7.37) is easy to perform using the explicit expressions for the pulsation field dependencies on the 'fast' time. This yields

$$\frac{\partial^2 \overline{\mathbf{v}}_0}{\partial \xi^2} = \nabla \overline{\Phi}_0 - \overline{\mathbf{f}}_0 + Re_\mathrm{p}\left[\mathbf{W}\frac{\partial \mathbf{V}^*}{\partial \xi} + \mathbf{W}^* \frac{\partial \mathbf{V}}{\partial \xi} + (\mathbf{V}\nabla)\mathbf{V}^* + (\mathbf{V}^*\nabla)\mathbf{V}\right] \tag{7.38}$$

Substituting ξ-dependencies (7.22) and (7.23) into (7.38) and integrating over ξ, satisfying both the boundary conditions and the minimal singularity principle, we obtain the explicit expression for the average velocity distribution in the boundary layer

$$\begin{aligned}\overline{\mathbf{v}}_0 = {} & Re_\mathrm{p}(\mathbf{U}\cdot\nabla)\mathbf{U}^*\left(\mathrm{i}\mathrm{e}^{-\alpha\xi} - \mathrm{i}\mathrm{e}^{-\alpha^*\xi} + \tfrac{1}{2}\mathrm{i}\mathrm{e}^{-(\alpha+\alpha^*)\xi} - \tfrac{1}{2}\right) \\ & + Re_\mathrm{p}\mathbf{U}\ \mathrm{div}\ \mathbf{U}^*\left(2\mathrm{i}\mathrm{e}^{-\alpha\xi} + \mathrm{e}^{-\alpha\xi} - \alpha^*\xi\mathrm{e}^{-\alpha^*\xi} - \tfrac{1}{2}\mathrm{i}\mathrm{e}^{-(\alpha+\alpha^*)\xi} - \tfrac{3}{2}\mathrm{i} - 1\right) + \mathrm{c.c.}\end{aligned} \tag{7.39}$$

It can be seen that generation of the average velocity is essentially concerned with the non-uniformity of the pulsation tangential velocity. Actually, there are two different sources of the average velocity. The first one is connected with inertial effects. The second one is related to the exchange of fluid between the boundary layer and external flow regions due to the pulsation process that takes place at non-zero div $\mathbf{U}$. Recall that here we deal with a two-dimensional divergence of a vector field, so that div $\mathbf{U}$ may be non-zero even in an incompressible fluid.

In Fig. 155 the spatial dependence of the contribution of the inertial term to (7.39) (curve 1) is presented together with the real and imaginary parts of the expression responsible for the divergent contribution (curves 2 and 3, respectively). Note that if the pulsation flow has a uniform phase, then vector $\mathbf{U}$ may be treated as real. This means that curve 3 describes the contribution related to the phase non-uniformity. As one may see from the figure, curve 1 changes its sign at the distance of the order of the skin layer thickness from the wall. Therefore, the average flow in the vicinity of the wall (inside the boundary layer) and that outside the boundary layer may have different directions.

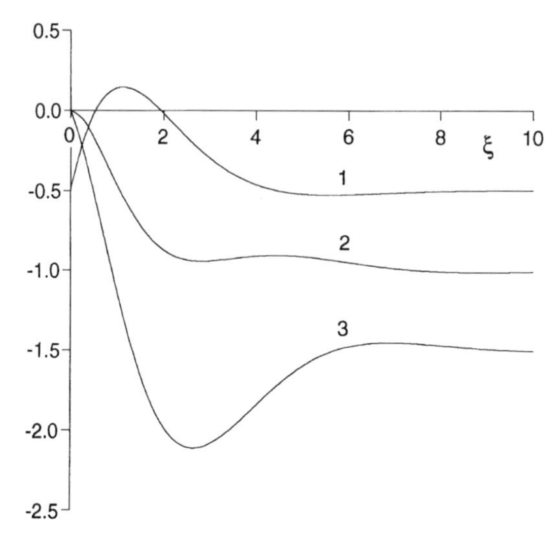

Fig. 155 Different parts of the average velocity in the boundary layer: 1, the inertial part; 2, 3, the real and imaginary components of the divergent term

1.5 Boundary Conditions for the External Flow

The limiting distribution of the velocity in the boundary layer at $\xi \to \infty$ in fact yields the effective boundary condition for the external (with respect to the boundary layer) flow. Indeed, the idea of the matched asymptotic expansion method is to construct the external expansion and to match it with the intrinsic one. As a matter of fact, the equations describing the external flow have been already constructed in Section 3 in Chapter 6. At the moment, we do not need their explicit form; it is enough to know that the corresponding solutions are regular functions of z. Like the case of pulsation velocity equations discussed above, to find the matching conditions one needs to take the leading term of the external expansion, rewrite it in terms of the intrinsic coordinate $z = \varepsilon \cdot \xi$ and write down a one-term expansion with respect to ε. Obviously, it will lead to $\overline{\mathbf{v}}_0$ and $\overline{w}_0$ at $z = 0$. After that one should take the one-term expansion given by (7.36) and (7.39), rewrite it in terms of the external coordinate z and find a one-term expansion with respect to ε. Comparison of the obtained expressions yields the matching condition that is sought. As a result, we come to the boundary conditions for the solution of the averaged equation in the external region

$$\text{at } z = 0: \qquad \overline{w} = 0, \qquad \overline{\mathbf{v}} = -\frac{\mathrm{Re_p}}{2}(\mathbf{U}\nabla)\mathbf{U}^* - Re_\mathrm{p}\left(1 + \frac{3}{2}\mathrm{i}\right)\mathbf{U}\,\mathrm{div}\,\mathbf{U}^* + \mathrm{c.c.} \tag{7.40}$$

For applications it might be more convenient to rewrite this formula in terms of real variables. Let the tangential component of the pulsation velocity at the rigid surface have the form

$$\widetilde{\mathbf{v}} = \mathbf{V}_1 \cos \tau + \mathbf{V}_2 \sin \tau \tag{7.41}$$

i.e. $\mathbf{U} = \frac{1}{2}(\mathbf{V}_1 - \mathrm{i}\mathbf{V}_2)$. By substituting this into (7.40), we obtain

$$\begin{aligned}\overline{\mathbf{v}}|_{z=0} = &-\frac{Re_{\mathrm{p}}}{4}[2\mathbf{V}_1 \operatorname{div} \mathbf{V}_1 + 2\mathbf{V}_2 \operatorname{div} \mathbf{V}_2 + (\mathbf{V}_1 \cdot \nabla)\mathbf{V}_1 \\ &+ (\mathbf{V}_2 \cdot \nabla)\mathbf{V}_2 + 3\mathbf{V}_2 \operatorname{div} \mathbf{V}_1 - 3\mathbf{V}_1 \operatorname{div} \mathbf{V}_2].\end{aligned} \tag{7.42}$$

Known formulas for the particular cases may be derived from (7.40) (or from (7.42)). In the simplest case of one-dimensional pulsations with a uniform phase ($\mathbf{V}_1 = V_1 j$, $\mathbf{V}_2 = 0$, $\mathbf{j}$ is the unit vector in the direction of the x axis), (7.42) turns into the well-known Schlichting formula [2]

$$\overline{v}|_{z=0} = -\frac{3}{8} Re_{\mathrm{p}} \frac{\partial V_1^2}{\partial x} \tag{7.43}$$

If pulsations are one-dimensional but their phase is non-uniform, (7.42) yields

$$\overline{v}|_{z=0} = -\frac{3}{4} Re_p \left(V_1 \frac{\partial V_1}{\partial x} + V_2 \frac{\partial V_2}{\partial x} + V_2 \frac{\partial V_1}{\partial x} - V_1 \frac{\partial V_2}{\partial x} \right). \tag{7.44}$$

It is convenient to rewrite the latter expression introducing the pulsation phase explicitly:

$$V_1 = A \cos \varphi, \qquad \mathrm{V}_2 = \mathrm{A} \sin \varphi \tag{7.45}$$

where A is the local amplitude of the pulsation velocity (in $b\Omega$ units) and φ is the phase. Substituting (7.45) into (7.44), we obtain [3, 7]

$$\overline{v}|_{z=0} = -\frac{3}{4} Re_{\mathrm{p}} \left(A \frac{\partial A}{\partial x} - A^2 \frac{\partial \varphi}{\partial x} \right) \tag{7.46}$$

Finally, in the particular case of non-one-dimensional pulsations with a uniform phase ($\mathbf{V}_2 = 0$), from (7.42) we have [4]

$$\overline{\mathbf{v}}|_{z=0} = -\frac{1}{4} Re_{\mathrm{p}}(2\mathbf{V}_1 \operatorname{div} \mathbf{V}_1 + (\mathbf{V}_1 \nabla)\mathbf{V}_1). \tag{7.47}$$

Formula (7.42) allows the average tangential velocity on a rigid surface to be calculated. As for the average normal velocity, one should be careful when applying formula (7.36). The point is that the study presented in this section was carried out for the reference frame in which the considered boundary surface domain was at rest. On the other hand, the equations for the average flow

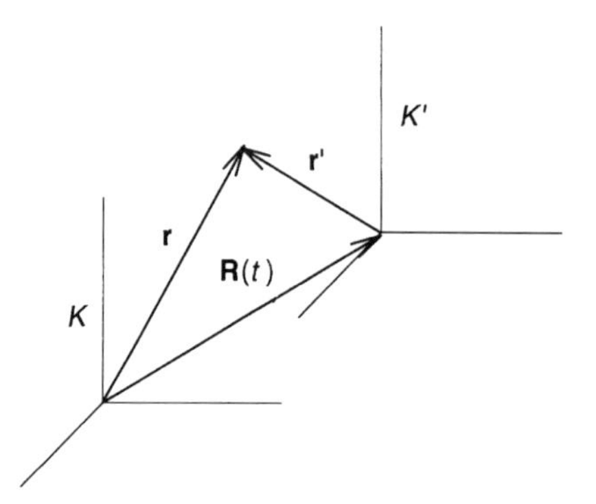

Fig. 156 The laboratory (K) and the oscillating (K') reference frames

obtained in Section 5 in Chapter 6 are formulated for the laboratory reference frame. Since the observation point in which the averaging is performed moves in space during oscillations of the reference frame, the result of averaging, generally, depends on the reference frame choice.

Let us discuss this point in more detail. Consider two different reference frames: K is the laboratory one and K' is moving (Fig. 156). Let $\mathbf{R}(t)$ be the coordinates of the origin of K' with respect to K and $\mathbf{r}$ and $\mathbf{r}'$ the coordinates of some point in K and K', respectively. Then, apparently,

$$\mathbf{r} = \mathbf{r}' + \mathbf{R} \tag{7.48}$$

and for the velocities the Galilei transformation holds

$$\mathbf{v}(\mathbf{r}, t) = \mathbf{v}'(\mathbf{r}', t) + \frac{\mathrm{d}\mathbf{R}}{\mathrm{d}t} \tag{7.49}$$

where $\mathbf{r}$ and $\mathbf{r}'$ are connected by (7.48).

Assume that the reference frame K' is subject to a small-amplitude oscillatory motion with respect to K. In this case it is possible to expand the right-hand side of (7.49) into a power series with respect to $\mathbf{R}$, eliminating $\mathbf{r}'$. Retaining the terms up to the first order we obtain

$$\mathbf{v} = \mathbf{v}' - (\mathbf{R}\nabla)\mathbf{v}' + \frac{\mathrm{d}\mathbf{R}}{\mathrm{d}t} \tag{7.50}$$

Then we decompose all the fields in both reference frames into the average and pulsation parts and perform the averaging in (7.50). We obtain

$$\overline{\mathbf{v}} = \overline{\mathbf{v}}' - \overline{(\mathbf{R}\nabla)\widetilde{\mathbf{v}}'} \tag{7.51}$$

Extracting the pulsation part from (7.50) we find that

$$\widetilde{\mathbf{v}} = \widetilde{\mathbf{v}}' + \frac{\mathrm{d}\mathbf{R}}{\mathrm{d}t} + \cdots \tag{7.52}$$

where small corrections are not written explicitly. Formulas (7.51) and (7.52) give the law of transformation for the average and pulsation velocity components upon transiting from the laboratory reference frame to the oscillating one. Let us apply (7.51) to recalculate the average normal velocity component on the rigid boundary:

$$\overline{w} = \overline{w}' - \overline{(\mathbf{R}\nabla)\widetilde{w}'} \tag{7.53}$$

Since $\widetilde{w}'$ vanishes at the boundary, (7.53) in fact includes differentiation only with respect to the normal coordinate:

$$\overline{w} = \overline{w}' - \overline{R_n \frac{\partial \widetilde{w}'}{\partial z}} \tag{7.54}$$

where R_n is the normal component of vector $\mathbf{R}$. It follows from (7.52) that

$$\frac{dR_n}{dt} = \widetilde{w}$$

and

$$\frac{\partial \widetilde{w}'}{\partial z} = \frac{\partial \widetilde{w}}{\partial z}$$

Then formula (7.54) takes the form

$$\overline{w} = \overline{w}' - S_n \tag{7.55}$$

where S_n is the normal component of the pulsational transport vector introduced in Section 3 in Chapter 6.

Thus, the condition of the vanishing normal component of the average velocity in the proper (moving) reference frame, when rewritten in the laboratory reference frame, assumes the form

$$\overline{w} + S_n = 0 \tag{7.56}$$

Hence, the boundary conditions for the average velocity on a rigid surface are given by formulas (7.40) where, in the case of a moving boundary, $\mathbf{U}$ should be treated as the amplitude of the relative pulsation velocity and the values of the average velocity should be corrected in accordance with (7.51). Finally, the boundary conditions for the average velocity on a rigid surface in the laboratory reference frame are presented as

$$\overline{w} = -S_n \tag{7.57}$$

$$\overline{\mathbf{v}} = -\frac{Re_p}{2}(\mathbf{U}_\tau \nabla)\mathbf{U}_\tau^* - Re_p\left(1 + \frac{3}{2}\mathrm{i}\right)\mathbf{U}_\tau \operatorname{div} \mathbf{U}_\tau^* + \mathrm{c.c.} - Re_P\,\overline{(\mathbf{R}\nabla)\widetilde{\mathbf{V}}} \tag{7.58}$$

where $\mathbf{U}_\tau$ is the complex amplitude of the relative tangential component of the pulsation velocity. In the case of vibrations with a uniform phase, the boundary conditions (7.57) to (7.58) simplify and take the form of the Niborg conditions:

$$\overline{w} = 0$$
$$\overline{\mathbf{v}} = -Re_{\rm p}(\mathbf{U}_\tau \nabla)\mathbf{U}_\tau - 2\, Re_{\rm p}\mathbf{U}_\tau \text{ div } \mathbf{U}_\tau \tag{7.59}$$

In conclusion, we remark that in the case of uniform vibrations in the reference frame of the vessel the pulsation velocity is proportional to the non-isothermality parameter (and identically absent in an isothermal fluid) and formulas (7.57) to (7.59) acquire the factor $(\beta\theta)^2$ in the right-hand sides. At the same time the term with the vibration force in the averaged equation of motion takes in the factor $(\beta\theta)$. This means that for uniform vibrations in the limiting case $\beta\theta \to 0$ we need to impose on the average velocity of a rigid surface the ordinary no-slip condition.

2 Boundary Conditions for the Pulsation Field on a Free Surface

In the general case on a free fluid surface the familiar boundary conditions of the normal and tangential stress balance and of coincidence of the normal velocities of the surface and fluid (the kinematic condition) must be satisfied. As emphasized in the preceding sections, the no-slip condition for the complete velocity leads to the formation of a well-pronounced boundary layer. For a free surface the condition on the tangential stress contains only the velocity derivatives. This means that the boundary layer near the free surface is less pronounced [8] and would manifest itself only in higher spatial derivatives of the velocity field. Therefore, the vorticity generated near the free surface must be small. It is then natural to assume that in the high-frequency limit for the pulsation field we do not have to require that the condition for the tangential stresses should be satisfied and in the condition for the normal stress balance we do not have to take the viscous stresses into account. As for the surface tension, since for any choice of units it induces an independent dimensionless parameter, various limiting situations may be considered.

2.1 Simplest Variants of the Boundary Conditions

In the simplest case, assuming the surface tension coefficient to be finite and considering the high-frequency limit, we have to retain in the boundary condition only the term with the pulsation pressure. Since the leading part of the pulsation velocity field is potential (as shown in Section 2 in Chapter 6), this means that the condition of the potential constancy or, in terms of velocities, the condition of the vanishing tangential velocity component is imposed at the free surface. Such a condition is quite acceptable for studying quasi-equilibrium shapes of a free surface under the action of vibrations and was applied in a number of works [9]. These boundary conditions being adopted, one obtains the following problem for the pulsation field:

$$\widetilde{\mathbf{v}} = \nabla\Phi\,, \qquad \Delta\Phi = 0 \tag{7.60}$$

with the following boundary conditions on the rigid boundaries:

$$\frac{\partial\Phi}{\partial n} = f_{\mathrm{n}} \tag{7.61}$$

where f_{n} is the normal component of the velocity of the boundary motion in the laboratory reference frame, and on the free boundary

$$\Phi = 0 \tag{7.62}$$

(for simplicity, we restrict ourselves to the case of a simply connected boundary).

The solution of the problem (7.60) to (7.62) determines the pulsation velocity field in a unique manner. Moreover, the tangential component of the pulsation velocity vanishes on the free surface, while the normal component may be non-zero.

The kinematic condition is not used for calculating the pulsation velocity field but it can be applied to determine the pulsations of the free surface shape. If the surface is described by the equation $z = \zeta(x, y, t)$, the kinematic condition may be written as

$$\frac{\partial\zeta}{\partial t} = \widetilde{w} \tag{7.63}$$

where $\widetilde{w}$ is the projection of the pulsation velocity on to the z axis. Integrating (7.63), one obtains pulsations of the free surface shape.

Note that there is no necessity to impose the requirement of the volume conservation on ζ. From equations (7.60) it follows that this condition is satisfied automatically.

2.2 Impossibility of Suppression of Deformations

One may expect that a strong gravity or a large surface tension will turn out to be capable of preventing deformations of the free surface. This assumption was used in [10]. In order to verify such a possibility, let us do the relevant estimations. We rewrite the complete expression of the normal stresses balance:

$$-p + \rho g\zeta + \sigma_{\mathrm{nn}} - \alpha\Delta_2\zeta = 0 \tag{7.64}$$

Here we suppose that gravity is directed along the negative direction of the z axis, σ_{nn} is the normal component of the viscous stress tensor and α is the surface tension coefficient. The capillary term is written assuming smallness of the free surface deviations from a flat horizontal and the hydrostatic part of the pressure is eliminated.

Let us estimate the terms in (7.65) for the case of pulsations. It follows from the equation of motion that the pulsation pressure is of the order of $\rho\Omega L\tilde{v}$, where L is the reference size. The reference pulsation velocity is $\tilde{v} \sim b\Omega$ for the general case of arbitrary vibrations and $\tilde{v} \sim b\Omega\beta\theta$ for the case of uniform vibrations in a weakly non-isothermal fluid.

To estimate the second term, we shall consider the kinematic condition (7.63), from which follows that $\zeta \sim \tilde{v}/\Omega$. Thus, the gravitational contribution to the normal stress is of the order of $\rho g\tilde{v}/\Omega$ and its ratio to that of the pulsation pressure is given by the dimensionless parameter $g/(\Omega^2 L)$. To prevent the pulsational deformations of the free surface by means of gravity it is necessary that

$$g/\Omega^2 L \gg 1 \tag{7.65}$$

Actually, however, in all the situations of practical importance the required frequencies would be so low that with respect to them the notion of the high-frequency approximation completely loses its meaning. Moreover, as shown in Section 5 in Chapter 6, the intensity of the thermovibrational convection in the general case is determined by the parameter $Gr_{\mathrm{v}} = b^2\Omega^2 L^2\beta\theta/\nu^2$, while that of the gravitational convection is determined by the ordinary Grashof number $Gr = g\beta\theta L^3/\nu^2$, so that

$$Gr = Gr_{\mathrm{v}} \frac{g}{\Omega^2 L}\left(\frac{L}{b}\right)^2 \tag{7.66}$$

It follows from (7.66) that even if the parameter $g/(\Omega^2 L)$ assumes some finite values (and more if these values are large) one obtains, due to the inequality $L \gg b$,

$$Gr \gg Gr_{\mathrm{v}} \tag{7.67}$$

In other words, if the gravity is strong enough to suppress pulsational deformations of the free surface, then the gravitational convection would be far greater than the thermovibrational one.

The latter estimation is based on equality (7.66) and should be modified for the case of uniform vibrations. If the fluid-filled vessel vibrations are performed precisely along the vertical direction, the leading (isothermal) part of the pulsation field is uniform. Consequently, the thermovibrational convection is described by the second-order equations (see Section 5 in Chapter 6) and the convection intensity is determined by the modified vibrational Grashof number $\widetilde{Gr}_v = b^2\Omega^2L^2(\beta\theta)^2/\nu^2$. In addition, the condition of suppression of pulsational surface deformations (7.65) by gravity remains valid. Rewriting this in terms of the Grashof numbers we obtain

$$\left(\frac{b}{L}\right)^2 \beta\theta\, Gr \gg Gr_v \tag{7.68}$$

At finite Gr this inequality may be satisfied only if the vibration amplitude greatly exceeds the vessel dimensions. Under such conditions, a fluid with a free surface could not be kept in a vessel since the vessel acceleration is greater that the gravitational one.

We have implemented the estimations of a possible contribution of the gravity forces to the normal stress balance. Let us discuss the contribution of the capillary forces. As can be seen from (7.64), the ratio of the reference value of the capillary forces to the normal pressure forces on the free surface is determined by the dimensionless parameter $(\alpha/\Omega^2\rho L^3) = 1/We$, where We is the Weber number. Since surface tension does not enter any other relations, the above parameter is independent and one may discuss just its particular numerical values in typical situations.

For example, the Weber number is of the order of 10^4 for water and the majority of liquid metals for frequencies of about 100 Hz and reference sizes of the order of 1 cm. This means that the capillary forces would not be able to prevent the pulsational deformations of the free surface. Only for free surfaces of very small dimensions would the Weber numbers become small enough for the boundary condition of the absence of the normal component of the pulsation velocity to acquire some meaning. At the same frequency of 100 Hz, the capillary forces would be comparable to those of the normal pressure only for dimensions in the order of a millimeter.

Thus, the high-frequency approximation for the description of the thermovibrational convection proves to be inconsistent with the assumption of the pulsational indeformability of the free surface.

2.3 Validity of the Simplest Boundary Conditions

The estimations made above do not mean that the capillary effects may be neglected entirely. The point is that in the context of considering the validity of the indeformability condition it was natural to estimate the capillary terms in respect to the size of the free surface. However, as soon as the conditions allowing the surface to be deformed are adopted, one should recall that the capillary waves can propagate along the surface. Here the spatial scale is not at all connected with the external size and is determined by the vibration frequency. Hence, high Weber numbers mean not the smallness of the capillary term but a comparative smallness of the capillary wavelength. Introduction of the capillary waves complicates the theory significantly. Therefore we begin by discussing those situations where the allowance for this factor is not obligatory.

At sufficiently high vibration frequencies the wavelength of the excited capillary waves is so small that these waves are effectively damped by viscosity. Moreover, the thickness of the wave zone, i.e. the region adjacent to the surface, where the motion essentially depends on the surface waves, is itself on the order of the wavelength. Therefore, at high frequencies it cannot affect the pulsation field in the interior fluid. It should be noted, however, that established requirements may impose more rigid restrictions on the vibration frequency than those discussed earlier. Let us perform the necessary estimations.

As known [8], the dispersion relation for the capillary waves on a flat infinite surface has the following form:

$$\Omega^2 = \frac{\alpha}{\rho} k^3 \tag{7.69}$$

where k is the wavenumber. Taking into account the smallness of the length in comparison with all the reference sizes, one may use relation (7.69) in the case of a curved surface as well. If the wavelength is required to be much smaller than the reference size, it is necessary that

$$kL \gg 1 \tag{7.70}$$

Substituting (7.69) into inequality (7.70), we obtain the frequency bound

$$\Omega^2 \gg \frac{\alpha}{\rho L^3} \tag{7.71}$$

i.e. as noted above, the Weber number must be higher than unity. Thus, at high *We* the thickness of the wave zone is small and the pulsation field is not sensible to the surface waves in most of the volume.

Let us now discuss the role of viscosity in the damping of short waves. In order for capillary waves to be absent on most of the surface it is neccessary for the decay length of the wave (i.e. the distance on which the wave amplitude decreases by a factor e) to be much smaller than the reference size. The decay length may easily be estimated by means of relation (7.69) using the well-known result for the damping decrement [8]:

$$\lambda_\Omega = 2\nu k^2 \tag{7.72}$$

Taking into account formula (7.72), one may write the general dispersion relation for a weak damping in the form

$$\lambda = \mathrm{i}\Omega_0(k) + 2\nu k^2 \tag{7.73}$$

where $\Omega_0(k)$ is determined by the dispersion relation (7.69). Assuming that the wave does not decrease with time but has an amplitude varying in space, we may set $\lambda = \mathrm{i}\Omega$ in formula (7.72), where Ω is the vibration frequency, and treat the wavenumber k as a complex one:

$$k = k_0 + \mathrm{i}k' \tag{7.74}$$

It is the value k' that determines the spatial decrement of damping. Assuming that $k' \ll k_0$, we obtain, from (7.73),

$$k' = \frac{4}{3}\frac{\nu k^3}{\Omega} \tag{7.75}$$

Now the condition of the small propagation length may be written as follows:

$$k'L \gg 1 \tag{7.76}$$

i.e.

$$\Omega \gg \Omega^*, \quad \Omega^* = \frac{\alpha}{\rho\nu L} \tag{7.77}$$

Condition (7.77) is relatively strong and requires high vibration frequencies in real situations. Thus, for example, for water at $L = 1\,\mathrm{cm}$ we get, from (7.77), $\Omega^* \approx 5 \times 10^3\,\mathrm{s}^{-1}$.

It is convenient to rewrite condition (7.77) in terms of the non-dimensional frequency $\omega = \Omega L^2/\nu$:

$$We \gg \omega$$

Since the high-frequency limit condition requires $\omega \gg 1$, this means that (7.77) is stronger than (7.71). Nevertheless, it might be satisfied, and this means that there indeed occur certain situations where the surface capillary waves are practically absent.

Therefore, if the vibration frequency is so large that conditions (7.71) and (7.77) are satisfied, the pulsation field may be found from the solution of the Laplace equation for the potential with the simple boundary condition (7.62).

2.4 Accounting for Capillary Waves

Now we discuss the situations where capillary waves are to be taken into account. There are several factors leading to this necessity. First, the frequency range in which the conditions of validity of the averaged approach are still satisfied but inequalities (7.71) and (7.77) are violated is interesting and important for many applications. Second, the waves propagating along the surface may generate the average flow not only in the wave zone but also in the whole fluid-filled volume. The mechanism of such generation and means for its description will be discussed in detail in the next section. Third, as the calculations show, sites of contact between the free surface and the rigid boundaries are the locations where the pulsation energy concentrates. If the angle between the free surface and rigid boundary is greater than $\pi/2$, this concentration may cause a singularity of the solution if the capillary waves are not taken into account. Capillary waves can be an effective mechanism of removal of excessive energy from the region of concentration.

Consideration of the capillary waves cannot be accomplished by just adding a capillary term to the normal stress balance condition. The point is that when the resonance conditions for the surface waves are satisfied the wave amplitude will increase. Moreover, in the absence of damping, this increase is unbounded. This implies that when considering the capillary waves the viscous effects must be taken into account. Since we assume, as before, the skin layer thickness to be much smaller than the reference size, the pulsation flow is (as discussed earlier) potential in the interior. Then the question under discussion is one of modification of the boundary conditions. It is important to emphasize that since the normal component of the fluid velocity entering the kinematic condition incorporates both potential and vortex components, it does not suffice just to take the viscous term into account in the boundary condition for the normal stress balance. The vortex component must be found in order to close the set of boundary conditions. The relevant calculation was done by Ruvinsky and Freidman [11, 12] while developing a theory which we are going to recall. Following mainly the ideas of these papers, we shall perform calculations in a different way, and not only for plane waves.

Let us present the pulsation field as the sum of potential and vortex components:

$$\widetilde{\mathbf{v}} = \nabla\Phi + \mathbf{u} \tag{7.78}$$

$$\widetilde{w} = \frac{\partial \Phi}{\partial z} + U \tag{7.79}$$

Here $\widetilde{\mathbf{v}}$ and $\widetilde{w}$ are the tangential and normal velocity components as in Section 1 and $\mathbf{u}$ and U are the tangential and normal components of the vortex part. Suppose that a local system of coordinates is chosen so that in the given point on the surface the z axis is directed along the vector normal to this surface. The potential part of the velocity is determined by the requirement

$$\rho\frac{\partial \Phi}{\partial t} = -\widetilde{P} \tag{7.80}$$

where $\widetilde{P}$ is the pulsation pressure. Since inside the vortex layer the linearized equation of motion reads

$$\rho \frac{\partial \widetilde{\mathbf{v}}}{\partial t} = -\nabla \widetilde{P} + \eta \Delta \widetilde{\mathbf{v}} \tag{7.81}$$

$$\rho \frac{\partial \widetilde{w}}{\partial t} = -\frac{\partial \widetilde{P}}{\partial z} + \eta \Delta \widetilde{w} \tag{7.82}$$

$$\frac{\partial \widetilde{w}}{\partial z} + \operatorname{div} \widetilde{\mathbf{v}} = 0 \tag{7.83}$$

then

$$\Delta \widetilde{P} = 0 \tag{7.84}$$

Relation (7.80) determines Φ with the accuracy of an arbitrary coordinate function. Thence one may assume that this potential also satisfies the Laplace equation:

$$\Delta \Phi = 0 \tag{7.85}$$

Let us write down the boundary conditions at $z = 0$ imposed on the solutions of equations (7.81) to (7.83). They are the condition of normal stress balance:

$$-\widetilde{P} + 2\eta \frac{\partial \widetilde{w}}{\partial z} - \alpha \widetilde{K} = 0 \tag{7.86}$$

the condition of vanishing tangential stress:

$$\frac{\partial \widetilde{\mathbf{v}}}{\partial z} = -\nabla \widetilde{w} \tag{7.87}$$

and the kinematic condition:

$$\frac{\partial \zeta}{\partial t} = \widetilde{w} \tag{7.88}$$

Here K denotes the pulsations of the surface curvature.

Using relation (7.80) and equation (7.85) we rewrite the equations of motion in terms of the vortex component of the velocity

$$\rho \frac{\partial \mathbf{u}}{\partial t} = \eta \Delta \mathbf{u} \tag{7.89}$$

$$\rho \frac{\partial U}{\partial t} = \eta \Delta U \tag{7.90}$$

$$\frac{\partial U}{\partial z} + \operatorname{div} \mathbf{u} = 0 \tag{7.91}$$

Having applied the operation div_2 to equation (7.87) and expressing $\mathrm{div}\,\widetilde{\mathbf{v}}$ from the continuity equation (7.83) we obtain:

$$\frac{\partial^2 \widetilde{w}}{\partial z^2} = \Delta_2 \widetilde{w} \qquad \text{at } z = 0 \tag{7.92}$$

Substituting here relation (7.79), we arrive at the boundary condition:

$$\frac{\partial^2 U}{\partial z^2} + \frac{\partial^3 \Phi}{\partial z^3} = \Delta_2 U + \Delta_2 \frac{\partial \Phi}{\partial z} \tag{7.93}$$

Using once more the equation for the potential we obtain

$$\frac{\partial^2 U}{\partial z^2} = \Delta_2 U + 2\Delta_2 \frac{\partial \Phi}{\partial z} \tag{7.94}$$

Writing equation (7.70) for the free surface and using (7.94), we find that

$$\frac{\partial U}{\partial t} = 2\nu \Delta_2 \left(U + \frac{\partial \Phi}{\partial z} \right) \tag{7.95}$$

It can be seen from (7.95) that the ratio of the vortex normal component of the velocity to the potential component is of the order δ^2/L^2. Since the right-hand side of (7.95) already contains the small factor ν, one may neglect U in comparison to $\partial U/\partial z$ and instead of (7.95) write

$$\frac{\partial U}{\partial t} = 2\,\nu \Delta_2 \frac{\partial \Phi}{\partial z} \tag{7.96}$$

On the same basis one may approximately replace $\partial \widetilde{w}/\partial z$ by

$$\frac{\partial^2 \Phi}{\partial z^2} = -\Delta_2 \Phi$$

in (7.86).

Collecting the obtained relations, we arrive at the complete set of boundary conditions for the pulsation field on the free surface:

$$\frac{\partial \Phi}{\partial t} - 2\,\nu \Delta_2 \Phi - \frac{\alpha}{\rho} \widetilde{K} = 0 \tag{7.97}$$

$$\frac{\partial \zeta}{\partial t} = \frac{\partial \Phi}{\partial z} + U \tag{7.98}$$

$$\frac{\partial U}{\partial t} = 2\nu \Delta_2 \frac{\partial \Phi}{\partial z} \tag{7.99}$$

These relations, together with the Laplace equation for the potential and boundary conditions on the rigid surfaces, determine the pulsation velocity field. It should be emphasized that we retain the terms containing viscosity in relations (7.97) to (7.99) while omitting the viscous terms in equations describing the pulsation velocity field in the interior. This is not a contradiction. As has already been noted, at high vibration frequencies the vorticity does not in practice spread from the boundary layer. Due to this, neglecting dissipation in the interior simply means that exponentially small terms are dropped.

3 Boundary Conditions for the Average Velocity on a Free Surface

It has already been mentioned that near a free surface there exists a dynamic boundary layer in which the vorticity of the pulsation flow differs from zero. Non-linear interaction of vorticity and velocity pulsations may generate the average flow. From this viewpoint, the phenomena near a free surface are similar to those near a rigid surface, but there are some important distinctions. The main one is that a rigid wall, contrary to a free surface, singles out a reference frame in which it is at rest. Because of that, one is able to express the average velocity on the external border of the boundary layer through the characteristics of the pulsation field (formula (7.58)). Near a free surface a uniform flow with an arbitrary velocity is always possible, so that one may hope just to determine the average velocity non-uniformities. In short, one sets the velocity on a rigid wall (the no-slip condition) and determines it on the external border of the skin layer. On a free surface one sets the shear rate (the condition of the tangential stress vanishing) and therefore has to find the shear rate outside the skin layer. The shear rate of the average flow was calculated by Longuet-Higgins [3] for the case of a monochromatic plane wave propagating along an infinite fluid surface. As it turned out, the effective shear rate, like the case of the Schlichting effect, is independent of viscosity. Usually, a one-dimensional consideration does not suffice to calculate the fluid entrainment by surface waves in the problems of thermovibrational convection. Investigation of a more general situation is required, which is the subject of the present section.

3.1 Deformed Coordinates

We are interested in the situations where the vibration amplitude (and consequently the amplitude of surface waves) is not necessarily small as compared to the skin layer thickness. Therefore, the boundary conditions on the true fluid surface cannot be shifted to the non-perturbed surface and one needs to use the deformed coordinate method in some form. However, certain difficulties arise concerning recalculation of the average values from the deformed coordinates back to ordinary ones. To avoid this, at the beginning we will not decompose the equations for pulsation and average components as we did in the case of the rigid boundary. To obtain the solution we are going to use the asymptotic matched expansion method.

Let us write down the complete set of equations and boundary conditions describing the fluid behavior in the layer near the free surface:

$$\frac{\partial \mathbf{v}}{\partial \tau} + \varepsilon S\left(\mathbf{v} \cdot \nabla v + w\frac{\partial \mathbf{v}}{\partial z}\right) = -\nabla p + \varepsilon^2 \frac{\partial^2 \mathbf{v}}{\partial z^2} + \varepsilon^2 \Delta \mathbf{v} \tag{7.100}$$

$$\frac{\partial w}{\partial \tau} + \varepsilon S\left(\mathbf{v} \cdot \nabla w + w\frac{\partial w}{\partial z}\right) = -\frac{\partial p}{\partial z} + \varepsilon^2 \frac{\partial^2 w}{\partial z^2} + \varepsilon^2 \Delta w \tag{7.101}$$

$$\frac{\partial w}{\partial z} + \operatorname{div} \mathbf{v} = 0 \tag{7.102}$$

On the free surface $z = \varepsilon S\zeta(x, y, \tau)$:

$$-p + \varepsilon^2 \sigma_{\mathrm{nn}} + Ca\, K = 0 \tag{7.103}$$

$$\sigma_{\mathrm{n}\tau} = 0 \tag{7.104}$$

$$\frac{\partial \zeta}{\partial \tau} + \varepsilon S\mathbf{v} \cdot \nabla \zeta = w \tag{7.105}$$

The equations and the boundary conditions are written in the dimensionless form. The following quantities are used to scale for, respectively, the velocity, time, pressure, length and displacement of the free surface: $b\Omega$, Ω^{-1}, $b\Omega^2\rho L$, L and b. The problem is characterized by the dimensionless parameters $\varepsilon = \delta/L$ (the skin layer thickness), $S = b/\delta = \sqrt{Re_{\mathrm{p}}}$ and the capillary parameter $Ca = \alpha/(b\Omega^2\rho L^2)$. The following notations are used: $\mathbf{v}$ for the projection of the velocity vector on the xy plane and w for the velocity projection on the z axis. The differentiation operators ∇ and the Laplacian act in the xy plane. The expressions for the normal and tangential components σ_{nn} and $\sigma_{\mathrm{n}\tau}$ of the viscous stress tensor in terms of $\mathbf{v}$, w and the operator ∇ will be derived below. In (7.103), K stands for the surface curvature, but in this item we would not need its explicit expression.

As known, the viscous stress tensor σ with the adopted units and in three-dimensional vector notation is written as

$$\sigma = \nabla \mathbf{v} + (\nabla \mathbf{v})^{\mathrm{T}} \tag{7.106}$$

Here the superscript T means the operation of tensor transposition, tensor products being written without indices. Separating explicitly the horizontal and vertical velocity components and the corresponding differentiation operators, as was done while presenting the set of equations, we can rewrite (7.106) as

$$\sigma = \nabla \mathbf{v} + (\nabla \mathbf{v})^{\mathrm{T}} + \nabla w \mathbf{j} + \mathbf{j} \nabla w + \mathbf{j} \frac{\partial \mathbf{v}}{\partial z} + \frac{\partial \mathbf{v}}{\partial z} \mathbf{j} + 2 \frac{\partial w}{\partial z} \mathbf{jj} \tag{7.107}$$

where $\mathbf{j}$ is the unit vector of the z axis.

The force acting upon a surface element with the normal vector $\mathbf{n}$ is $\sigma \cdot \mathbf{n}$. The vector normal to the surface determined by the equation $z = \varepsilon S \zeta$ is given by the formula

$$\mathbf{n} = \frac{\mathbf{j} - \varepsilon S \nabla \zeta}{\sqrt{1 + \varepsilon^2 S^2 (\nabla \zeta)^2}} \tag{7.108}$$

Restricting ourselves to the terms linear in $\nabla \zeta$ (one does not need higher accuracy) we obtain from (7.107) and (7.108) the following expression for the force acting on the free surface element:

$$\sigma \cdot \mathbf{n} = \nabla w + \frac{\partial \mathbf{v}}{\partial z} + 2 \frac{\partial w}{\partial z} \mathbf{j} - \varepsilon S \left(\nabla \mathbf{v} \cdot \nabla \zeta + \nabla \zeta \cdot \nabla \mathbf{v} + \mathbf{j} \nabla w \cdot \nabla \zeta + \mathbf{j} \frac{\partial \mathbf{v}}{\partial z} \cdot \nabla \zeta \right) \tag{7.109}$$

The normal stress may be calculated from (7.109) by contraction with the normal vector:

$$\sigma_{\mathrm{nn}} = \sigma : \mathbf{nn} = 2 \frac{\partial w}{\partial z} - 2 \varepsilon S \left(\nabla w \cdot \nabla \zeta - \frac{\partial \mathbf{v}}{\partial z} \cdot \nabla \zeta \right) \tag{7.110}$$

Multiplying (7.110) by $\mathbf{n}$ and subtracting it from (7.109) one finds the tangential stress vector:

$$\sigma_{\mathrm{n}\tau} = \nabla w + \frac{\partial \mathbf{v}}{\partial z} - \varepsilon S \left(\nabla \mathbf{v} \cdot \nabla \zeta + \nabla \zeta \cdot \nabla \mathbf{v} - 2 \nabla \zeta \frac{\partial w}{\partial z} - \mathbf{j} \nabla w \cdot \nabla \zeta - \mathbf{j} \frac{\partial \mathbf{v}}{\partial z} \cdot \nabla \zeta \right) \tag{7.111}$$

At the free surface the tangential stress vanishes. Moreover, it is sufficient to impose this condition only on the horizontal component $\sigma_{\mathrm{n}\tau}$. This yields the following form of the boundary conditions (7.104):

$$\nabla w + \frac{\partial \mathbf{v}}{\partial z} - \varepsilon S \left(\nabla \mathbf{v} \cdot \nabla \zeta + \nabla \zeta \cdot \nabla \mathbf{v} - 2 \nabla \zeta \frac{\partial w}{\partial z} \right) = 0 \tag{7.112}$$

Now we proceed to construct the solution inside the boundary layer. First of all, we shall introduce a new intrinsic variable instead of the z coordinate:

$$\xi = \frac{z - \varepsilon S \zeta(x, y, \tau)}{\varepsilon} \tag{7.113}$$

so that the boundary conditions are now to be imposed at $\xi = 0$. The variable ξ changes by a finite value upon receding from the boundary to the distance of the order of the skin layer thickness.

Since transformation (7.113) incorporates the horizontal coordinates and time, the differentiation operators in the equations and boundary conditions are to be changed accordingly:

$$\frac{\partial}{\partial z} = \frac{1}{\varepsilon}\frac{\partial}{\partial \xi}, \qquad \frac{\partial}{\partial \tau} \mapsto \frac{\partial}{\partial \tau} - S\frac{\partial \zeta}{\partial \tau}\frac{\partial}{\partial \xi}, \qquad \nabla \mapsto \nabla - S\nabla\zeta\frac{\partial}{\partial \xi} \tag{7.114}$$

This yields the following form of equations (7.100) to (7.102):

$$\frac{\partial \mathbf{v}}{\partial \tau} - S\frac{\partial \zeta}{\partial \tau}\frac{\partial \mathbf{v}}{\partial \xi} + S\left(\varepsilon \mathbf{v}\cdot\nabla\mathbf{v} + w\frac{\partial \mathbf{v}}{\partial \xi} - \varepsilon S\mathbf{v}\cdot\nabla\zeta\frac{\partial \mathbf{v}}{\partial \xi}\right)$$
$$= -\nabla p + S\nabla\zeta\frac{\partial p}{\partial \xi} + \frac{\partial^2 \mathbf{v}}{\partial \xi^2} + \varepsilon^2\widehat{\Delta}\mathbf{v} \tag{7.115}$$

$$\frac{\partial w}{\partial \tau} - S\frac{\partial \zeta}{\partial \tau}\frac{\partial w}{\partial \xi} + S\left(\varepsilon \mathbf{v}\cdot\nabla w + w\frac{\partial w}{\partial \xi} - \varepsilon S\mathbf{v}\cdot\nabla\zeta\frac{\partial w}{\partial \xi}\right)$$
$$= -\frac{1}{\varepsilon}\frac{\partial p}{\partial \xi} + \frac{\partial^2 w}{\partial \xi^2} + \varepsilon^2\widehat{\Delta}w \tag{7.116}$$

$$\frac{\partial w}{\partial \xi} + \varepsilon \operatorname{div} \mathbf{v} - \varepsilon S\nabla\zeta\cdot\frac{\partial \mathbf{v}}{\partial \xi} = 0 \tag{7.117}$$

where $\widehat{\Delta}$ is the Laplace operator transformed according to (7.114):

$$\widehat{\Delta} = \Delta - S\Delta\zeta\frac{\partial}{\partial \xi} - 2S\nabla\zeta\cdot\nabla\frac{\partial}{\partial \xi} + S^2(\nabla\zeta)^2\frac{\partial^2}{\partial \xi^2} \tag{7.118}$$

It is convenient to introduce a new W function by means of the relation

$$W = w - \varepsilon S\nabla\zeta\cdot\mathbf{v} \tag{7.119}$$

Thence the continuity equation (7.117) takes the form

$$\frac{\partial W}{\partial \xi} + \varepsilon \operatorname{div} \mathbf{v} = 0 \tag{7.120}$$

In the transformed coordinates, the boundary condition of the vanishing tangential stress (7.112) and the kinematic condition (7.105) look as follows:

$$\frac{1}{\varepsilon}\frac{\partial \mathbf{v}}{\partial \xi} + \nabla W + S\nabla\zeta\frac{\partial W}{\partial \xi} - \varepsilon S(\nabla\zeta\cdot\nabla\mathbf{v} - \nabla\nabla\zeta\cdot\mathbf{v}) = 0 \tag{7.121}$$

$$\frac{\partial \zeta}{\partial \tau} = W \tag{7.122}$$

3.2 Intrinsic Expansion

We seek the solution in the form of expansions with respect to the small parameter ε:

$$\begin{aligned} \mathbf{v} &= \mathbf{v}_0 + \varepsilon \mathbf{v}_1 + \cdots \\ w &= w_0 + \varepsilon w_1 + \cdots \\ \zeta &= \zeta_0 + \varepsilon \zeta_1 + \cdots \\ p &= p_0 + \varepsilon p_1 + \cdots \end{aligned} \tag{7.123}$$

It follows from equations (7.116) and (7.117) that w_0 and p_0 do not depend on ξ:

$$\frac{\partial w_0}{\partial \xi} = 0, \qquad \frac{\partial p_0}{\partial \xi} = 0 \tag{7.124}$$

Then the boundary condition (7.122) yields

$$w_0 = W_0 = \frac{\partial \zeta_0}{\partial \tau} \tag{7.125}$$

Substituting expansions (7.123) into (7.115) and using (7.124) and (7.125), one gets the equation for $\mathbf{v}_0$:

$$\frac{\partial \mathbf{v}_0}{\partial \tau} = -\nabla p_0 + \frac{\partial^2 \mathbf{v}_0}{\partial \xi^2} \tag{7.126}$$

with the boundary condition at $\xi = 0$:

$$\frac{\partial \mathbf{v}_0}{\partial \xi} = 0 \tag{7.127}$$

From (7.126) and (7.127) it follows that $\mathbf{v}_0$ does not depend on ξ as well. Assuming vibrations to be monochromatic, we present $\mathbf{v}_0$, ζ_0 and p_0 in the form

$$\begin{aligned} \mathbf{v}_0 &= \mathbf{U} e^{i\tau} + \mathbf{U}^* e^{-i\tau} \\ \zeta_0 &= H e^{i\tau} + H^* e^{-i\tau} \\ p_0 &= \Pi e^{i\tau} + \Pi^* e^{-i\tau} \end{aligned} \tag{7.128}$$

Moreover,

$$\mathbf{U} = -\nabla \Pi \tag{7.129}$$

In the next order we have the following equations for p_1 and W_1:

$$\frac{\partial W_1}{\partial \xi} + \text{div } \mathbf{v}_0 = 0 \tag{7.130}$$

$$\frac{\partial p_1}{\partial \xi} = -\frac{\partial w_0}{\partial \tau} \tag{7.131}$$

and for the horizontal velocity:

$$\frac{\partial \mathbf{v}_1}{\partial \tau} + S\mathbf{v}_0 \cdot \nabla \mathbf{v}_0 = -\nabla p_1 + S\nabla\zeta_0 \frac{\partial p_1}{\partial \xi} + \frac{\partial^2 \mathbf{v}_1}{\partial \xi^2} \tag{7.132}$$

with the boundary condition at $\xi = 0$:

$$\frac{\partial \mathbf{v}_1}{\partial \xi} + \nabla W_0 = 0 \tag{7.133}$$

Presenting p_1 in the form

$$p_1 = -\frac{\partial w_0}{\partial \tau}\xi - \frac{1}{2}S\mathbf{v}_0^2 + \frac{1}{2}S(\nabla\zeta_0)^2 + q_1 \tag{7.134}$$

where q_1 is a new unknown function independent of ξ, we rewrite (7.132) as

$$\frac{\partial \mathbf{v}_1}{\partial \tau} = -\nabla q_1 + \xi\nabla \frac{\partial w_0}{\partial \tau} + \frac{\partial^2 \mathbf{v}_1}{\partial \xi^2} \tag{7.135}$$

The solution of (7.135) satisfying the boundary condition (7.133) is given by the formula

$$\mathbf{v}_1 = \nabla H(\mathrm{i}\xi - 2\alpha \mathrm{e}^{\alpha\xi})\mathrm{e}^{\mathrm{i}\tau} - \nabla H^*(\mathrm{i}\xi + 2\alpha^* \mathrm{e}^{\alpha^*\xi})\mathrm{e}^{-\mathrm{i}\tau} \tag{7.136}$$

where $\alpha = (1 + \mathrm{i})/\sqrt{2}$.

Note that equation (7.135) and condition (7.133) do not determine $\mathbf{v}_1$ uniquely. To the solution (7.136) an arbitrary term depending on time and independent of ξ might be added. This uncertainty cannot be removed and reflects the uncertainty of the complete velocity mentioned at the beginning of this section. However, it is not at all essential for what follows.

Since in this section we are interested only in the average velocity, in the next order it is sufficient to write down the equations for the part of $\mathbf{v}_2$ independent of time:

$$\frac{\partial^2 \mathbf{v}_2}{\partial \xi^2} = -S\xi \overline{\frac{\partial \mathbf{v}_1}{\partial \xi} \text{ div } \mathbf{v}_0} + S\,\overline{\mathbf{v}_0 \cdot \nabla \mathbf{v}_1} + S\,\mathbf{v}_1 \cdot \nabla \mathbf{v}_0 + \nabla \overline{p}_2 - S\overline{\nabla\zeta_0 \frac{\partial p_2}{\partial \xi}} \tag{7.137}$$

Equation (7.116) yields

$$\frac{\partial p_2}{\partial \xi} = -\frac{\partial w_1}{\partial \tau} - S\mathbf{v}_0 \cdot \nabla w_0 \tag{7.138}$$

It can easily be seen that while substituting (7.138) into (7.137) all the terms that are linear in ξ cancel, i.e. the Van Dyke principle of minimal singularity [6] is satisfied. Otherwise it would be impossible to match the intrinsic expansion with the external one obtained under the assumption of a regular coordinate dependence of the solution in the exterior region.

The terms in (7.137) independent of ξ yield the terms in $\mathbf{v}_2$ quadratic with respect to ξ that do not affect the shear rate. Thus, in the right-hand side it is sufficient to retain the terms exponential in ξ:

$$\frac{\partial^2 \mathbf{v}_2}{\partial \xi^2} = 2\mathrm{i}S\nabla H \text{ div } \mathbf{U}^* \xi \mathrm{e}^{\alpha\xi} - 2\alpha S\nabla(\nabla H \cdot \mathbf{U}^*)\mathrm{e}^{\alpha\xi} + \text{c.c.} \tag{7.139}$$

from which, by integration over ξ, one gets

$$\frac{\partial \mathbf{v}_2}{\partial \xi} = 2\mathrm{i}S\nabla H \text{ div } \mathbf{U}^*(\alpha\xi - 1)\mathrm{e}^{\alpha\xi} - 2S\nabla(\nabla H \cdot \mathbf{U}^*)\mathrm{e}^{\alpha\xi} + \text{c.c.} + \mathbf{c} \tag{7.140}$$

Here $\mathbf{c}$ is the vector constant of integration that must be determined from the boundary condition following from (7.121) after time averaging:

$$\frac{\partial \mathbf{v}_2}{\partial \xi} = -S\overline{\nabla\zeta_0 \text{ div } \mathbf{v}_0} + S\overline{\nabla\zeta_0 \cdot \nabla\mathbf{v}_0} - S\overline{\nabla\nabla\zeta_0 \cdot \nabla\mathbf{v}_0} \tag{7.141}$$

Setting $\xi = 0$ in (7.140) and comparing with equation (7.141), we find that

$$\mathbf{c} = 3S\big(\overline{\nabla\zeta_0 \text{ div } \mathbf{v}_0} + \overline{\nabla\zeta_0 \cdot \nabla\mathbf{v}_0}\big) + S\overline{\mathbf{v}_0 \cdot \nabla\nabla\zeta_0} \tag{7.142}$$

3.3 Matching Conditions

Now we can commence matching the external and intrinsic expansions. The general scheme of this procedure is as follows. Let some values $y(z, \varepsilon)$ be presented by the external expansion

$$y^{\mathrm{o}} = Y_0(z) + \varepsilon Y_1(z) + \varepsilon^2 Y_2(z) + \cdots \tag{7.143}$$

and by the intrinsic expansion

$$y^{\mathrm{i}} = y_0(\xi) + \varepsilon y_1(\xi) + \varepsilon^2 y_2(\xi) + \cdots \tag{7.144}$$

where $\xi = (z - \varepsilon\eta)/\varepsilon$ and η is a function of time and the rest of the spatial coordinates.

We suppose the external expansion to be regular, so that $Y_0(z)$ and the others are differentiable with respect to z. The functions $y_0(\xi)$, $y_1(\xi), \ldots$ contain exponentially decreasing terms and polynomial terms whose power does not exceed the sequential number of the expansion term (the minimal singularity principle). Let us rewrite the two-term external expansion (7.143) in terms of the intrinsic coordinate, replacing z by $\varepsilon(\xi + \eta)$. Then we expand the result with respect to ε up to the second order, assuming ξ to be fixed. We shall obtain what is called 'the two-term intrinsic expansion of the two-term external expansion' [5]. It is convenient to write this result in terms of z:

$$(y^{\mathrm{o}})^{\mathrm{i}} = Y_0(0) + zY_0{}'(0) + \tfrac{1}{2}z^2 Y_0{}''(0) + \varepsilon[Y_1(0) + zY_1{}'(0)] + \varepsilon^2 Y_2(0) \quad (7.145)$$

Now we take the two-term intrinsic expansion (7.144), rewrite it in the terms of the external z coordinate and expand it with respect to ε up to the second order under fixed z. In this way we obtain 'the two-term external expansion of the two-term intrinsic expansion', which we denote as $(y^{\mathrm{i}})^{\mathrm{o}}$:

$$\begin{aligned}(y^{\mathrm{i}})^{\mathrm{o}} &= y_{00} + \varepsilon y_{10} + (z - \varepsilon\eta)y_{11} + \varepsilon^2 y_{20} + \varepsilon(z - \varepsilon\eta)y_{21} + (z - \varepsilon\eta)^2 y_{22} \\ &= y_{00} + zy_{11} + z^2 y_{22} + \varepsilon(y_{10} - \eta y_{11} + zy_{21} - 2\eta z y_{22}) \\ &\quad + \varepsilon^2(y_{20} - \eta y_{21} + \eta^2 y_{22}) \end{aligned} \quad (7.146)$$

where y_{jk} is the coefficient at the kth power in the polynomial part of $y_{\mathrm{i}}(\xi)$.

The Van Dyke principle of matching requires that $(y^{\mathrm{i}})^{\mathrm{o}}$ coincides with $(y^{\mathrm{o}})^{\mathrm{i}}$. Comparing the coefficients in (7.145) and (7.146) we obtain the matching conditions:

$$\begin{gathered} Y_0(0) = y_{00}, \qquad Y_0'(0) = y_{11}, \qquad Y_0''(0) = 2y_{22} \\ Y_1(0) = y_{10} - \eta y_{11}, \qquad Y_0'(0) = y_{21} - 2\eta y_{22}, \quad Y_2(0) = \eta^2 y_{22} - \eta y_{21} + y_{20} \end{gathered} \quad (7.147)$$

The polynomial parts of the intrinsic expansion terms being known, one may find the limiting values of the external expansion terms and their derivatives.

3.4 Boundary Conditions for the External Solution

In our case the role of the zeroth term of the external expansion is played by the pulsation field, and the first-order terms contain the average flow we are interested in. First, we apply the matching conditions (7.147) to find the boundary condition for the normal component of the average velocity. On the basis of (7.119) and (7.130) one may write

$$w_1 = S\nabla\zeta_0 \cdot \mathbf{v}_0 - \xi \ \mathrm{div}\ \mathbf{v}_0 \quad (7.148)$$

so that for the z component of the velocity we have

$$y_{10} = S\nabla\zeta_0 \cdot \mathbf{v}_0, \qquad y_{11} = -\operatorname{div} \mathbf{v}_0 \tag{7.149}$$

Substituting (7.149) into (7.147), where η should be taken as $S\zeta_0$, and averaging over time, we see that the average velocity component $\mathbf{w}$ normal to the free surface satisfies the boundary condition

$$\overline{w}|_{z=0} = \varepsilon S \operatorname{div} \overline{(\zeta_0 \mathbf{v}_0)} \tag{7.150}$$

The deviation of the average normal velocity from zero in the general case where the boundary is at rest on the average has a simple meaning. As was shown in Section 3 in Chapter 6, it is the sum of the average velocity and the pulsational transport vector that plays the role of specific density for the mass flux. One may easily prove that as a direct consequence of (7.150), the average mass flux towards the free surface vanishes.

Proceeding to consideration of the tangential components of the average velocity, we recall that one cannot calculate the tangential velocity value itself by means of the pulsation characteristics since for the given pulsation field the tangential velocity may be arbitrary. However, we have all the necessary information to calculate the average shear rate. Indeed, in the matching condition (7.147) the vector $\mathbf{c}$ determined by formula (7.142) acts as y_{21}, and the value of y_{21} may be found using the external solution for the pulsation field:

$$2y_{22} = \left.\frac{\partial^2 \mathbf{v}_0}{\partial z^2}\right|_{z=0} = \nabla \left.\frac{\partial w}{\partial z}\right|_{z=0} = -\nabla \operatorname{div} \mathbf{v}_0 \tag{7.151}$$

where we have taken into account potentiality and solenoidality of the pulsation field.

Finally we arrive at the following expression for the average shear rate:

$$\left.\frac{\partial \overline{\mathbf{v}}}{\partial z}\right|_{z=0} = \varepsilon S\left(3\overline{\nabla\zeta_0 \operatorname{div} \mathbf{v}_0} + 3\overline{\nabla\zeta_0 \cdot \nabla\mathbf{v}_0} + \overline{\mathbf{v}_0 \cdot \nabla\nabla\zeta_0} + \overline{\zeta_0 \nabla \operatorname{div} \mathbf{v}_0}\right) \tag{7.152}$$

Let us rewrite once more the formulas for the average normal velocity and the shear rate going back to the dimensional values and omitting the index '0':

$$\overline{w} = \operatorname{div} \left(\overline{\zeta \mathbf{v}}\right) \tag{7.153}$$

$$\frac{\partial \overline{\mathbf{v}}}{\partial z} = 3\overline{\nabla\zeta \operatorname{div} \mathbf{v}} + 3\overline{\nabla\zeta \cdot \nabla\mathbf{v}} + \overline{\mathbf{v} \cdot \nabla\nabla\zeta} + \overline{\zeta\nabla \operatorname{div} \mathbf{v}} \tag{7.154}$$

These formulas simplify significantly for the case of two-dimensional flows. Using representation (7.128) and bearing in mind the fact that $\mathbf{U}$ has only one component, we obtain, for the average vorticity $\overline{\varphi}$,

$$\overline{\varphi} = \frac{\partial \overline{v}}{\partial z} - \frac{\partial \overline{w}}{\partial x} = 8 \operatorname{Re}\left(\frac{\partial H}{\partial x}\frac{\partial U^*}{\partial x}\right) \qquad (7.155)$$

which coincides with the results of Longuet-Higgins [3].

It is interesting to note that these formulas do not contain any parameters at all, contrary to the formula for the average velocity on the rigid wall which includes frequency. This enables formulas (7.153) and (7.154) to be used in the case of non-monochromatic vibrations.

References

1. Lord Rayleigh. On the circulation of air observed in Kundt's tubes, and on some allied acoustical problems, *Phil. Trans. Roy. Soc. London*, 1883, **A175**, 1–21.
2. Schlichting, H. Berechnung ebener periodischer Grenzschichtstromungen., *Z. Phys.*, 1932, **33**, 327.
3. Longuet-Higgins, M.S. Mass transport in water waves, *Phil. Trans. Roy. Soc. London*, 1953, **A245**, 535–81.
4. Nyborg, W.L. Acoustic streaming, in *Physical Acoustics* (ed. W.P. Mason), Academic Press, 1965.
5. Nayfeh, A.H. *Introduction to Perturbation Techniques*, John Wiley, Chichester, 1981.
6. Van Dyke, M. *Perturbation Methods in Fluid Mechanics*, Academic Press, 1964.
7. Batchelor, G.K. *An Introduction to Fluid Dynamics*, Cambridge University Press, 1970.
8. Landau, L.D. and E.M. Lifshitz. *Fluid Mechanics*, Pergamon Press, Oxford, 1987.
9. Lyubimov, D.V., M.V. Savvina and A.A. Cherepanov. On quasi-equilibrium shape of a free fluid surface in modulated gravity field, in *Problems of Hydrodynamics and Heat/Mass Transfer with Free Boundaries*, Novosibirsk, 1987, pp. 9–17.
10. Briskman, V.A. Vibration-thermocapillary convection and stability, in *Proc. of the First Int. Symp. on Hydromechanics and Heat/Mass Transfer in Microgravity*, Perm, Moscow, 1991, Gordon and Breach, London, 1992, pp. 111–19.
11. Ruvinsky, K.D. and G.I. Freidman. The fine structure of strong gravity-capillary waves, in *Nonlinear Waves: Structures and Bifurcations* (eds. A.V.Gaponov-Grekhov and M.I. Rabinovich), Nauka, Moscow, 1987, pp. 304–26.
12. Ruvinsky, K.D., F.I. Feldstein and G.I. Freidman. Numerical simulations of the quasi-stationary stage of ripple excitation by steep gravity-capillary waves, *J. Fluid Mech.*, 1991, **30**, 339–53.

8 The Second-Order Effects

It has already been mentioned in Chapter 6 that in cases where the isothermal pulsation field (*I-field*) is uniform, to detect the effect of vibrations on the average behavior one needs to take into account the part of the pulsation field that is produced by non-uniform heating (*T-field*). In this chapter we analyze these problems in more detail, paying particular attention to the validity ranges of different limiting cases and to the boundary conditions to be used.

While deriving the equations for thermovibrational convection, in the case of arbitrary vibrations (Sections 2 and 3 in Chapter 6) it was shown that in a general case the major contribution to the vibrational force is due to the interaction between the density non-uniformities and the non-uniformities of the isothermal field of the pulsation velocity. However, there exists a number of situations where the pulsation field, having been evaluated disregarding weak density non-uniformities, turns out to be uniform itself. In these cases the thermovibrational effect of first order vanishes and one needs to consider the effects of second order related to the influence of non-isothermality on the pulsation field.

The governing equations for thermovibrational convection in the case of translational vibrations of a vessel completely filled with a fluid were discussed in Chapter 1. In Section 3 in Chapter 6 it was shown how these equations may be derived from the general equations for the average flow of a non-uniform fluid under arbitrary vibrations. Here we discuss the possible existence of intermediate cases when both linear and quadratic (with respect to the non-isothermality parameter effect) terms work simultaneously, and then analyze the situations where the first-order effects prove to be small.

Therefore we consider the situation where the isothermal pulsation field is nearly uniform, so that the effects of the first order are small, and it is necessary to take into account the second-order effects. General equations describing the average flow of an isothermally incompressible non-uniform fluid (6.62) and (6.63) remain valid as well as the equations for the pulsation field (6.41) to (6.43).

1 Closed Cavity: Linear Vibrations

Let us first discuss the most important particular case where a closed cavity completely filled with a fluid performs vibrations in a fixed direction without changing its own shape and orientation. In this case the I-field is uniform and is determined by the following expression:

$$\widetilde{\mathbf{v}}_0 = b\Omega \cos \Omega t \mathbf{k} \tag{8.1}$$

where $\mathbf{k}$ is the unit vector along the axis of vibrations. Taking into account a weak non-isothermality yields, as shown in Section 2 in Chapter 6, the equations for the T-field in the form

$$\begin{aligned} \operatorname{curl} \mathbf{w} &= \beta \nabla T \times \mathbf{k} \\ \operatorname{div} \mathbf{w} &= 0 \end{aligned} \tag{8.2}$$

where it is assumed that the T-field is written as

$$\widetilde{\mathbf{v}}_1 = b\Omega \mathbf{W} \cos \Omega t \tag{8.3}$$

On the rigid boundaries of the cavity, the normal component of the vector $\mathbf{W}$ should vanish:

$$W_{\mathrm{n}} = 0 \tag{8.4}$$

It follows from (8.2) that the field $\mathbf{W}$ is of order $\beta\theta$, so that the T-field is much smaller than the vibration velocity. The T-field has a simple physical meaning—it is the pulsation field in the reference frame of the oscillating cavity. The smallness of the T-field means that, to find it, the conventional Boussinesq approximation is sufficient. As a matter of fact, equations (8.2) were obtained for the first time [1] right in the framework of the Boussinesq approximation.

Let us discuss the equations for the average velocity and temperature. Note, first, that the vibrational force in the equation of fluid motion takes the form found in Section 3 in Chapter 6:

$$-\tfrac{1}{2}\beta b^2 \Omega^2 (\mathbf{W} \cdot \mathbf{k}) \nabla T \tag{8.5}$$

Consider now the pulsational transport phenomena. As shown in Section 3 in Chapter 6, pulsational transport is described by the vector field $\mathbf{S}$, which is defined in the reference frame of the cavity by the following expression:

$$\mathbf{S} = \overline{\mathbf{q} \cdot \nabla \widetilde{\mathbf{v}}_1}, \qquad \frac{\partial \mathbf{q}}{\partial t} = \widetilde{\mathbf{v}}_1 \tag{8.6}$$

For linear vibrations (8.3), the phase of oscillations is uniform, and from (8.6) it follows that $\mathbf{S}$ vanishes.

If oscillations are translational, i.e. the orientation of the vessel does not change during the period of vibrations, but not linear (for instance, circular polarization of vibrations), the phase is non-uniform and one needs to estimate the contribution of pulsational transport to the total transport of heat and average vorticity. From (8.6) the order of magnitude of S may be estimated as

$$\frac{\tilde{v}^2}{\Omega L} \sim \frac{b^2 \Omega \beta^2 \theta^2}{L} \tag{8.7}$$

Comparing (8.7) with the reference viscous scale of the average velocity ν/L, we obtain the dimensionless parameter characterizing the pulsation transport contribution:

$$R_s = \frac{b^2 \Omega \beta^2 \theta^2}{\nu} \tag{8.8}$$

At the same time, the governing parameter for thermovibrational convection is

$$Gr_v = \frac{b^2 \Omega^2 \beta^2 \theta^2 L^2}{\nu^2} \tag{8.9}$$

The ratio of these parameters,

$$\frac{R_s}{Gr_v} = \frac{\nu}{\Omega L^2} = \left(\frac{\delta}{L}\right)^2 \ll 1 \tag{8.10}$$

is small due to the assumed smallness of the skin-layer thickness. This means that the pulsation transport contribution to the total average transport should be neglected. Thus, the equations of motion and heat transfer take the same form as those obtained in the Boussinesq approximation.

As for the boundary conditions, according to the general formula (7.42), the tangential component may be estimated using (8.7). Then, due to (8.10), for the average velocity the ordinary no-slip conditions may be accepted.

Besides the cases where the non-uniformity of the I-field is comparable with the vibration velocity (the first-order theory with respect to the Boussinesq parameter $\varepsilon = \beta\theta$) and the cases when the non-uniformity of the I-field identically vanishes (the theory of the second order with respect to ε), there are, of course, situations where the non-uniformities of the I-field differ from zero, but are small.

Particular examples of such situations are considered in the two following sections. Here we will put forward some general remarks.

Let the non-uniformity of the I-field be related to some additional small parameter ε_i, so that if both ε and ε_i equal zero, the pulsation field is uniform. We present its amplitude $\mathbf{V}$ in the form

$$\mathbf{V} = \mathbf{k} + \varepsilon \mathbf{W}_T + \varepsilon_i \mathbf{W}_i \tag{8.11}$$

Substituting (8.11) into general equations of the pulsation field obtained in Section 2 in Chapter 6, we have

$$\operatorname{curl} \mathbf{W}_{\mathrm{T}} = \nabla\vartheta \times \mathbf{k}, \qquad \operatorname{div} \mathbf{W}_{\mathrm{T}} = 0 \tag{8.12}$$

$$\operatorname{curl} \mathbf{W}_{\mathrm{i}} = 0, \qquad \operatorname{div} \mathbf{W}_{\mathrm{i}} = 0 \tag{8.13}$$

with the boundary conditions uniform for $\mathbf{W}_{\mathrm{T}}$ and non-uniform for $\mathbf{W}_{\mathrm{i}}$. The actual form of the boundary conditions is determined by the vibration law and is insignificant here.

The thermovibrational force in the general equations (Section 3 in Chapter 6) has the form

$$\tfrac{1}{4} b^2 \Omega^2 \mathbf{V}^2 \nabla\rho \tag{8.14}$$

Substituting $\mathbf{V}$ from (8.11) and the equation of state in the form $\rho = 1 - \varepsilon\vartheta$ into (8.14) and omitting the gradient terms, with second-order accuracy one obtains the thermovibrational force:

$$-\tfrac{1}{2} b^2 \Omega^2 [\varepsilon^2 (\mathbf{k} \cdot \mathbf{W}_{\mathrm{T}}) + \varepsilon\varepsilon_{\mathrm{i}} (\mathbf{k} \cdot \mathbf{W}_{\mathrm{i}})] \nabla\vartheta \tag{8.15}$$

Thus, the thermovibrational force comprises two terms being of the second order with respect to the small parameters. The first one, proportional to ε^2, is the ordinary thermovibrational force of the second-order theory. The second one describes the interaction between the non-uniformities of the I-field and temperature non-uniformities. Note that the second term is linear in the Boussinesq parameter, which leads to the change of symmetry type of equation in comparison with the case $\varepsilon_{\mathrm{i}} = 0$.

The two above-discussed terms in the thermovibrational force (8.15) cause two dimensionless parameters. One may choose, for example, $Gr_{\mathrm{v}} = b^2\Omega^2\beta^2\theta^2L^2/(2\nu^2)$ and $Gr_{\mathrm{vi}} = b^2\Omega^2\beta\theta\varepsilon_{\mathrm{i}}L^2/(2\nu^2)$.

2 Closed Cavity: Swing Vibrations

In this section we consider the behavior of a fluid in a closed rigid-boundary cavity subject to angular swings about a fixed z axis so that each point of the boundary with the coordinates $\mathbf{R}(t)$ moves according to

$$\frac{\mathrm{d}\mathbf{R}}{\mathrm{d}t} = \omega(t)\mathbf{k} \times \mathbf{R} \tag{8.16}$$

Here $\mathbf{k}$ is the unit vector of the swing axis. The origin of the coordinate system lies on the axis $\mathbf{k}$ and $\omega(t) = \alpha\Omega\cos\Omega t$, where α is the swing amplitude (see Fig. 157). In a general case, the vibration law (8.16) gives rise to a non-uniform I-field. For example, in [2] an exact solution for the I-field is found for the case

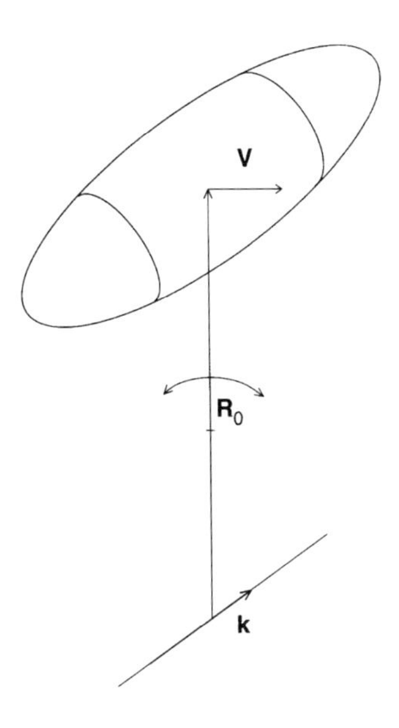

Fig. 157 Relative position of the fluid-filled cavity and the axis of the flow **k**. **V** is the pulsation velocity in the case of a body of revolution

of swing vibrations of a cylinder (its cross-section being an ellipse or an equilateral triangle) where the swing axis coincides with the symmetry axis of the cylinder. To describe thermovibrational convection in similar situations, one needs to use the general approach developed in Chapters 6 and 7. Moreover, the I-field is sufficient to describe the pulsations.

There are, however, two important cases when the I-field is uniform or nearly uniform under swing vibrations, so that it is necessary to evaluate the T-field. The first case is concerned with vanishing of the I-field non-uniformity for a special cavity shape. The second one is the case of a large arm of swings, so that the vibrations are nearly linear. Let us look at these cases in more detail, beginning with the first one.

2.1 I-Field Uniformity for a Special Cavity Shape

First we shall find the possible cavity shapes for which the I-field is uniform under swing vibrations of small amplitude. We seek the pulsation field in the form

$$\widetilde{\mathbf{v}} = \omega(t)\mathbf{V} \tag{8.17}$$

where $\mathbf{V}$ is a constant vector. On the cavity boundary $\mathbf{V}$ must satisfy the condition

$$\mathbf{V} \cdot \mathbf{n} = \mathbf{n} \cdot (\mathbf{k} \times \mathbf{R}) \tag{8.18}$$

where $\mathbf{n}$ is the surface normal vector.

It is convenient to split the vectors $\mathbf{R}$ and $\mathbf{V}$ into components parallel and orthogonal to the $\mathbf{k}$ axis:

$$\begin{aligned} \mathbf{R} &= z\mathbf{k} + \mathbf{r}, \qquad \mathbf{r} \cdot \mathbf{k} = 0 \\ \mathbf{V} &= V\mathbf{k} + \mathbf{v}, \qquad \mathbf{v} \cdot \mathbf{k} = 0 \end{aligned} \tag{8.19}$$

The surface bounding the region will be sought in the form

$$z = \zeta(\mathbf{r}) \tag{8.20}$$

The normal vector for the surface (8.20) may be written with the accuracy of an insignificant factor in the form

$$\mathbf{n} = \mathbf{k} - \nabla\zeta \tag{8.21}$$

Substituting (8.19) to (8.21) into (8.18) we obtain the equation for the surface:

$$V - \mathbf{v} \cdot \nabla\zeta = -\nabla\zeta \cdot (\mathbf{k} \times \mathbf{r}) \tag{8.22}$$

To simplify equation (8.22), we shift the origin of the coordinate system:

$$\mathbf{r} = \mathbf{r}' + \mathbf{v} \times \mathbf{k} \tag{8.23}$$

Then (8.22) takes the form

$$V = -\nabla\zeta \cdot (\mathbf{k} \times \mathbf{r}') = \mathbf{k} \cdot (\mathbf{r}' \times \nabla\zeta) \tag{8.24}$$

The operator $\mathbf{k} \cdot (\mathbf{r}' \times \nabla)$ is the operator of rotation (angular momentum) about the axis parallel to $\mathbf{k}$. Thus, in the polar coordinates r' and φ equation (8.24) reads as

$$V = \frac{\partial\zeta}{\partial\varphi}$$

For a non-zero V we obtain a spiral surface. Therefore, the closure condition will require $V = 0$. This means that ζ does not depend on φ, but is an arbitrary function of z.

Thus we have shown that a uniform I-field takes place under swing vibrations if a vessel has the shape of a rotation body with the axis parallel to the swing axis (Fig. 157). In this case the I-field is the same as under linear vibrations and is defined by the expression

$$\widetilde{\mathbf{v}} = \omega(t)\mathbf{k} \times \mathbf{R}_0$$

where $\mathbf{R}_0$ is the vector connecting the swing axis with the symmetry axis of the cavity. Under these conditions, the thermovibrational effect of the first order vanishes, and one should use the conventional equations with the boundary conditions from the second-order theory. If the cavity shape differs only weakly from the rotation body, then the I-field will be weakly non-uniform and one can simultaneously take into account the first- and second-order effects. This is done in the same way as considered in the preceding section.

2.2 The Case of a Large Arm of Swings

Let us consider now the case where the swing arm R_0 by far exceeds the size L of the fluid-filled cavity (Fig. 158). In this case the cavity moves almost straight forward, so that the I-field should be nearly uniform. Let us prove that and obtain the equations describing thermovibrational convection in this situation.

The general problem for the pulsation velocity field described in Chapter 6 takes the following form for the case of swing vibrations:

$$\operatorname{curl} \rho \mathbf{V} = 0 \tag{8.25}$$

$$\operatorname{div} \mathbf{V} = 0 \tag{8.26}$$

$$\mathbf{n} \cdot \mathbf{V}|_S = \mathbf{n} \cdot (\mathbf{k} \times \mathbf{R}) \tag{8.27}$$

where $\mathbf{V}$ is the pulsation velocity amplitude determined by the formula

$$\widetilde{\mathbf{v}} = \alpha_0 \, \Omega \mathbf{V} \cos \Omega t \tag{8.28}$$

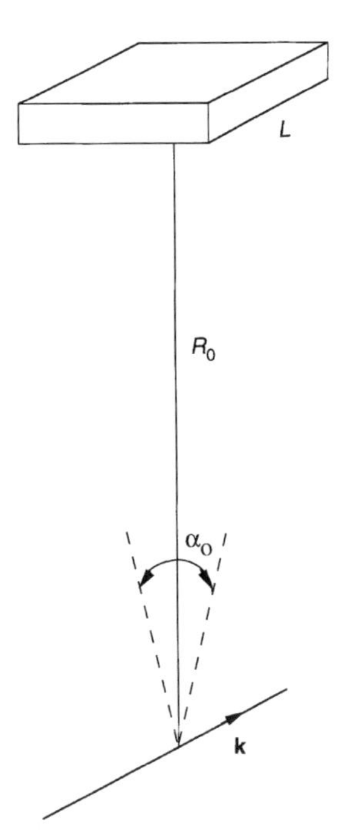

Fig. 158 Large arm of swings. The effects of the first order with respect to $\beta\theta$ are small to the extent of the smallness of the relative size of the domain

α_0 is the amplitude of angular swings, $\mathbf{n}$ is the normal vector at the surface bounding the fluid and $\mathbf{k}$ is the unit vector of the swing axis on which the origin of the coordinate system lies.

Let us shift the origin of the coordinate system to a point $\mathbf{R}_0$ situated somewhere in the vicinity of the oscillating cavity:

$$\mathbf{R} = \mathbf{R}_0 + \mathbf{r} \tag{8.29}$$

so that the value $r = |\mathbf{r}|$ is of the order of the cavity size L and $R_0 = |\mathbf{R}_0|$ is the swing arm. Now we extract from $\mathbf{V}$ the part $\mathbf{V}_0$ corresponding to linear vibrations with the amplitude $\alpha_0 R_0$:

$$\mathbf{V} = \mathbf{V}_0 + \mathbf{W}, \qquad \mathbf{V}_0 = \mathbf{k} \times \mathbf{R}_0 \tag{8.30}$$

The problem (8.25) to (8.27) transforms to

$$\nabla\rho \times \mathbf{V}_0 + \operatorname{curl} \rho\mathbf{W} = 0 \tag{8.31}$$

$$\operatorname{div} \mathbf{W} = 0 \tag{8.32}$$

$$\mathbf{n} \cdot \mathbf{W}|_S = \mathbf{n} \cdot (\mathbf{k} \times \mathbf{r}) \tag{8.33}$$

It can be seen from (8.31) to (8.33) that the isothermal part of $\mathbf{W}$ (the I-field) is of the order L and the part related to non-isothermality (the T-field) has the order $\beta\theta V_0$. Thus, in the case of a large swing arm, where $L \ll R_0 = |\mathbf{V}_0|$, $\mathbf{W}$ proves to be much smaller than the uniform part $\mathbf{V}_0$.

We linearize equation (8.31) with respect to $\beta\theta$ and the vibration force with respect to $\mathbf{W}$. Then we extract the I-field $\mathbf{W}_\mathrm{i}$ and T-field $\mathbf{W}_\mathrm{T}$ from $\mathbf{W}$. Thus we obtain the following set of equations to describe the thermovibrational convection under large-arm swing vibrations:

$$\frac{\partial \mathbf{v}}{\partial t} + \mathbf{v} \cdot \nabla\mathbf{v} - (G_\mathrm{v}\mathbf{W}_\mathrm{i} + G_\mathrm{k}\mathbf{W}_\mathrm{T}) \cdot \mathbf{j}\nabla T = -\nabla p + \Delta\mathbf{v} + Gr\, T\mathbf{j} \tag{8.34}$$

$$\frac{\partial T}{\partial t} + \mathbf{v} \cdot \nabla T = \frac{1}{Pr}\Delta T \tag{8.35}$$

$$\operatorname{div} \mathbf{v} = 0 \tag{8.36}$$

$$\operatorname{curl} \mathbf{W}_\mathrm{i} = 0, \qquad \operatorname{div} \mathbf{W}_\mathrm{i} = 0 \tag{8.37}$$

$$\operatorname{curl} \mathbf{W}_\mathrm{T} = \nabla T \times \mathbf{j}, \qquad \operatorname{div} \mathbf{W}_\mathrm{T} = 0 \tag{8.38}$$

At the cavity boundaries the following conditions for the pulsation fields are satisfied:

$$\mathbf{n} \cdot \mathbf{W}_\mathrm{i} = \mathbf{n} \cdot (\mathbf{k} \times \mathbf{r}) \tag{8.39}$$

$$\mathbf{n} \cdot \mathbf{W}_\mathrm{T} = 0 \tag{8.40}$$

Here ν/L, L^2/ν, θ, L, $\rho_0\nu^2/L$, L and $R_0\beta\theta$ are taken as the scales for the average velocity, time, temperature, length, pressure, W_{i} and W_{T}, respectively, and the notation $\mathbf{j} = \mathbf{k} \times \mathbf{R}_0/R_0 = \mathbf{V}_0/R_0$ is introduced. Besides the Grashof number Gr and the Prandtl number Pr, the set of equations (8.34) to (8.38) contains the dimensionless parameters

$$G_{\mathrm{v}} = \frac{b^2\Omega^2L^2\beta\theta}{\nu^2}\frac{L}{R_0} \qquad \text{and} \qquad G_{\mathrm{k}} = \frac{b^2\Omega^2L^2\beta^2\theta^2}{\nu^2} \tag{8.41}$$

where $b = \alpha R_0$.

As is usual in the treatment of second-order effects, the set (8.34) to (8.40) may be derived by applying the averaging method to the equations of convection in the Boussinesq approximation written in the reference frame of the swinging cavity. The corresponding calculations were performed in [3] where the obtained set differs from (8.34) to (8.40) only in its form. In the same paper the particular case of a cavity shape—an infinite plane layer orthogonal to $\mathbf{R}_0$—was considered. Of course, in this case it was implied that all the dimensions of the cavity are much smaller than R_0. This is the assumption that the size of the cavity in one direction is much smaller than in the two others, which allows one to call it a layer. In [3] the linear problem of the quasi-equilibrium stability in such a layer was shown to be equivalent to the problem of linear vibrations under a static gravity field. Let us show that such equivalence holds for an arbitrary orientation of the layer and for a non-linear problem as well.

Consider a layer formed by two parallel planes with the normal vector $\mathbf{n}$. One may prove by direct substitution that the solution of the problem presented by (8.37) and (8.39) is

$$\mathbf{W}_{\mathrm{i}} = \mathbf{q}(\mathbf{r}\cdot\mathbf{n}) + \mathbf{n}(\mathbf{r}\cdot\mathbf{q}) \tag{8.42}$$

where $\mathbf{q} = \mathbf{n} \times \mathbf{k}$. The I-field contribution to the vibration force is determined by the expression

$$-G_{\mathrm{v}}(\mathbf{W}_{\mathrm{i}}\cdot\mathbf{j})\nabla T = G_{\mathrm{v}}T\nabla(\mathbf{W}_{\mathrm{i}}\cdot\mathbf{j}) - G_{\mathrm{v}}\nabla(T\mathbf{W}_{\mathrm{i}}\cdot\mathbf{j}) \tag{8.43}$$

After omitting insignificant gradient terms, we find that the vibrational force determined by the I-field is equivalent to the buoyancy force in a static gravity field, the strength of which (in appropriate units) equals

$$\nabla(\mathbf{W}_{\mathrm{i}}\cdot\mathbf{j}) \tag{8.44}$$

Since the field $\mathbf{W}_{\mathrm{i}}$ determined by (8.42) is linear with respect to coordinates, expression (8.44) does not contain the coordinate dependence, i.e. the effective gravitational field proves to be uniform.

It should be emphasized that this conclusion does not depend on the layer orientation and does not concern the non-linear terms in the equation of motion. Thus, in cases where the first-order effects vanish under swing vibrations (i.e. when a fluid-filled cavity has the shape of a rotation body with the axis parallel to the swing axis or when the cavity dimensions are small in comparison to the swing arm) one should take into account the second-order effects with respect to the Boussinesq parameter $\beta\theta$ and with respect to an additional small parameter which might be, for example, the ratio of the cavity dimensions to the swing arm. As in other cases of the second-order theory, the governing equations may be obtained either from the general equations of Chapter 6 or by averaging the Boussinesq equations written in the proper reference frame of the cavity.

3 Weakly Deformable Free Surface

Let the oscillating cavity be filled just partially. With a free surface, even under linear vibrations, the I-field, generally speaking, will be non-uniform, due to the free surface deformability. In this section we consider the possible situations where the surface deformations lead only to a weak non-uniformity of the I-field. The first case is where vibrations are orthogonal to the free surface. It is obvious that for this situation in the reference frame of the oscillating vessel the isothermal fluid may remain quiescent with respect to the vessel. Of course, such a state may turn out to be unstable with respect to the parametric excitation of waves on the free surface. If, however, the vibration frequency does not satisfy the parametric resonance conditions, the I-field will be uniform and the first-order effects will vanish. In this case the equations of the second-order theory are valid. Interaction of the corresponding second-order thermovibrational convection with the thermocapillary convection generated by the Marangoni effect on the free surface is studied in papers [4] and [5].

3.1 Suppression of Pulsation Deformations by Gravity

Let us consider the possibility of suppression of the I-field non-uniformities in the situations where the condition of orthogonality of vibrations to the free surface is not satisfied. First, we discuss the possibility of suppressing the free surface deformations by the gravitational field. Consider a free surface displacement from the horizontal position which is described by the equation $z = \zeta(x, y)$. Extracting the hydrostatic part from the pressure, we present the gravity force contribution to the normal stress on the free surface in the form $\rho g \zeta$. Comparing this to the pulsation pressure which has the order of $b\Omega^2 L$, (L being the reference size), we obtain the estimation for the order of magnitude of the free surface deformations under the joint action of the gravity field and vibrations

$$\zeta \sim \frac{b\Omega^2 L}{g} \tag{8.45}$$

Hence, for the normal velocity of the free surface points,

$$\frac{\partial \zeta}{\partial t} \sim \frac{b\Omega^3 L}{g} \tag{8.46}$$

To be able to neglect deformations, the value determined by (8.46) must be much smaller than the reference vibration velocity $b\Omega$. This is so if

$$\frac{\Omega^2 L}{g} \ll 1 \tag{8.47}$$

If the condition (8.47) holds, the normal pulsation velocities on the free surface are small and may be neglected in comparison with the pulsation velocities in the fluid interior. This enables the impermeability condition to be imposed on the pulsation field on the free surface.

The physical meaning of condition (8.47) is that the vibration frequency is much smaller than the eigenfrequency of the surface oscillations in the gravity field. In this sense, condition (8.47) is similar to the non-acoustic limit condition adopted in this book (the acoustic wavelength at the vibration frequency is much greater than the reference size or, otherwise, the vibration frequency is small compared to that of the acoustic oscillations with the wavelength $\sim L$). Whereas the non-acoustic approximation is justified in a majority of practically important situations, this is not the case for condition (8.47). As we are going to show, condition (8.47) implies a gravity field that is so strong that the thermal buoyancy convection completely suppresses the thermovibrational one for all the vibration amplitudes that allow the averaged description.

We have already mentioned that the intensity of the thermovibrational convection is determined by the dimensionless parameter $Gr_v = b^2\Omega^2 L^2 \beta\theta/\nu^2$ for the case of the first-order effects and by $\widetilde{Gr_v} = Gr_v \beta\theta$ for the case of the second-order effects. The governing parameter for the gravitational convection is the Grashof number $Gr = g\beta\theta L^3/\nu^2$. Apparently, their ratio is

$$\frac{Gr_v}{Gr} = \frac{b^2\Omega^2}{gL} = \frac{\Omega^2 L}{g}\left(\frac{b}{L}\right)^2 \tag{8.48}$$

On the other hand, one of the validity conditions for the averaged approach is the possibility to linearize the equations for the pulsation field for which one needs $b \ll L$. This means that inequality (8.47) yields the inequality $Gr_v \ll Gr$, and the thermovibrational convection should be entirely excluded from consideration.

If the vibration conditions are such that the first-order effects are suppressed due to translational vibrations and, as we hope, to non-deformability of the free surface, the above-given estimations should be made more precise. The point is that in this case the condition allowing the pulsation equation to linearize becomes less strong. One should simply require that $b\beta\theta \ll L$. (We remark that in any case the linearization condition means that the pulsation velocities in an appropriate reference frame were small in comparison with ΩL, i.e. a particle displacement relative to a vessel during the vibration period was much smaller than the vessel dimensions.) However, condition (8.47) is not sufficient to neglect the I-field as compared to the T-field. Since now the T-field has the order not of $b\Omega$ but of $b\Omega\beta\theta$, instead of inequality (8.47) we have

$$\frac{\Omega^2 L}{\beta\theta g} \ll 1 \tag{8.49}$$

This results in the following relation between the gravitational and vibrational Grashof numbers:

$$\frac{\widetilde{Gr_{\text{v}}}}{Gr} = \frac{\Omega^2 L}{\beta\theta g}\left(\frac{b\beta\theta}{L}\right)^2 \ll 1 \tag{8.50}$$

where the inequality is provided by the product of two small parameters.

Thus, for the cases of both small and large vibration amplitudes, the gravity field capable of suppressing the surface deformations suppresses the thermo-vibrational convection as well. Luckily, inequalities (8.47) and, more so, (8.49) usually do not hold. This enables one to investigate the thermovibrational convection in the presence of a free surface, being allowed to use the averaged approach as well.

One needs only to take into account correctly the free surface deformability, which certainly poses its own difficulties; namely, replacing inequality (8.47) with the approximate equality condition means the occurrence of resonant excitation of large-scale gravitational waves, so that the fluid would be thrown out of the vessel. To avoid that, one should instead require not the inequality (8.49) but the inversed one. Indeed, the resonance is not very dangerous for short waves—they damp made strongly under viscosity. As to the large-scale oscillations that generate the large-scale I-field, they would not be resonant any longer. Moreover, as their amplitudes are determined by estimation (8.45), it is not difficult to make them smaller.

3.2 Suppression of Deformations by Surface Tension

Consider now the second factor that could suppress pulsation deformations of a free surface, i.e. the surface tension. Unlike the gravity field which, besides suppressing deformations, by itself is a source of thermal convection, the surface tension has no such effect. Consequently, one needs just to consider the availability of the parameters for which the pulsation suppression effect takes place.

To make it simpler, we take the case of a flat non-perturbed surface. The order of magnitude of the pulsation pressure is, as usual, determined by the expression $\widetilde{p} \sim \rho b\Omega^2 L$. Estimation of the surface displacement ζ through the capillary term yields

$$\zeta \sim \frac{\rho b \Omega^2 L^3}{\alpha} \tag{8.51}$$

where α is the surface tension coefficient.

Taking the kinematic condition into account, we obtain for the order of magnitude of the I-field velocities

$$\widetilde{\mathbf{v}} \sim b\Omega \, We \tag{8.52}$$

where We is the Weber number, $We = \rho\Omega^2 L^3/\alpha$. As before, the T-field has the order $\beta\theta b\Omega$, so that the ratio of the reference values of the I- and T-fields is equal to

$$\frac{We}{\beta\theta} \ll 1 \tag{8.53}$$

One may neglect the first-order effects and assume the surface to be pulsationally indeformable if the I-field is much smaller than the T-field, i.e. as follows from estimation (8.53), We $\ll \beta\theta$. It is convenient to rewrite this condition in terms of the dimensionless frequency $\omega = \Omega L^2/\nu$:

$$\omega^2 \ll \frac{\beta\theta}{On^2} \tag{8.54}$$

where $On^2 = \rho\nu^2/\alpha L$ and On is the Onezorge number. Since a large value of ω is the validity condition for the high-frequency approach (experiments and calculations in real time show that ω should be $\geqslant 10^3 \div 10^5$), for the frequency the two-sided inequality is obtained:

$$10^6 \div 10^8 \leqslant \omega^2 \ll \frac{\beta\theta}{On^2} \tag{8.55}$$

This means that the surface tension must be very large to suppress the surface deformations. At any rate, it must be

$$On^2 \ll \beta\theta \times \left(10^{-6} \div 10^{-8}\right) \tag{8.56}$$

i.e.

$$\alpha \gg \frac{\rho\nu^2}{\beta\theta L} \times \left(10^6 \div 10^8\right) \tag{8.57}$$

In practice, inequalities (8.56) and (8.57) are never satisfied. Therefore, it should not be considered that deformations are indeed completely suppressed by capillary forces.

3.3 Approximation of Weak Deformability

However, in some situations where On^2 does not satisfy inequality (8.56), there is no opposite inequality. Such a situation takes place, for example, in liquid metals if the cavity size is not very small. This justifies the use of the weak deformability approximation, i.e. the second-order theory that takes into account simultaneously the I- and T-fields.

We perform the corresponding calculations for the case of a free surface that is flat in the average. Let us choose as the reference frame that of one of the fluid-filled vessels. Extracting from the pulsation pressure the leading part determined by the motion of the fluid as a whole, one finds that

$$\tilde{p} = -\rho b \Omega^2 \cos \Omega t \cdot (\mathbf{k} \cdot \mathbf{r}) + \cdots \tag{8.58}$$

where vibrations are supposed to be linear, monochromatic and performed along the axis $\mathbf{k}$. Retaining the leading terms in the dynamic boundary condition for the free surface (assuming small Weber numbers), we obtain the equations for the amplitude of the surface-shaped pulsations:

$$\Delta_2 \zeta = (\mathbf{k} \cdot \mathbf{r}) \tag{8.59}$$

and after taking into account the kinematic condition,

$$\Delta_2 w = (\mathbf{k} \cdot \mathbf{r}) \tag{8.60}$$

where w is the amplitude of the component of the pulsation velocity normal to the free surface.

Together with the equations for the pulsation velocity amplitude,

$$\text{curl } \mathbf{V}_\text{I} = 0, \qquad \text{div } \mathbf{V}_\text{I} = 0 \tag{8.61}$$

and the conditions of impermeability on the rigid boundaries, we obtain the problem for the I-field. The problem for the T-field has the ordinary form inherent to the second-order theory:

$$\text{curl } \mathbf{V}_\text{T} = \mathbf{k} \times \nabla T, \qquad \text{div } \mathbf{V}_\text{T} = 0 \tag{8.62}$$

with the conditions of impermeability on all the boundaries including the free one.

The solution of the problems for the I- and T-fields must be substituted into the general expression for the second-order vibrational force obtained in Section 1. The contribution of the I- and T-fields to the vibrational convection is determined by two independent dimensionless parameters, which read

$$G_{\mathrm{vT}} = \frac{b^2\Omega^2\beta^2\theta^2L^2}{\nu^2} \tag{8.63}$$

$$G_{\mathrm{vi}} = \frac{b^2\Omega^2\beta\theta\, We\, L^2}{\nu^2} = \frac{b^2\Omega^4L^5\rho\beta\theta}{\alpha\nu^2} \tag{8.64}$$

Thus, if the frequency is high enough for the dynamic skin-layer thickness to be much smaller than the reference size ($\Omega > 10^3$), but also low enough that the capillary wavelength $> L$ (i.e. $We \ll 1$), one should take into account the free surface deformations caused by the I-field but may neglect those of T-field origin.

Let us write down the complete set of equations and boundary conditions of thermovibrational convection for the case of high surface tension of the free surface. Combining the I- and T-fields, we obtain from (8.59) to (8.62) and the general formulas of Section 1,

$$\frac{\partial \mathbf{v}}{\partial t} + \mathbf{v}\cdot\nabla\mathbf{v} - \frac{1}{2}b^2\Omega^2\beta(\mathbf{k}\cdot\mathbf{W})\nabla T = -\frac{1}{\rho}\nabla p + \nu\Delta\mathbf{v} \tag{8.65}$$

$$\frac{\partial T}{\partial t} + \mathbf{v}\cdot\nabla T = \chi\Delta T \tag{8.66}$$

$$\operatorname{div}\,\mathbf{v} = 0 \tag{8.67}$$

$$\operatorname{curl}\,\mathbf{W} = \beta\mathbf{k}\times\nabla T \tag{8.68}$$

$$\operatorname{div}\,\mathbf{W} = 0 \tag{8.69}$$

with the boundary conditions on the free surface:

$$\alpha\Delta_2(\mathbf{w}\cdot\mathbf{n}) = \rho\Omega^2\mathbf{k}\cdot\mathbf{r} \tag{8.70}$$

and on the rigid boundaries:

$$\mathbf{W}\cdot\mathbf{n} = 0 \tag{8.71}$$

The boundary conditions for the average fields are determined by the particular configuration of the problem and may vary.

Once more we note that condition (8.70) may be used only for small values of the Weber number. If this does not hold, a more correct approach is to neglect the effect of gravity and surface tension on the pulsational deformations of the free surface, i.e. to use the boundary conditions obtained in [6] to [8] (see Section 2 in Chapter 7). A comparative study of two variants of the boundary conditions performed in [5] shows that even for the case of vibrations normal to the free surface, the use of the impermeability condition instead of condition (7.62) leads to results that are significantly different.

References

1. Zenkovskaya, S.M. and I.B. Simomenko. On the influence of high frequency vibrations on the onset of convection, *Izv. AN SSSR, Mekhanika Zhidkosti i Gaza*, 1966, **5**, 51–5.
2. Povitskii, A.S. and L.Ya. Lyubin. *Foundatious of Dynamics and Heat/Mass Transfer in Fields and Gases in Weightlessness*, Machinostroenie, Moscow, 1972.
3. Kozlov, V.G. On vibrational thermal convection in a cavity subject to high frequency vibrational swings, *Izv. AN SSSR, Mekhanika Zhidkosti i Gaza*, 1988, **3**, 138–44.
4. Briskman, V.A. Vibration-thermocapillary convection and stability, in *Proc. of the First Int. Symp. on Hydrodynamics and Heat/Mass Transfer in Microgravity*, Perm, Moscow, 1991, Gordon and Breach, London, 1992, pp. 111–19.
5. Birikh, R.V., B.A. Briskman, A.L. Zuev, V.I. Chernatynskii and V.I. Yakushin. Interaction between thermogravitational and thermocapillary convection mechanisms, *Izv. RAN, Mekhanika Zhidkosti i Gaza*, 1994, **5**, 107–21.
6. Lyubimov, D.V. and A.A. Cherepanov. On the appearance of stationary relief on a fluid interface under vibrational field, *Izv. AN SSSR, Mekhanika Zhidkosti i Gaza*, 1986, **6**, 8–13.
7. Lyubimov, D.V., M.V. Savvina and A.A. Cherepanov. On quasi-equilibrium shape of a free fluid surface in modulated gravity field, in *Problems of Hydrodynamics and Heat/Mass Transfer with Free Boundaries*, Novosibirsk, 1987, pp. 97–105.
8. Lyubimov, D.V. Thermovibrational flows in a fluid with a free surface, *Microgravity Quart.*, 1994, **4**(2), 107–12.

9 Some Particular Problems

In this chapter the approach developed in Chapters 6 to 8 is used to study a number of particular problems. At first, we consider the flows induced by oscillations of a heated sphere. In this case the isothermal pulsation field is non-uniform, so that thermovibrational convection of the first order with respect to $\beta\theta$ and an average flow of the Schlichting type would be generated. Then the flow in an annular gap between two cylinders is studied for the case where the inner cylinder performs circularly polarized oscillations. Here, a significant non-uniformity of the oscillation phase makes the pulsational transport of vorticity and heat essential. Further, vibrational flows and heat transfer in a liquid zone when excited in the axial direction are analyzed. The presence of a free surface along which the capillary waves can propagate makes it necessary to take into account the surface-wave mechanism of the average flow generation. Finally, thermovibrational convection in a rectangular cavity subject to swing oscillations is considered.

1 Flows Induced by a Heated Oscillating Sphere

In this section we consider the vibrational flows induced by linearly polarized vibrations of a heated sphere immersed in a fluid. Both the case of a sphere in an infinite fluid and the case of a rigid quiescent spherical envelope coaxial with the inner sphere (Fig. 159) are analyzed. If the envelope had performed oscillations according to the same law as the inner sphere, then in the isothermal case the fluid would have moved as a solid body together with the spheres, thus realizing the case of uniform vibrations. In the situation considered herein, the amplitude of oscillations of the fluid is non-uniform. However, the vibrations are performed along a fixed direction, i.e. the motion of the sphere is described by a single function of time. Therefore, in terms of Chapter 6, the case considered here is an example of non-uniform vibrations with a uniform phase. Due to that, the isothermal mechanisms of the average flow generation as well as the

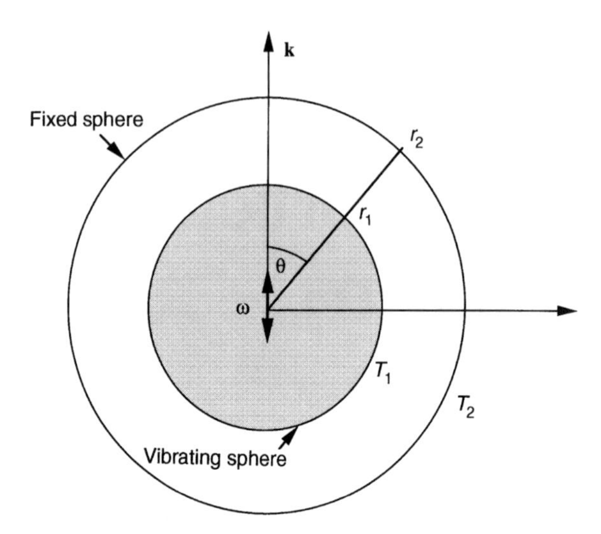

Fig. 159 Thermovibrational convection between spheres and geometry of the problem

first-order volumic thermovibrational mechanism are in action, but the pulsational transport effect is absent.

The flows near oscillating solids, particularly near a vibrating sphere, have been widely investigated. In [1] to [4] the matched asymptotic expansion method is applied to study secondary flows induced by slow oscillations of a sphere in a viscous incompressible fluid quiescent at infinity. In [5] the results of [1] and [2] are extended to finite frequencies. It is shown that using the assumption on smallness of the oscillation amplitude, it is possible to construct a uniformly valid expansion without using the matching procedure. The problem of the flow near a sphere is characterized by two dimensionless parameters:

$$\varepsilon = \frac{b}{R}, \qquad M = R\left(\frac{\Omega}{\nu}\right)^{1/2} \tag{9.1}$$

where R is the sphere radius and b is the oscillation amplitude. Apparently, the pulsational Reynolds number Re_{p} introduced in Section 5 in Chapter 6 is connected to the parameters ε and M by the relations

$$Re_{\mathrm{p}} = \varepsilon^2 M^2 \tag{9.2}$$

Expansion with respect to ε at fixed M yields the pulsation flow in the zeroth order:

$$\Psi_0 = \tfrac{1}{2}\left[f_0(r)\mathrm{e}^{\mathrm{i}t} + f_0^*(r)\mathrm{e}^{-\mathrm{i}t}\right]\sin^2\theta \tag{9.3}$$

where Ψ is the stream function. The quantities Ω^{-1}, R and $b\Omega$ are chosen as the scales for the time, length and velocity, respectively; the function $f_0(r)$ is given by the expression

$$f_0(r) = \tfrac{1}{2}r^2 + Ar^{-1} + B(1+\rho^{-1})\mathrm{e}^{-\rho} \tag{9.4}$$

where

$$\rho = Mr\sqrt{\mathrm{i}}, \qquad A = -\frac{1}{2}\left(1 + \frac{3}{M\sqrt{\mathrm{i}}} + \frac{3}{M^2\mathrm{i}}\right), \qquad B = \frac{3}{2M\sqrt{\mathrm{i}}}\,\mathrm{e}^{M\sqrt{\mathrm{i}}}$$

For $M \to \infty$ and finite r the solution of (9.3) turns into the known solution for a potential flow of an inviscid fluid around a sphere. In [5] the solution of the next order, i.e. first, with respect to ε has been found.

The stationary component of the solution is given by the expression

$$\Psi_1 = f_{10}(r)\sin^2\theta\cos\theta \tag{9.5}$$

where $f_{10}(r)$ is expressed through the gamma function. For the limiting case of small frequencies and large distances from the sphere, the asymptotic expressions are obtained. At $M \to 0$ and fixed $\sigma = Mr\sqrt{2}$,

$$\begin{aligned} f_{10} = -\frac{9}{2}\Bigg\{\sigma^{-2} - \Bigg[\left(\sigma^{-2} + \frac{1}{2}\sigma^{-1}\right)\cos\frac{1}{2}\sigma \\ + \left(\frac{1}{2}\sigma^{-1} + \frac{1}{6}\right)\sin\frac{1}{2}\sigma\Bigg]\mathrm{e}^{-\delta/2}\Bigg\} + O(M) \end{aligned} \tag{9.6}$$

For large σ the asymptote is

$$f_{10} = P_2 + Q_2\sigma^{-2} \tag{9.7}$$

where

$$P_2 = -\frac{27}{160}M\sqrt{2} + \frac{27}{360}M^2 - \frac{1}{64}M^3\sqrt{2} + O(M^4\log M)$$

$$Q_2 = -\frac{9}{2} - \frac{1359}{560}M\sqrt{2} - \frac{99}{280}M^2 + O(M^3)$$

Similar to the case (9.6), in the case (9.7) at M small enough, the flow on the vibration axis is directed towards the oscillating sphere. On the other hand, analysis of the solution at large M shows that in this case the average flow on the vibration axis is directed from the sphere. This means that there must exist a critical value of M at which a stagnation point emerges on the vibration axis.

In [6] the heat transfer from a heated oscillating sphere was studied. The case of small amplitudes and high frequencies of vibrations was examined which allows the boundary layer approximation to be applied. It was assumed that the major contribution to the heat transport came from the Schlichting flow. The heat transfer law was found from the creeping flow approximation. However, neither in [6] nor in [5] was the effect of density non-uniformities on the average flow (thermovibrational convection) taken into account. This step has been done in [7] and makes the subject of the present section.

1.1 Governing Equations and Boundary Conditions

We shall consider the flows induced by oscillations of a solid sphere placed in a quiescent rigid spherical envelope so that the average position of the center of the inner sphere coincides with the center of the envelope. The temperatures of the inner and outer spheres are different, so that the fluid, completely filling the gap between the spheres, is non-isothermal. Let us analyze the case of zero gravity. Let the inner sphere vibrate with amplitude b and frequency Ω according to a sinusoidal law along the axis given by the unit vector $\mathbf{k}$. We assume that the vibrations are of high frequency and small amplitude so that the conditions

$$\frac{\nu}{\Omega L^2} \ll 1 \tag{9.8}$$

and

$$\frac{b}{L} \ll 1 \tag{9.9}$$

hold. Here ν is the kinematic viscosity and L is the reference size. We assume that inequality (9.9) is satisfied at $L = \min(r_1,\ r_2 - r_1)$, where r_1 is the radius of the inner sphere. As shown in Section 3 in Chapter 6, if conditions (9.8) and (9.9) take place, the average flow is described by the following equations:

$$\frac{\partial \mathbf{v}}{\partial t} + (\mathbf{v} \cdot \nabla)\mathbf{v} = -\nabla p + \Delta \mathbf{v} + Gr_{\mathrm{v}}(\nabla \Phi)^2 \nabla T \tag{9.10}$$

$$\operatorname{div} \mathbf{v} = 0 \tag{9.11}$$

$$\frac{\partial T}{\partial t} + \mathbf{v} \cdot \nabla T = \frac{1}{Pr} \Delta T \tag{9.12}$$

Here $\mathbf{v}$ is the average velocity, T and p are the fluid temperature and pressure and Φ is the potential of the pulsation velocity amplitude which obeys the Laplace equation

$$\Delta \Phi = 0 \tag{9.13}$$

Equations (9.10) to (9.12) are written in a dimensionless form. The radius r_1 of the inner sphere, the quantity ν/r_1 and the temperature difference Θ between the inner and the outer spheres are chosen as the length, velocity and temperature scales. The equations contain two dimensionless parameters, viz. the Prandtl number $Pr = \nu/\chi$ and the vibrational Grashof number $Gr_v = \beta\Theta b^2\Omega^2 r_1^2/\nu^2$, ν being the kinematic viscosity, χ the thermal diffusivity and β the thermal expansion coefficient.

On the pulsation velocity component the impermeability conditions are imposed, i.e. on the vibrating sphere at $r = 1$:

$$\frac{\partial \Phi}{\partial r} = \cos\theta \tag{9.14}$$

and on the envelope at $r = R$:

$$\frac{\partial \Phi}{\partial r} = 0 \tag{9.15}$$

where $R = r_2/r_1$ is the reduced radius of the envelope. Conditions (9.14) and (9.15) are written in a spherical coordinate system, r being the radius and θ the polar angle. The polar axis is directed along the vibration axis **k**. Condition (9.14) is assumed to hold at the mean position of the inner sphere. This is justified by the smallness of the amplitude.

The use of the impermeability condition instead of the no-slip one means that the set (9.13) to (9.15) describes the field of the pulsation velocity correctly everywhere except for the viscous boundary skin layers near rigid surfaces. However, the skin layer thicknesses are rather small due to relation (9.8).

Thermal conductivities of the solid spheres are assumed to be high, so that we may impose the isothermality conditions on their surfaces:

$$T = 0 \qquad \text{at } r = R \tag{9.16}$$

and

$$T = 1 \qquad \text{at } r = 1 \tag{9.17}$$

The temperature of the envelope is chosen as the reference one.

The pulsation part of velocity is not potential in the viscous skin layers which leads to generation of the average vortex flow. As shown in Section 1 in Chapter 7, this effect may be described by means of the effective boundary condition for the average velocity. For the configuration considered herein, this boundary condition is

$$\mathbf{v}_\tau = -Re_\mathrm{p}\left[2\mathbf{U}_\tau(\nabla_\tau \mathbf{U}_\tau) + \tfrac{1}{2}\nabla_\tau \mathbf{U}_\tau^2\right] \tag{9.18}$$

on the envelope $r = R$ and

$$\mathbf{v}_\tau = -Re_\mathrm{p}\left[2\mathbf{W}_\tau(\nabla_\tau \mathbf{W}_\tau) + \tfrac{1}{2}\nabla_\tau \mathbf{W}_\tau^2\right] \tag{9.19}$$

on the vibrating sphere surface at $r = 1$. Here $\mathbf{U} = \nabla\Phi$ and $\mathbf{W} = \mathbf{U} - \mathbf{k}$ is the pulsation velocity amplitude in the reference frame of the vibrating sphere. The subscript τ denotes the tangential components of the vectors; the operator ∇_τ acts along the surfaces of the rigid spheres. Boundary conditions (9.18) and (9.19) include an extra dimensionless parameter $Re_\mathrm{p} = b^2\Omega/\nu$, which is responsible for the generation of the average vortex flows in the skin layers.

We restrict ourselves to axisymmetrical solutions where all the quantities are azimuth-independent. Then conditions (9.18) and (9.19) take the form

$$v_\theta = -Re_\mathrm{p}\left\{3\frac{\partial\Phi}{\partial\theta}\frac{\partial^2\Phi}{\partial\theta^2} + 2\cot\theta\left[\frac{\partial\Phi}{\partial\theta}\right]^2\right\} \quad \text{at} \quad r = R \tag{9.20}$$

and

$$v_\theta = -Re_\mathrm{p}\left\{3\frac{\partial\Phi_1}{\partial\theta}\frac{\partial^2\Phi_1}{\partial\theta^2} + 2\cot\theta\left[\frac{\partial\Phi_1}{\partial\theta}\right]^2\right\} \quad \text{at} \quad r = 1 \tag{9.21}$$

where $\Phi_1 = \Phi - \cos\theta$. The normal components of the average velocities satisfy the impermeability condition:

$$v_r = 0 \qquad \text{at} \quad r = 1 \text{ and at } \quad r = R \tag{9.22}$$

The problem (9.13) to (9.15) has the solution

$$\Phi = \frac{\cos\theta}{1 - R^3}\left(\frac{R^3}{2r^2} + r\right) \tag{9.23}$$

By substituting (9.23) in (9.10), (9.20) and (9.21), we obtain the closed problem for the average velocity $\mathbf{v}$ and temperature T. An axisymmetrical case allows the stream function to be introduced:

$$v_r = \frac{1}{r\sin\theta}\frac{\partial}{\partial\theta}(\Psi\sin\theta) \tag{9.24}$$

$$v_\theta = -\frac{1}{r}\frac{\partial(r\Psi)}{\partial r} \tag{9.25}$$

For the solution (9.24) and (9.25) the continuity equation (9.11) is satisfied identically whereas for the stationary flow equations (9.10) and (9.12) and boundary conditions (9.20) to (9.22) the following form is taken:

$$\frac{1}{r}\left(\frac{\partial\Psi}{\partial r}\cot\theta - \frac{1}{r}\frac{\partial\Psi}{\partial\theta}\right)\zeta + M(\Psi,\zeta) + \frac{Gr_\mathrm{v}}{r}\left(\frac{\partial E}{\partial r}\frac{\partial T}{\partial\theta} - \frac{\partial E}{\partial\theta}\frac{\partial T}{\partial r}\right) = L\zeta \tag{9.26}$$

$$\zeta = -L\Psi \tag{9.27}$$

$$M(\Psi, T) = \frac{1}{Pr}\Delta T \tag{9.28}$$

$$\Psi = 0, \qquad \frac{\partial \Psi}{\partial r} = \frac{45}{8} Re_{\mathrm{p}} \frac{1}{R(1-R^3)^2} \sin 2\theta \qquad \text{at } r = R \tag{9.29}$$

$$\Psi = 0, \qquad \frac{\partial \Psi}{\partial r} = \frac{45}{8} Re_{\mathrm{p}} \frac{R^6}{(1-R^3)^2} \sin 2\theta \qquad \text{at } r = 1 \tag{9.30}$$

The operators M and L are

$$M(a,b) = \frac{1}{r\sin\theta}\left(\frac{\partial}{\partial\theta}(a\sin\theta)\frac{\partial b}{\partial r} - \frac{1}{r}\frac{\partial}{\partial r}(ar)\frac{\partial b}{\partial r}\right)$$

$$L = \Delta - \frac{1}{r^2\sin^2\theta}$$

with

$$\Delta = \frac{1}{r^2}\frac{\partial}{\partial r}\left(r^2\frac{\partial}{\partial r}\right) + \frac{1}{r^2\sin\theta}\frac{\partial}{\partial\theta}\left(\sin\theta\frac{\partial}{\partial\theta}\right) \tag{9.31}$$

The new variable E is proportional to the average density of the pulsation flow energy:

$$E = \left(\frac{\partial\Phi}{\partial r}\right)^2 + \frac{1}{r^2}\left(\frac{\partial\Phi}{\partial\theta}\right)^2 \tag{9.32}$$

1.2 Creeping Flows

At small values of the parameters Gr_{v} and Re_{p} it is possible to neglect non-linear terms in the motion and heat transfer equations of the fluid. Then we arrive at the equations

$$\frac{Gr_{\mathrm{v}}}{r}\left(\frac{\partial E}{\partial r}\frac{\partial T}{\partial\theta} - \frac{\partial E}{\partial\theta}\frac{\partial T}{\partial r}\right) = L\zeta \tag{9.33}$$

$$\zeta = -L\Psi \tag{9.34}$$

$$\Delta T = 0 \tag{9.35}$$

The solution of equation (9.35) with the boundary conditions (9.16) and (9.17) is

$$T = \frac{1}{R-1}\left(\frac{R}{r} - 1\right) \tag{9.36}$$

For a slow creeping flow, we obtain, by substituting (9.23) and (9.36) into (9.33),

$$L\zeta = -\frac{3R^4\,Gr_{\mathrm{v}}}{4(R-1)(1-R^3)^2}\frac{R^3-4r^3}{r^9}\sin 2\theta \tag{9.37}$$

It follows from (9.37) and (9.29) and (9.30) that the solution for Ψ is proportional to $\sin 2\theta$. Introducing a new variable ψ by means of the relation

$$\Psi = \frac{45}{8}\frac{R^6}{(1-R^3)^2}\psi\, Re_{\mathrm{p}} \sin 2\theta$$

one gets, from (9.37) and (9.34),

$$N\xi = -B\frac{R^3-4r^3}{r^9} \tag{9.38}$$

$$\xi = -N\psi \tag{9.39}$$

where

$$N = \frac{\mathrm{d}^2}{\mathrm{d}r^2} + \frac{2}{r}\frac{\mathrm{d}}{\mathrm{d}r} - \frac{6}{r^2} \tag{9.40}$$

with the boundary conditions

$$\psi = 0, \qquad \frac{\mathrm{d}\psi}{\mathrm{d}r} = -\frac{1}{R^7} \qquad \text{at } r = R \tag{9.41}$$

$$\psi = 0, \qquad \frac{\mathrm{d}\psi}{\mathrm{d}r} = -1 \qquad \text{at } r = 1 \tag{9.42}$$

where we introduced the notation

$$B = \frac{2}{15R^2(R-1)}\frac{Gr_{\mathrm{v}}}{Re_{\mathrm{p}}} \tag{9.43}$$

Thus, for the case of a weak linear convection, the flow structure is not determined by the two parameters Gr_{v} and Re_{p} separately, but by their ratio $Gr_{\mathrm{v}}/Re_{\mathrm{p}}$.

1.2.1 Limiting case: $R \to \infty$

Before studying the convection in a container with an arbitrary ratio of the radii R, let us investigate the simpler case where the sphere vibrates in an infinite fluid, i.e. $R \to \infty$. In this case (9.38) to (9.42) take the form

$$N\xi = -\frac{b}{r^9}, \qquad \xi = -N\psi \tag{9.44}$$

and

$$\psi = 0, \qquad \frac{\mathrm{d}\psi}{\mathrm{d}r} = -1 \qquad \text{at } r = 1 \tag{9.45}$$

$$\psi = 0, \qquad \frac{\mathrm{d}\psi}{\mathrm{d}r} = 0 \qquad \text{at } r \to \infty \tag{9.46}$$

where

$$b = \frac{2}{15}\frac{Gr_\mathrm{v}}{Re_\mathrm{p}}$$

The solution of the problem (9.44) to (9.46) is

$$\psi = \frac{1}{r}\left[\frac{b}{504}\left(\frac{1}{r^2} - 1\right)^2 + \frac{1}{2}\left(\frac{1}{r^2} - 1\right)\right] \tag{9.47}$$

or, going back to the stream function, Ψ:

$$\Psi = \frac{3}{16r}\left[\frac{Gr_\mathrm{v}}{126}\left(\frac{1}{r^2} - 1\right)^2 + 15\,Re_\mathrm{p}\left(\frac{1}{r^2} - 1\right)\right] \sin 2\theta \tag{9.48}$$

It can be seen that the resulting creeping flow is a superposition of the thermoconvective and Schlichting flows. This flow is symmetrical with respect to the equatorial plane.

The pulsation Reynolds number Re_p is always positive. Thus, the flow described by the second term in (9.48) is of the same structure at any Re_p. It is a jet-type flow going away from the sphere poles and directed towards the sphere in the equatorial plane.

The parameter Gr_v may be of either sign. If the value of Gr_v is negative, i.e. the vibrating sphere is colder than the surrounding fluid, the flows induced by the thermovibrational and Schlichting mechanisms are of the same direction. This means that the structure of the resulting flow is the same as that of the pure Schlichting flow. In this case the stream lines are not closed, the liquid particles move outward to infinity at the poles and from infinity to the sphere near the equatorial plane.

The situation is essentially different when the vibrating sphere is warmer than the surrounding fluid. In this case the flows induced by the two mechanisms are of opposing directions. Close to the vibrating sphere the circulation of the flow is defined by the boundary condition (9.45) and, hence, the flow in this region is always directed from the equator to the poles. At arbitrary distances, the direction of the flow is defined by the competition of the two mechanisms. If Gr_v is positive and small enough, the last term in (9.47) prevails. Then the direction of circulation is the same as in the case discussed above, but the stream lines are closed. If Gr_v is higher than some critical value, the flow structure becomes a 'two-storied' one. There appears a certain spherical separation surface of radius r^*. Below it, the stream lines are closed and the fluid particles do not penetrate the outer space. The flow direction at $r < r^*$ is determined by (9.45). To the outside of the separation surface, the stream lines are not closed and the flow direction is opposite, i.e. the fluid particles approach the sphere along the polar axis and move from it in the equatorial plane. Since at $r = r^*$ the value of ψ turns to zero, we obtain from (9.48)

$$r^{2*} = \frac{b}{b - 252} \tag{9.49}$$

It follows from (9.49) that the 'two-storied' structure is possible if b is positive and larger than the critical value equal to 252. In terms of non-dimensional parameters Gr_v and Re_p the existence condition for the two-storied structure is

$$\frac{Gr_v}{Re_p} > 1890 \tag{9.50}$$

The spherical separation surface emerges at infinity at the critical value of the ratio Gr_v/Re_p and approaches the sphere when this ratio increases.

Note that the parameter

$$b = \tfrac{2}{15}\nu\beta\Theta\Omega r_1^2$$

does not depend on the vibration amplitude.

1.2.2 General case: finite *R*

At a finite radius of the envelope, a 'one-storied' flow is impossible since the boundary conditions (9.41) and (9.42) induce the flow of the same direction near both rigid surfaces. One might find the analytical solution of the problem at arbitrary values of R as well, but it is too cumbersome to present it here.

The above-defined stream function Ψ has a simple meaning: it is the azimuthal component of the velocity vector potential. However, it is not the best mean to present the results since the surfaces of constant value of this function (except for the surface $\Psi = 0$) are not the integral surfaces for the velocity field. The necessary properties possess the stream function $\widehat{\Psi} = \Psi \sin\theta$. Therefore we express the results in terms of the radial part of the stream function $\widehat{\Psi}$ which is defined as $r\psi(r)$.

The data obtained by tabulation are shown in Figs. 160 to 163 for different values of the envelope radius R and the ratio $k = Gr_v/Re_p$. Here the vertical axis corresponds to the product $r\psi(r)$, while the horizontal axis corresponds to

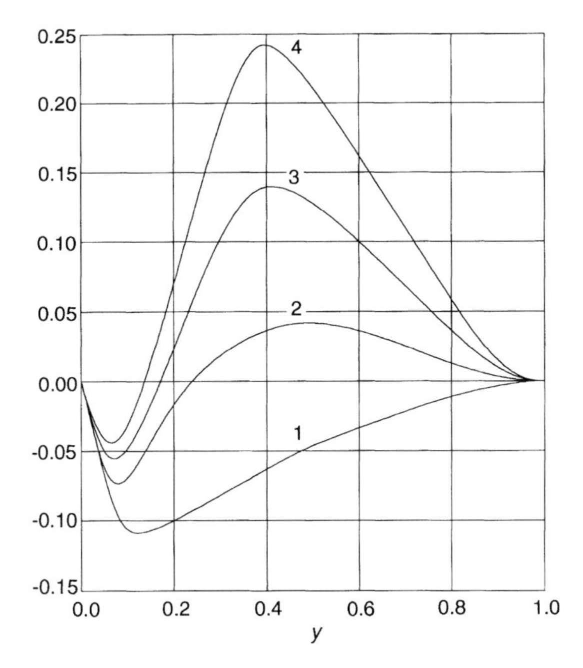

Fig. 160 Radial distribution of the stream function at $R = 4$ and various Gr_v/Re_p

the reduced radial coordinate $y = (r - 1)/(R - 1)$, with $y = 0$ on the inner (vibrating) sphere and $y = 1$ on the envelope.

It can be seen that, when the distance between the spheres is large enough (Fig. 160, $R = 4$), the flow structure is similar to that obtained for the surrounding infinite fluid. If the temperature difference between the spheres is not too high (curve 1, $k = 2000$), then the structure of the resulting flow is defined by the Schlichting mechanism. The space between the spheres is occupied by a single vortex. The direction of its circulation corresponds to the Schlichting flow induced in a viscous skin layer near the inner sphere. Evidently, due to the boundary conditions (9.41), there exists a vortex in the opposite direction close to the fixed sphere. However, its intensity is extremely low and it occupies so small a part of the layer that in a graph describing the whole gap this vortex is not resolved. Although this vortex exists at any value of the parameters, hereafter we would not discuss it because, due to its small size and low intensity, its importance is minor.

With the growth of the vibrational Grashof number, the influence of the thermovibrational flow increases, and a vortex of the opposite direction appears (curve 2, $k = 3000$). With further growth of Gr_v, the intensity of the thermovibrational flow increases while the intensity of the Schlichting flow goes down. In addition, the area that it occupies near the inner sphere (curves 3 and 4,

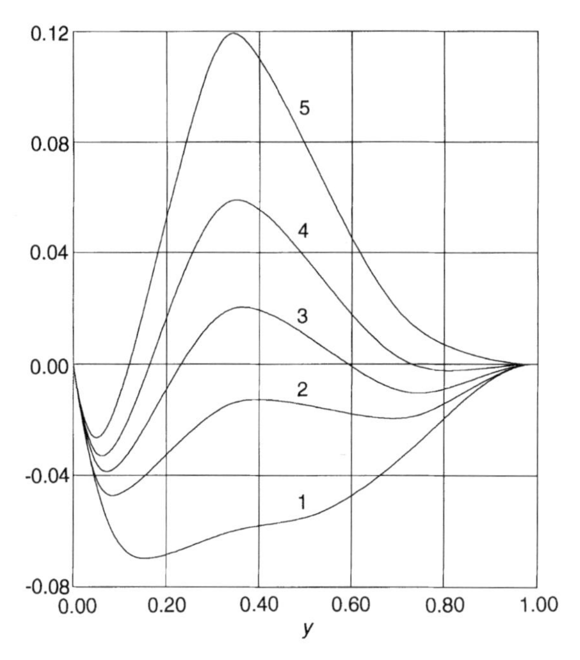

Fig. 161 Radial distribution of the stream function at $R = 2.5$ and various Gr_v/Re_p

$k = 4000$ and 5000) is reduced. In comparison with the case of an infinite fluid, the presence of the envelope results in larger critical values of the parameters at which the two-storied structure appears.

The decrease in the envelope radius leads to the growth of the critical parameters for the transition from a one-vortex structure to a multivortex one (Fig. 161, $R = 2.5$). The flow becomes more complicated. At comparatively low values of Gr_v (curve 1, $k = 6000$), a one-vortex flow takes place. At higher temperatures of the inner sphere, the major one-vortex flow is sustained but inside its nuclei two weak vortices in the same direction develop (curve 2, $k = 8333$). When k exceeds a certain critical value, in the central part of the spherical layer there appears a vortex in the opposite direction (curve 3, $k = 10\,000$). Upon Gr_v growth, its intensity and size become larger whereas the outer vortices are pushed away towards the inner sphere and the envelope (curves 4 and 5, $k = 12\,000$ and $15\,000$).

Figure 162 corresponds to $R = 2.1$. In this case the effect of the Schlichting flow is rather strong, and up to sufficiently high values of k (curve 1, $k = 50\,000$) it determines the structure of the resulting flow. Only at very high k do the vortices in the opposite direction appear (curves 2, 3 and 4, $k = 80\,000$, $100\,000$ and $150\,000$). We note that at these layer thicknesses a purely thermovibrational flow has a two-vortex structure. The corresponding

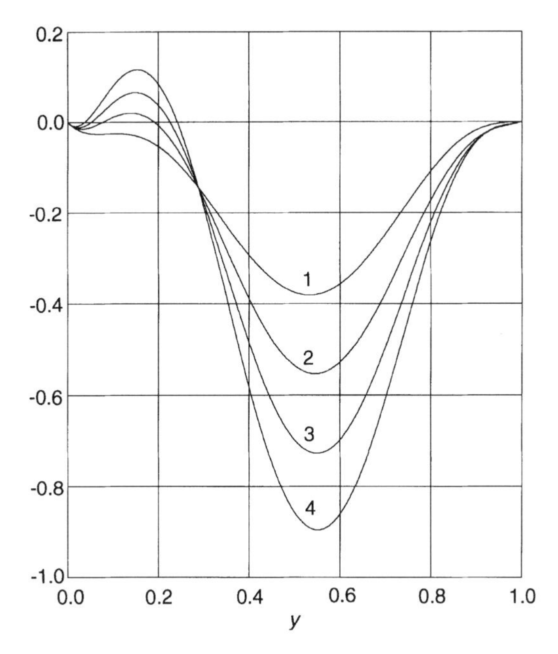

Fig. 162 Radial distribution of the stream function at $R = 2.1$, $Gr_v > 0$ and various Gr_v/Re_p

part of the stream function vanishes at $x \approx 0.28$, and due to that all the lines cross in the same point.

If the inner sphere is cooled (Gr_v and k are negative) and the layer thickness is large, the flow structure does not change qualitatively with the change in the inner sphere temperature. The flow structure is rather simple: it comprises one vortex, the direction of which corresponds to the Schlichting flow. At smaller values of the layer thickness, a complicated flow structure may occur (Fig. 163, $R = 2.1$) due to a complicated structure of a purely thermovibrational flow. With the increase of the absolute value of k, the net one-vortex flow (curve 1, $k = -6000$) changes to a two-vortex one (curves 2 to 5, $k = -10\,000$, $-15\,000$, $-20\,000$ and $-25\,000$).

Thus, an analytical investigation performed for small values of the governing parameters Re_p and Gr_v in a creeping flow approximation shows that the vibration flow essentially depends on the layer thickness and in the case of small thicknesses may be rather complicated. For a fixed layer thickness, the flow structure is determined by a single parameter which is the ratio of the vibrational Grashof number to the pulsational Reynolds number.

In conclusion, we would like to emphasize once more the qualitative difference between the case considered in this section of an oscillating sphere surrounded by a quiescent rigid envelope and the case where both the inner

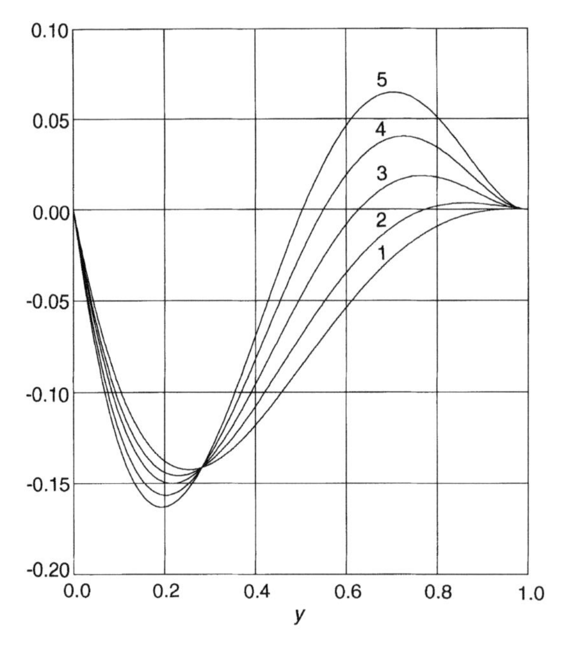

Fig. 163 Radial distribution of the stream function at $R = 2.1$, $Gr_v < 0$ and various Gr_v/Re_p

sphere and the envelope oscillate according to the same law. In the latter case the first-order effects vanish, and one should apply the equations and boundary conditions of the second-order theory. As a result, first, the flow structure does not any longer depend on the sign of the imposed temperature difference and, second, the absence of the Schlichting generation makes it less complicated. Such a behavior takes place in the case of a gap between two coaxial cylinders [8, 9]: the structure of the flow induced by synchronous vibrations of the cylinders does not depend on the sign of the imposed temperature difference.

2 Thermovibrational Convection between Two Coaxial Cylinders: The Case of Circularly Polarized Vibrations

The goal of this section is to consider the thermovibrational flow in an annular gap between two infinite cylinders. An inner cylinder of radius R_1 is subject to small oscillations of circular polarization, so that each point of this cylinder moves along a circle of radius b. The plane of oscillations is perpendicular to the cylinder axis. The outer cylinder of radius R_2 stays at rest. The cylinders are maintained at different fixed temperatures: T_1 for the inner cylinder and T_2 for the outer one (Fig. 164).

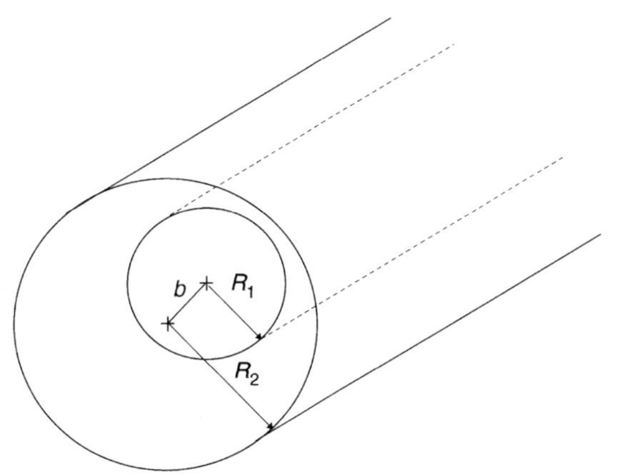

Fig. 164 Geometry of the problem

Oscillatory motion of the inner cylinder generates a pulsation flow inside the gap. Here both the pulsation velocity field and its phase are non-uniform. Under these conditions one should expect both the Schlichting generation of an average flow as well as the effect of pulsational transport. The latter was absent for the case considered in the previous section where the oscillation phase was uniform.

The symmetry of the problem allows us to assume that the basic average flow is axisymmetrical. Since the temperature distribution in a thermoconductive regime is also axisymmetrical, in the basic state the gradients of density and of the kinetic energy of pulsations are parallel. This means that thermovibrational convection does not affect the basic state. Thus, the basic flow in itself is independent of heating. However, its stability and secondary flows depend both on the value and sign of the temperature difference between the two cylinders.

2.1 Pulsation Field

We begin the analysis from the study of the pulsation velocity field. Let us write down the oscillation law for the inner cylinder as

$$\mathbf{R}(t) = \mathbf{R}_0 + b(\cos\Omega t\mathbf{i} + \sin\Omega t\mathbf{j}) \tag{9.51}$$

where $\mathbf{i}$ and $\mathbf{j}$ are unit vectors orthogonal to the cylinder axis, $\mathbf{R}(t)$ renders the instantaneous coordinates of an arbitrary point of the cylinder and $\mathbf{R}_0$ describes the average values of those coordinates. Differentiating (9.51), we obtain the velocity $\mathbf{V}(t)$ of an arbitrary point of the inner cylinder:

$$\mathbf{V}(t) = b\Omega(-\sin\Omega t\mathbf{i} + \cos\Omega t\mathbf{j}) \tag{9.52}$$

Taking into account the impermeability boundary condition, we arrive at the following problem for the pulsation velocity potential $\Phi(t)$:

$$\Delta\Phi = 0 \tag{9.53}$$

$$\frac{\partial\Phi}{\partial r} = 0 \qquad \text{at } r = R_2 \tag{9.54}$$

$$\frac{\partial\Phi}{\partial r} = \mathbf{V}(t)\cdot\mathbf{e}_r \qquad \text{at } r = R_1 \tag{9.55}$$

where r is the radius in the cylindrical coordinate system (r, φ, z), $\mathbf{e}_r = \nabla r$.
We seek the solution of the problem (9.53) to (9.55) in the form

$$\Phi = b\Omega F(r)\sin(\varphi - \Omega t) \tag{9.56}$$

Substituting (9.56) into (9.53), we obtain for $F(r)$:

$$F(r) = Ar + \frac{B}{r} \tag{9.57}$$

Boundary conditions (9.54) and (9.55) are satisfied at

$$A = \frac{R_2^2}{R_2^2 - R_1^2}, \qquad B = R_1^2 A \tag{9.58}$$

This means that the pulsation velocity field $\widetilde{\mathbf{v}}$ has the form

$$\widetilde{\mathbf{v}} = b\Omega A\left[\left(1 - \frac{R_1^2}{r^2}\right)\sin(\varphi - \Omega t)\mathbf{e}_r + \left(1 + \frac{R_1^2}{r^2}\right)\cos(\varphi - \Omega t)\mathbf{e}_\varphi\right] \tag{9.59}$$

and the pulsation displacements field $\mathbf{q}$ may be written as

$$\mathbf{q} = bA\left[\left(1 - \frac{R_1^2}{r^2}\right)\cos(\varphi - \Omega t)\mathbf{e}_r - \left(1 + \frac{R_1^2}{r^2}\right)\sin(\varphi - \Omega t)\mathbf{e}_\varphi\right] \tag{9.60}$$

Substitution of (9.59) and (9.60) into formula (6.61) for the pulsational transport vector yields

$$\mathbf{S} = -2b^2\Omega A^2\frac{R_1^4}{r^5}\mathbf{e}_\varphi \tag{9.61}$$

2.2 The Problem of the Average Flow

The general problem of the average flow and heat transfer, accounting for the volumic thermovibrational effect, generation of vorticity in the boundary layers and pulsating transport effect, has been formulated in Chapter 6. Let us specify it for the configuration considered here. Let the quantities R_1, R_1^2/ν, ν/R_1,

$\rho\nu^2/R_1^2$ and θ be the scales for length, time, average velocity, average pressure and temperature, respectively:

$$\frac{\partial \mathbf{u}}{\partial t} + (\mathbf{u}\cdot\nabla)\mathbf{u} + Re_{\mathrm{p}}\,S(r)\boldsymbol{\zeta}\times\mathbf{e}_\varphi = -\nabla p + \Delta\mathbf{u} + Gr_{\mathrm{v}}\,S(r)\vartheta\mathbf{e}_r \tag{9.62}$$

$$\frac{\partial \vartheta}{\partial t} + \mathbf{u}\cdot\nabla\vartheta + Re_{\mathrm{p}}\,S(r)\mathbf{e}_\varphi\cdot\nabla\vartheta = Pr^{-1}\Delta\vartheta \tag{9.63}$$

$$\mathrm{div}\ \mathbf{u} = 0 \tag{9.64}$$

$$\mathbf{u}\cdot\mathbf{e}_r = 0, \qquad \mathbf{u}\cdot\mathbf{e}_\varphi = Re_{\mathrm{p}}\frac{3R^4}{(R^2-1)^2}, \qquad \vartheta = 0 \quad \text{at} \quad r = 1 \tag{9.65}$$

$$\mathbf{u}\cdot\mathbf{e}_r = 0, \qquad \mathbf{u}\cdot\mathbf{e}_\varphi = Re_{\mathrm{p}}\frac{3}{R(R^2-1)^2}, \qquad \vartheta = 1 \quad \text{at} \quad r = R \tag{9.66}$$

Here we have introduced the notations:

$$\boldsymbol{\zeta} = \mathrm{curl}\,\mathbf{u}, \qquad S(r) = \frac{2r^4}{(R^2-1)^2}\frac{1}{r^5} \tag{9.67}$$

The problem includes the following dimensionless parameters: the pulsational Reynolds number $Re_{\mathrm{p}} = b^2\Omega/\nu$, vibrational Grashof number $Gr_{\mathrm{v}} = b^2\Omega^2\beta\theta R_1^2/\nu^2$, the Prandtl number $Pr = \nu/\chi$ and the geometrical parameter $R = R_2/R_1$. Positive values of Gr_{v} correspond to the heated outer cylinder and negative ones to the heated inner one.

2.3 The Basic State

The problem (9.62) to (9.66) admits a stationary axisymmetrical solution with a zero axial velocity uniform in the axial direction. Obviously, at small Re_{p} and Gr_{v} this solution will be unique.

Let us seek the velocity of the basic flow in the form

$$\mathbf{u} = Re_{\mathrm{p}}\,u(r)\mathbf{e}_\varphi \tag{9.68}$$

From (9.62) it follows that

$$u(r) = A_1 r + \frac{B_1}{r} \tag{9.69}$$

Constants A_1 and B_1 are found from the boundary conditions (9.65) and (9.66):

$$A_1 = -3\frac{R^2+1}{(R^2-1)^2}, \qquad B_1 = 3\frac{R^4+R^2+1}{(R^2-1)^2} \tag{9.70}$$

We have obtained the same velocity distribution as for the flow between two rotating cylinders in the case where the angular velocities of the inner and outer cylinders rotation are related as R^6. Such coincidence of flows stems from the fact that for the considered case (axial symmetry and uniformity in the z direction) the pulsational transport and non-isothermality do not affect the basic flow structure and contribute only to the pressure distribution. On the occurrence of a spontaneous symmetry breakdown owing to the loss of stability of the basic state, the above effects would manifest themselves.

The stationary temperature distribution in the basic state is given by the solution to the Laplace equation. Taking into account the boundary conditions, it follows that

$$\vartheta = \frac{\ln r}{\ln R} \tag{9.71}$$

2.4 Linear Stability Problem

Now we formulate the problem of the behavior of small disturbances of the basic state. Linearizing equations (9.62) to (9.64) about (9.68) and (9.71) and retaining the same notations for the velocity, temperature and pressure perturbations, we obtain

$$\begin{aligned}\frac{\partial u_r}{\partial t} + Re_{\mathrm{p}}\left[\frac{u(r)}{r}\frac{\partial u_r}{\partial \varphi} - 2\frac{u(r)}{r}u_\varphi - S(r)\left(\frac{\partial u_\varphi}{\partial r} + \frac{u_\varphi}{r} - \frac{1}{r}\frac{\partial u_r}{\partial \varphi}\right)\right] \\ = -\frac{\partial p}{\partial r} + \Delta u_r - \frac{u_r}{r^2} - \frac{2}{r^2}\frac{\partial u_\varphi}{\partial \varphi} + Gr_{\mathrm{v}}\, S(r)\vartheta\end{aligned} \tag{9.72}$$

$$\frac{\partial u_\varphi}{\partial t} + Re_{\mathrm{p}}\frac{u(r)}{r}\frac{\partial u_\varphi}{\partial \varphi} + 2A_1\, Re_{\mathrm{p}}\, u_r = -\frac{1}{r}\frac{\partial p}{\partial \varphi} + \Delta u_\varphi - \frac{u_\varphi}{r^2} + \frac{2}{r^2}\frac{\partial u_r}{\partial \varphi} \tag{9.73}$$

$$\frac{\partial u_z}{\partial t} + Re_{\mathrm{p}}\left[\frac{u(r)}{r}\frac{\partial u_z}{\partial \varphi} + S(r)\frac{1}{r}\frac{\partial u_z}{\partial \varphi} - S(r)\frac{\partial u_\varphi}{\partial z}\right] = -\frac{\partial p}{\partial z} + \Delta u_z \tag{9.74}$$

$$\frac{\partial \vartheta}{\partial t} + Re_{\mathrm{p}}\left[\frac{u(r) + S(r)}{r}\frac{\partial \vartheta}{\partial \varphi} + \frac{u_r}{r\ln R}\right] = \frac{1}{Pr}\Delta\vartheta \tag{9.75}$$

$$\frac{\partial u_r}{\partial r} + \frac{u_r}{r} + \frac{1}{r}\frac{\partial u_\varphi}{\partial \varphi} + \frac{\partial u_z}{\partial z} = 0 \tag{9.76}$$

Equations (9.72) to (9.76) are written in a cylindrical coordinate system. The boundary conditions are

$$u_\varphi = u_z = u_r = \vartheta = 0 \qquad \text{at } r = 1,\ R \tag{9.77}$$

The uniformity of the problem with respect to t, z and φ allows the analysis to be restricted to normal modes for which all the fields, i.e. u_r, u_z, u_φ, ϑ and p, are proportional to $\exp(\lambda t + \mathrm{i}kz + n\varphi)$. Here the axial and azimuthal wave-numbers k and n describe perturbations, and the increment λ is the eigenvalue of the problem. In terms of the normal modes, equations (9.72) to (9.76) take the form

$$\lambda u_r + Re_\mathrm{p}\left\{\frac{\mathrm{i}n}{r}u(r)u_r - \frac{2}{r}u(r)u_\varphi - S(r)\left(u_\varphi' + \frac{1}{r}u_\varphi - \frac{in}{r}u_r\right)\right]$$
$$= -p' + Du_r - \frac{u_r}{r^2} - 2\frac{\mathrm{i}n}{r^2}u_\varphi + Gr_\mathrm{v}\,S(r)\vartheta \tag{9.78}$$

$$\lambda u_\varphi + Re_\mathrm{p}\frac{\mathrm{i}n}{r}u(r)u_\varphi + 2A_1\,Re_\mathrm{p}\,u_r = -\frac{\mathrm{i}n}{r}p + Du_\varphi - \frac{u_\varphi}{r^2} + 2\frac{\mathrm{i}n}{r^2}u_r \tag{9.79}$$

$$\lambda u_z + \mathrm{i}\,Re_\mathrm{p}\left[\frac{n}{r}u(r)u_z + \frac{i}{r}S(r)u_z - kS(r)u_\varphi\right] = -\mathrm{i}kp + Du_z \tag{9.80}$$

$$\lambda\vartheta + \frac{\mathrm{i}n}{r}Re_\mathrm{p}[u(r) + S(r)]\vartheta + \frac{u_r}{r\ln R} = \frac{1}{Pr}D\vartheta \tag{9.81}$$

$$u_r' + \frac{1}{r}u_r + \frac{\mathrm{i}n}{r}u_\varphi + \mathrm{i}ku_z = 0 \tag{9.82}$$

Here

$$D = \frac{\partial^2}{\partial r^2} + \frac{2}{r}\frac{\partial}{\partial r} - \frac{n^2}{r^2} - k^2$$

and the prime denotes differentiation with respect to r.

2.5 A Narrow Gap

The set of equations (9.78) to (9.82) may be simplified significantly for the case of a narrow gap. Let us set $R = 1 + \varepsilon$ and stretch the radial coordinate

$$r = 1 + \varepsilon x \tag{9.83}$$

We also make the replacements

$$u_r = \varepsilon w, \qquad u_z = \varepsilon v, \qquad \vartheta = \varepsilon^2\,Pr\,\widetilde{\vartheta}, \qquad u_\varphi = 3\,Re_\mathrm{p}\,C, \qquad p = S(r)u_\varphi + \widetilde{p}$$

Then in the leading order in ε we obtain the set of equations

$$\widetilde{\lambda} w = -\widetilde{p}' + \widetilde{D} w + \widetilde{Ra}\,\widetilde{\vartheta} + \widetilde{Rd}\,C \tag{9.84}$$

$$\widetilde{\lambda} v = -\mathrm{i}\widetilde{k}\widetilde{p} + \widetilde{D} v \tag{9.85}$$

$$\widetilde{\lambda} C = \widetilde{D} C + w \tag{9.86}$$

$$\widetilde{\lambda}\, Pr\, \widetilde{\vartheta} = \widetilde{D}\vartheta + w \tag{9.87}$$

where the following notations are introduced:

$$\widetilde{\lambda} = \varepsilon^2 \lambda + \frac{5}{4}\mathrm{i} n Re_{\mathrm{p}}, \qquad \widetilde{k} = \varepsilon k, \quad \widetilde{D} = \frac{\partial^2}{\partial x^2} - \widetilde{k}^2,$$
$$\widetilde{Ra} = -\frac{1}{2} Pr\, \varepsilon\, Gr_{\mathrm{v}}, \qquad \widetilde{Rd} = \frac{27}{2} Re_{\mathrm{p}}^2 \tag{9.88}$$

The boundary conditions for (9.84) to (9.87) are

$$w = v = C = \widetilde{\vartheta} = 0 \qquad \text{at } x = 0 \ \text{ and } \ x = 1 \tag{9.89}$$

The formulated problem formally coincides with the stability problem for the equilibrium of a binary mixture in a horizontal layer under equilibrium vertical gradients of temperature and concentration. In such an interpretation, $\widetilde{Ra}$ and $\widetilde{Rd}$ are the thermal and concentrational Rayleigh numbers, C the concentration of the light component, w and v the vertical and horizontal components of the velocity and $\widetilde{\lambda}$ the increment of perturbations. The wavenumber transformation means that the reference size for perturbations is the thickness of the gap and not the cylinder radius. Due to the same reasons (the change of the reference size), the parameter Gr_{v} is multiplied by ε. The transformation of λ reflects that in the considered limiting case, where the axial wavenumbers tend to infinity and azimuthal ones remain finite, n affects just the disturbance frequency. The nature of this influence may be easily understood. In the limiting case of a narrow gap from (9.67) and (9.68) follows that

$$u(r) + S(r) = \frac{5}{4}\frac{1}{\varepsilon^2} \tag{9.90}$$

Thus, transformation (9.88) means that at non-zero n the perturbations have the form of spiral rolls. These rolls rotate with a velocity equal to the sum of the average flow velocity and the velocity of the pulsational transport, i.e. with a velocity equal to the average Lagrange one of the particles suspended in a flow. Certainly, this system of spiral rolls may additionally move along the cylinder axis that corresponds to the appearance of an imaginary part of $\widetilde{\lambda}$.

For neutral monotonic (in the sense that Im $\widetilde{\lambda} = 0$) perturbations the problem (9.84) to (9.87) simplifies even more. Introducing the function

$$\widetilde{T} = \frac{\widetilde{Rd}\, C + \widetilde{Ra}\, \widetilde{\vartheta}}{\widetilde{Rd} + \widetilde{Ra}} \tag{9.91}$$

we obtain the equations

$$-\widetilde{p}' + \widetilde{D}w + \widehat{Ra}\, \widetilde{T} = 0 \tag{9.92}$$

$$-\mathrm{i}\, \widetilde{k}\, \widetilde{p} + \widetilde{D}v = 0 \tag{9.93}$$

$$\widetilde{D}\, \widetilde{T} + w = 0 \tag{9.94}$$

with the boundary conditions

$$w = v = \widetilde{T} = 0 \qquad \text{at } \ x = 0 \ \text{ and } \ x = 1 \tag{9.95}$$

where we introduced the notation $\widehat{Ra} = \widetilde{Ra} + \widetilde{Rd}$.

The problem (9.92) to (9.95) coincides with the classic Rayleigh–Benard problem where the role of the Rayleigh number is now played by $\widehat{Ra}$. It is known that for the most dangerous perturbations $\widehat{Ra} = \widehat{Ra}^* \approx 1707.762$. Hence, with respect to monotonous (at $n = 0$) perturbations, we may write for the stability boundary

$$\frac{27}{2} Re_{\mathrm{p}}^2 - \frac{1}{2} \varepsilon\, Pr\, Gr_{\mathrm{v}} = \widehat{Ra}^* \tag{9.96}$$

It can be seen that the instability takes place in the absence of heating ($Gr_{\mathrm{v}} = 0$) as well. Moreover, the heating has a stabilizing effect if the outer cylinder is heated ($Gr_{\mathrm{v}} > 0$) and a destabilizing effect in the case of a heated inner cylinder. This complies well with the analogy noted in Section 3 in Chapter 6 between thermovibrational and thermal buoyancy convections, the role of the gravitational potential being played by the kinetic energy of pulsations. Indeed, since the pulsations are more intense near the inner cylinder, the effective gravitational field will be directed from the outer cylinder to the inner one. Thus, the heating of the outer cylinder corresponds to the 'heating from above' regime.

2.6 Numerical Results

The calculations were carried out by the Tau–Chebyshev method [10]. Detailed results for the isothermal case are reported in [11]. Here we give an example concerning the narrow gap at a non-zero temperature difference. In Fig. 165 the squared minimal values of Re_{p} versus Gr_{v} are plotted for $Pr = 6.75$, $R = 1.05$ and $n = 0$. For all the negative values of Gr_{v} (heated inner cylinder, the case

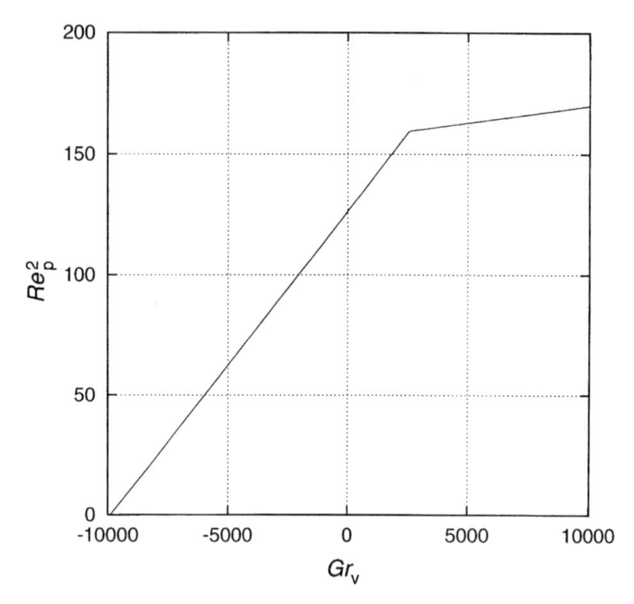

Fig. 165 The stability border in the plane $(Re_p^2,\ Gr_v)$ at $R = 1.05$

of 'heating from below'), the instability is monotonous and is well described by dependence (9.96). Everywhere in this parameter range heating leads to destabilizing of the system. At some value Gr_v^*, the critical value of Re_p turns to zero. It follows from (9.96) that Gr_v^* is determined by the expression

$$Gr_v^* = -2\,\frac{\widehat{Ra}^*}{\varepsilon\,Pr} \tag{9.97}$$

Substituting numerical values into (9.97), we obtain $Gr_v^* = -10\,120$, and the calculation leads to $Gr_v^* \approx -9885$. Formula (9.97) may be used as the leading term of an asymptotic expansion of Gr_v^* into a power series in ε. Since (9.97) includes ε^{-1}, one may anticipate that the next term of the expansion does not depend on ε and the difference between the asymptotic and numerical values of Gr_v^* for given R will be nearly the same for other R. The calculations confirm that. In Fig. 166 the values of Gr_v^* obtained numerically (solid line) and found from (9.97) are plotted for different R. It can be seen that the difference in value of Gr_v^* remains nearly the same up to $R = 1.1$.

At $Gr_v > 0$ the heating stabilizes the system and starting from some value Gr_v the monotonous instability is replaced by an oscillatory one. This corresponds well with the known result of the instability of a binary mixture [12] where the thermal Prandtl numbers are greater than the diffusional ones. Recall that in our case the effective diffusional Prandtl number is equal to unity.

$$\operatorname{div}\ \mathbf{u} = 0 \tag{9.64}$$

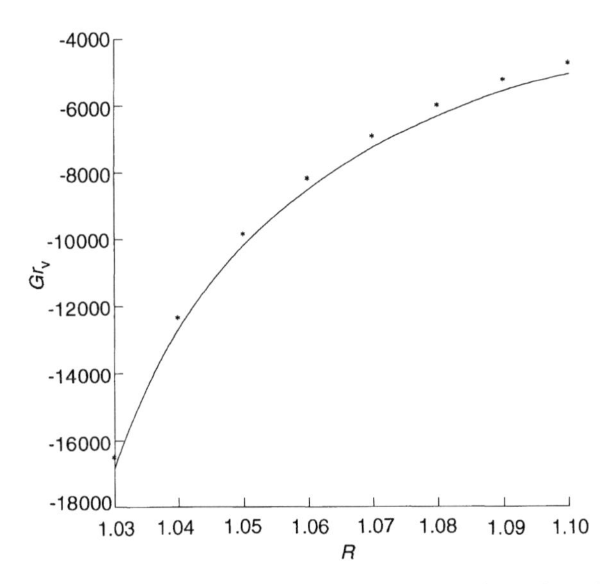

Fig. 166 Critical values of Gr_v at $Re_p = 0$ as a function of R. The points denote the numerical results; the line is the asymptotic result in the narrow-gap approximation

Transition to the oscillatory instability is accompanied by the occurrence of a break on the dependence of Re_p on Gr_v, which is related to the transition to a different leaf of the neutral surface.

3 Average Flows and Heat Transfer in a Liquid Zone Subject to Axial Vibrations

In this section we consider the average flows and heat transfer in a liquid zone when its end walls perform high-frequency axial vibrations. In this situation, the non-uniformities of both amplitude and phase of oscillations cause the free surface to deform, and one more vibrational effect emerges. It is different from the mechanisms working in the example considered in the previous section which are the Schlichting flow, first-order thermovibrational convection and pulsational transport of vorticity and heat. This extra effect is the generation of an average flow by waves propagating along the free surface from the vibrating rigid end walls. The account is based mainly on paper [13].

3.1 Averaged Flows in the Isothermal Case

We shall consider the situation where in the absence of vibration the zone is a cylinder of radius R and length $2L$ (Fig. 167). The average deformation of the

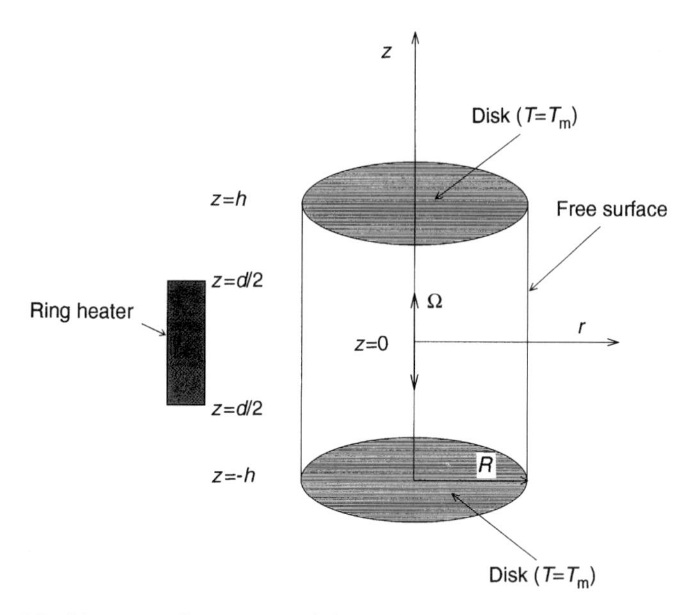

Fig. 167 Liquid zone and geometry of the problem

free surface is neglected. Both end walls perform synchronous high-frequency vibrations with the frequency Ω and amplitude b along the direction of the zone axis according to the sinusoidal law

$$\dot{z} = 2b\Omega \cos \Omega t \tag{9.98}$$

(z is the axial coordinate).

As shown in Section 2 in Chapter 7, under ordinary conditions ensuring the high-frequency approximation, the pulsation velocity field is potential so that the pulsation velocity $\mathbf{v}_\mathrm{p}$ in the laboratory reference frame is given by the following relations:

$$\mathbf{v}_p = b\Omega[\mathbf{W}\exp(\mathrm{i}\Omega t) + \mathbf{W}^*\exp(-\mathrm{i}\Omega t)] \tag{9.99}$$

$$\mathbf{W} = \nabla\Phi, \qquad \Delta\Phi = 0 \tag{9.100}$$

where $\mathbf{W}$ is the dimensionless complex amplitude of the pulsation velocity defined in $b\Omega$ units.

On the end walls, the pulsation velocity satisfies the impermeability condition:

$$\frac{\partial\Phi}{\partial z} = 1 \qquad \text{at } z = \pm L \tag{9.101}$$

On the free surface $r = R$ we assume that the boundary conditions by Ruvinsky and Freidman discussed in Section 2 in Chapter 7 hold. These conditions are the dynamic stress balance condition and the kinematic condition accounting for the viscous effects in the skin layer near the free surface. For the considered configuration, under the assumption of an axial symmetry they take the following form:

$$\mathrm{i}\Omega\Phi = 2\nu\frac{\partial^2\Phi}{\partial z^2} + \frac{\alpha}{\rho\Omega}\left(\frac{\zeta}{R^2} + \frac{\partial^2\zeta}{\partial z^2}\right) \tag{9.102}$$

$$\mathrm{i}\Omega\zeta = \Omega\frac{\partial\Phi}{\partial r} - 2\mathrm{i}\nu\frac{\partial^3\Phi}{\partial r\partial z^2} \tag{9.103}$$

where ζ is the amplitude of the pulsational deviation of the surface from the cylindrical one defined in the units b, r is the radial coordinate, ν is the fluid viscosity and α is the surface tension coefficient.

The equations for the average flow were discussed in Section 3 in Chapter 6. Axial vibrations of the end walls generate the surface waves propagating along the free surface from the end walls to the center of the zone. This results in a non-uniformity of the pulsation phase. Moreover, the phase of pulsations will be non-uniform not only on the free surface but in the interior as well. Thus, we need to apply the general equations taking into account the pulsational transport effect:

$$\frac{\partial\mathbf{u}}{\partial t} - (\mathbf{u} + \mathbf{S}) \times \operatorname{curl}\mathbf{u} = -\frac{1}{\rho}\nabla p + \nu\Delta\mathbf{u} \tag{9.104}$$

$$\operatorname{div}\mathbf{u} = 0 \tag{9.105}$$

where $\mathbf{u}$ is the average velocity, the Bernoulli pressure of the pulsation field is included into p and $\mathbf{S}$ is the pulsational transport vector introduced in Section 3 in Chapter 6 and given by the formulas

$$\mathbf{S} = \overline{(\mathbf{q}\nabla)\mathbf{v}_\mathrm{p}}, \qquad \frac{\partial\mathbf{q}}{\partial t} = \mathbf{v}_\mathrm{p}, \qquad \overline{\mathbf{q}} = 0 \tag{9.106}$$

Using (9.99) we may rewrite the expression for $\mathbf{S}$ in the form

$$\mathbf{S} = \tfrac{1}{2}b^2\Omega\,\mathrm{Im}\,[(\mathbf{W}\nabla)\mathbf{W}^*] \tag{9.107}$$

Let us restrict ourselves to the axisymmetrical flows in which only the radial u_r and axial u_z components of the average velocity are non-zero. In this case it is convenient to introduce the stream function ψ and the vorticity φ of the average flow:

$$u_r = -\frac{1}{r}\frac{\partial\psi}{\partial z}, \qquad u_z = \frac{1}{r}\frac{\partial\psi}{\partial r}, \qquad \varphi = -\frac{1}{r}\left(\frac{\partial^2\psi}{\partial r^2} - \frac{1}{r}\frac{\partial\psi}{\partial r} + \frac{\partial^2\psi}{\partial z^2}\right) \tag{9.108}$$

Let us also introduce the stream function ψ_s for the pulsational transport. Using (9.107), it is not difficult to obtain the explicit expression for ψ_s:

$$\psi_s = \frac{1}{2} b^2 \Omega \, \mathrm{Im}\left(r \frac{\partial \Phi}{\partial r} \frac{\partial \Phi^*}{\partial z} \right) \tag{9.109}$$

As shown in Section 3 in Chapter 6, the total mass transport vector equals $\rho(\mathbf{u} + \mathbf{S})$. Then the impermeability condition is to be imposed on the sum $\mathbf{u} + \mathbf{S}$, i.e. the condition

$$\psi + \psi_s = 0 \tag{9.110}$$

must hold on all the boundaries.

Taking into account the generation of the average vorticity in the dynamic skin layers near the rigid and free boundaries, we require that on the end walls the Schlichting-type condition is (see Section 1 in Chapter 7)

$$\text{at } z = \pm L: \qquad \frac{\partial \psi}{\partial z} = \frac{1}{4} b^2 \Omega r \left\{ 3\,\mathrm{Re}\left[(1+\mathrm{i}) \frac{\partial \Phi}{\partial r} \frac{\partial^2 \Phi^*}{\partial r^2} \right] + \frac{2}{r} \left| \frac{\partial \Phi}{\partial r} \right|^2 \right\} \tag{9.111}$$

and on the free surface the Longuet–Higgins condition is (see Section 3 in Chapter 7)

$$\text{at } \ r = 1: \qquad \varphi = 2b^2 \Omega \, \mathrm{Re}\left(\frac{\partial \zeta}{\partial z} \frac{\partial^2 \Phi^*}{\partial z^2} \right) \tag{9.112}$$

must be satisfied.

Let us write down the complete set of equations and boundary conditions in the dimensionless form using the values R, R^2/ν and νR as the scales for length, time and the stream function, respectively:

$$\frac{\partial \varphi}{\partial t} + \frac{1}{r}\left(\frac{\partial \psi_e}{\partial r} \frac{\partial \varphi}{\partial z} - \frac{\partial \psi_e}{\partial z} \frac{\partial \varphi}{\partial r} + \frac{\partial \psi_e}{\partial z} \frac{\varphi}{r} \right) = \frac{\partial^2 \varphi}{\partial r^2} + \frac{1}{r} \frac{\partial \varphi}{\partial r} - \frac{1}{r^2} \varphi + \frac{\partial^2 \varphi}{\partial z} \tag{9.113}$$

$$\varphi = -\frac{1}{r}\left(\frac{\partial^2 \psi}{\partial r^2} - \frac{1}{r} \frac{\partial \psi}{\partial r} + \frac{\partial^2 \psi}{\partial z^2} \right) \tag{9.114}$$

$$\psi_e = \psi + \psi_s, \qquad \psi_s = \frac{1}{2} Re_p \, \mathrm{Im}\left(r \frac{\partial \Phi}{\partial r} \frac{\partial \Phi^*}{\partial z} \right) \tag{9.115}$$

$$\Delta \Phi = 0 \tag{9.116}$$

at $z = \pm l$:

$$\psi_e = 0 \tag{9.117}$$

$$\frac{\partial \psi}{\partial z} = \frac{1}{4} Re_p \, r \left\{ 3\mathrm{Re}\left[(1+\mathrm{i}) \frac{\partial \Phi}{\partial r} \frac{\partial^2 \Phi^*}{\partial r^2} \right] + \frac{2}{r} \left| \frac{\partial \Phi}{\partial r} \right|^2 \right\} \tag{9.118}$$

$$\frac{\partial \Phi}{\partial z} = 1 \tag{9.119}$$

at $r = 1$:

$$\psi_e = 0 \tag{9.120}$$

$$\varphi = 2Re_p \, \mathrm{Re}\left(\frac{\partial \zeta}{\partial z} \frac{\partial^2 \Phi^*}{\partial z^2} \right) \tag{9.121}$$

$$\mathrm{i}\zeta = \frac{\partial \Phi}{\partial r} - 2\mathrm{i}\omega^{-1} \frac{\partial^3 \Phi}{\partial r \, \partial z^2} \tag{9.122}$$

$$\mathrm{i}\Phi = 2\omega^{-1} \frac{\partial^2 \Phi}{\partial z^2} + We^{-1} \left(\zeta + \frac{\partial^2 \zeta}{\partial z^2} \right) \tag{9.123}$$

For the pulsational displacement of a free surface on the contact line we accept the stuck-edge condition:

$$\zeta(\pm l) = 0 \tag{9.124}$$

The problem is characterized by four dimensionless parameters: the pulsational Reynolds number $Re_p = b^2 \Omega / \nu$, the Weber number $We = \rho \Omega^2 R^3 / \alpha$, the dimensionless frequency $\omega = \Omega R^2 / \nu$ and the geometrical parameter $l = L/R$.

3.2 The Pulsation Field

It can be seen from the complete set of equations and boundary conditions that the problem for the pulsation fields decouples and may be solved independently: equation (9.116) should be solved with the boundary conditions (9.119), (9.122) to (9.124), whereas the pulsation field structure is determined by the geometry and the vibration law. To find the solution, we present the potential Φ as a sum of a linear function of z and the axial eigenfunctions of the problem for a liquid zone that is infinite in the axial direction:

$$\Phi = z + \sum_{n=0}^{\infty} a_n I_0(k_n r) \sin k_n z \tag{9.125}$$

where $k_n = \pi(1+2n)/2l$ and I_0 is the modified Bessel function of the zeroth order.

Substituting (9.125) into the dynamic boundary condition (9.123) and taking into account the stuck-edge condition (9.124) we arrive at the expression for the displacement of the free surface from its equilibrium position:

$$\zeta = -\mathrm{i}\, We\left(F\frac{\sin z}{\sin l} - z - \sum_{n=0}^{\infty} b_{\mathrm{n}}(-1)^n \sin k_n z\right) \tag{9.126}$$

where the notation F is introduced:

$$F = l + \sum_{n=0}^{\infty} b_n \tag{9.127}$$

and the coefficients b_n are connected with a_n by the relation

$$a_n = (-1)^n b_n \frac{1-k_n^2}{I_0(k_n)(1-2\mathrm{i}\omega^{-1}k_n^2)} \tag{9.128}$$

Substituting expansions (9.125) and (9.126) into the kinematic condition (9.122) we get

$$b_n = \frac{2}{lG_n}\left(\frac{F\cot(l)}{k_n^2-1} - \frac{1}{k_n^2}\right) \tag{9.129}$$

where

$$G_n = 1 + We^{-1}\, k_n(1-k_n^2)\frac{I_1(k_n)}{I_0(k_n)}\frac{1+2\mathrm{i}\omega^{-1}k_n^2}{1-2\mathrm{i}\omega^{-1}k_n^2} \tag{9.130}$$

($I_1(k_n)$ is the modified Bessel function of the first order). Substituting (9.129) into (9.130) and resolving the obtained relation for F, we obtain

$$F = \frac{l + \sum_{n=0}^{\infty}\beta_n}{1+\sum_{n=0}^{\infty}\gamma_n}, \qquad \beta_n = -\frac{2}{lG_n k_n^2}, \qquad \gamma_n = -\frac{2\cot(l)}{lG_n(k_n^2-1)} \tag{9.131}$$

Formulas (9.128) to (9.131) together with (9.125) and (9.126) yield the exact solution of the problem for the pulsation velocity field and free surface pulsations.

3.3 Numerical Results

Sums of the series included in the formula for the pulsation fields are found numerically by direct summing up. The number of retained terms is determined by the needed accuracy. The average fields are found from (9.113) and (9.114) with the boundary conditions (9.117), (9.118), (9.120) and (9.121) by the finite difference method using the ADI scheme. The ratio of the zone length to the radius was fixed: $l = 1$.

At small values of Re_p, in the creeping flow regime, the symmetry of the equations and boundary conditions with respect to $z = 0$ leads to the symmetry of the solution. With the increase of Re_p, the relative role of the non-linear terms grows and one might expect a symmetry break through some bifurcation. However, test calculations done for several values of parameters from the investigated range have shown that the symmetry of the solution remains intact. This enabled calculations to be performed in half of the zone using the symmetry conditions at $z = 0$.

Before proceeding to the numerical calculation results proper, let us first discuss the effect of the vibration frequency. Since it is included both in ω, Re_p and We, it is convenient to introduce two independent combinations of the given parameters which do not contain frequency. As one of these parameters one may take the dimensionless vibration amplitude:

$$\frac{b}{R} = \sqrt{\frac{Re_p}{\omega}} \tag{9.132}$$

As the other, the Onezorge number

$$On = \nu\sqrt{\frac{\rho}{\alpha R}} = \frac{\sqrt{We}}{\omega}$$

may be chosen, since it does not contain any vibration characteristics.

Most of the calculations were carried out at $On = 0.22 \times 10^{-3}$. The chosen Onezorge number is inherent to liquid metals at the zone size approximately equal to one centimeter. The Weber number We has been taken as the parameter characterizing the frequency. Since the experimental data [14] testify that the high-frequency limit for a generation of average flows in the boundary layer is achieved only at the frequencies $\omega > 10^4$, one must take a Weber number large enough. For the chosen Onezorge numbers, We should be at least several units.

Actually one has to choose a Weber number even greater. The point is that at small We large-scale oscillations of the free surface, which are weakly damped by viscosity, may be excited in a resonance way. However, the non-linear limitation of the amplitude is beyond the scope of the present theory employed here. At large Weber numbers only high resonances may emerge. These resonances correspond to the motions with sufficiently small scales, so that the viscosity could effectively affect them.

Besides suppressing resonances, the viscosity will cause another important effect. At sufficiently high viscosity, the oscillations of the free surface do not form a set of standing waves but have a considerable component corresponding to the waves propagating from the end walls to the central part of the zone. This causes entrainment of the fluid. At very large viscosity, due to significant damping of the waves, the major part of the free surface will be at rest, and therefore the surface-flow effect will be suppressed as well. It is natural to assume that the optimal set of parameters to generate the average flow is the one where the damping length $l_{\rm d}$ of the surface waves has the same order of magnitude as the size of the zone. It is known that at weak dissipation $l_{\rm d}$ is equal to the ratio of the group velocity to the decrement. For short waves the curvature of the free surface cross-section is irrelevant. Thus, as a rough estimation, one may use the dispersion relation for the capillary waves on a plane surface. This leads to $l_{\rm d} \approx 3\omega/(4We)$, i.e. the damping length will significantly exceed the zone length up to very large *We* values. Consequently, the increase of *We* in general yields amplification of the near-surface flow.

The calculations show that the average flow formed by the coupled Schlichting and surface-wave mechanisms has a four-vortex structure. Two symmetrical vortices localized near the free surface are generated by the surface waves propagating along the free surface from the oscillating end walls. The direction of circulation in these vortices is such that the fluid moves from the end walls to the central part of the zone along the free surface. Two symmetrical vortices of the Schlichting type are situated near the two rigid end walls. In these vortices the fluid moves along the rigid surfaces towards the axis of the zone, i.e. it quits the regions of high pulsation energy.

Figures 168a to d, where the stream lines of the average mass transport (lines of constant values of $\psi_{\rm e}$) are presented, illustrate the change of the average flow structure with the increase in the vibration frequency. The corresponding calculations were carried out at fixed values of Onezorge number whereas the parameters *We*, ω and $Re_{\rm p}$ were varied compatibly. As the calculations show, at small values of *We* (Fig. 168a) the contribution of the Schlichting mechanism of the average flow generation is higher than that of the surface-wave flow (except for the narrow resonant ranges where a large amplitude of the surface waves leads to a sharp amplification of the near-surface vortices). At higher values of *We*, due to a significant effect of entrainment, the near-surface vortices play the dominating role in the ranges between the resonances as well (Figs. 168b to d).

Thus, the numerical results prove the existence of the parameter range in which the resulting flow has the structure of two dominating vortices located close to the free surface. Note that under ordinary conditions of liquid zone heating (a hot center and cold end walls) the flow induced by the normal thermocapillary effect near the free surface is directed from the zone center to its walls, whereas the surface-wave mechanism induces the motion in the opposite direction. Thus, the thermocapillary and surface-wave mechanisms are competing, and one may expect their mutual suppression under certain conditions.

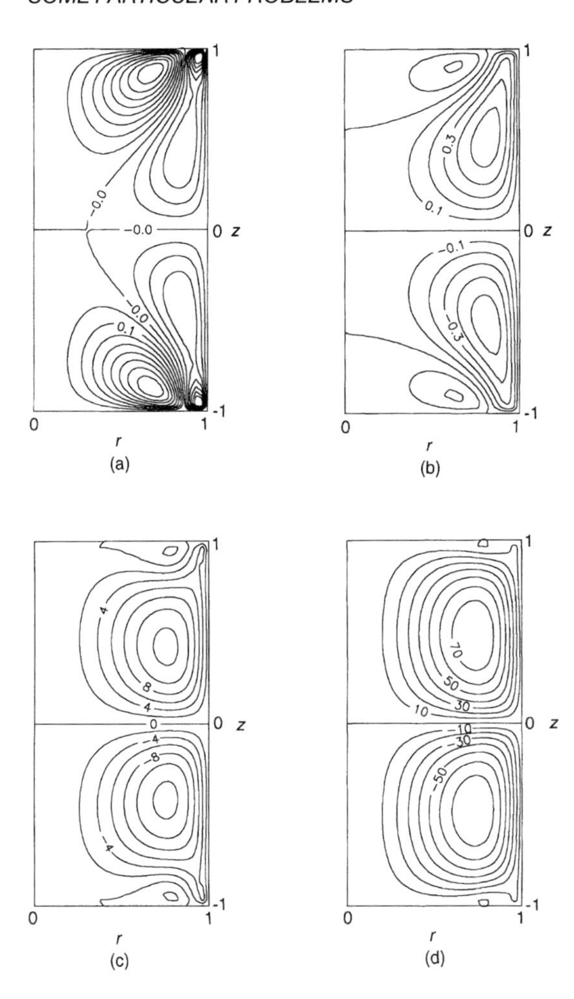

Fig. 168 Structure of the vibrational flow in an isothermal liquid zone ($Ra_v = 0$, $Ma = 0$, $On = 0.22 \times 10^{-3}$): (a) $We = 1000$, (b) $We = 1200$, (c) $We = 1800$, (d) $We = 3500$

3.4 Convective Flows in a Heated Liquid Zone

Let us consider the vibration flows and heat transfer in a heated liquid zone. Now, unlike the isothermal case discussed above, there appears to be a thermovibrational volumic mechanism of the average flow generation related to non-uniformity of heating. The average flow in a heated liquid zone subject to high-frequency vibrations was considered in [15] to [19]. For the first time the influence of axial high-frequency vibrations on the average flow and heat transfer in the floating zone was studied in [15] on the basis of the conventional approach. In that paper the possibility of partial suppression of a thermocapillary flow by vibrations was demonstrated and the orienting effect of vibrations on the temperature field was demonstrated. The same results were obtained on the basis of the same approach in [16]. Later on, the same problem was studied in the framework of the new approach (see [17] to [19]). The thermovibrational effect of the first order with respect to the Boussinesq parameter and the interaction of this mechanism with the Schlichting, wave-induced and thermocapillary flows were analyzed. Numerical investigation of the influence of the residual accelerations on the behavior of the heated liquid zone has been carried out in [20], taking into account the thermocapillary effect. In [21] the surface vibration-induced flow was studied experimentally for the half-zone configuration. The same situation was studied theoretically in [22] for low-frequency vibrations.

We consider the case of a ring heater positioned in the central part of the zone. The end walls are assumed to be at a constant temperature. In this case three various mechanisms of the average flow generation are set to work. Two of them are isothermal and related to the existence of the boundary layers near rigid and free surfaces. The third mechanism is of the thermovibrational origin. We are interested in the interaction of these mechanisms and the thermocapillary convection. We will now discuss the thermal boundary conditions. Let the temperature be constant on the end walls of the zone:

$$T = T_{\mathrm{m}} \qquad \text{at } z = \pm L \tag{9.133}$$

On the free surface the heat flux condition is adopted:

$$\frac{\partial T}{\partial r} = -\epsilon(T - T_{\mathrm{a}}) \tag{9.134}$$

where ϵ is the effective coefficient of the heat flux, T_{a} is the temperature of the exterior medium depending on the axial coordinate and modeling of the ring heater and

$$T_{\mathrm{a}}(z) = T_{\mathrm{m}} + \Delta T \exp\left(-\frac{z^2}{d^2}\right)$$

Here ΔT is the temperature difference between the heater and the end walls and d is the effective thickness of the heater.

As before, we consider the solutions with axial symmetry. Taking ΔT as the scale for temperature and choosing T_{m} as the initial temperature, we write the equations and the boundary conditions in the dimensionless form:

$$\begin{aligned}\frac{\partial \varphi}{\partial t} + \frac{1}{r}\left(\frac{\partial \psi_{\mathrm{e}}}{\partial r}\frac{\partial \varphi}{\partial z} - \frac{\partial \psi_{\mathrm{e}}}{\partial z}\frac{\partial \varphi}{\partial r} + \frac{1}{r}\frac{\partial \psi_{\mathrm{e}}}{\partial z}\varphi\right) + \frac{Ra_{\mathrm{v}}}{Pr}\left(\frac{\partial E}{\partial r}\frac{\partial T}{\partial z} - \frac{\partial E}{\partial z}\frac{\partial T}{\partial r}\right)& \\ = \frac{\partial^2 \varphi}{\partial r^2} + \frac{1}{r}\frac{\partial \varphi}{\partial r} - \frac{1}{r^2}\varphi + \frac{\partial^2 \varphi}{\partial z^2}&\end{aligned} \tag{9.135}$$

$$\frac{\partial T}{\partial t}+\frac{1}{r}\left(\frac{\partial \psi_{\mathrm{e}}}{\partial r}\frac{\partial T}{\partial z}-\frac{\partial \psi_{\mathrm{e}}}{\partial z}\frac{\partial T}{\partial r}\right)=\frac{1}{Pr}\Delta T \tag{9.136}$$

$$\Delta\Phi = 0 \tag{9.137}$$

$$z=\pm l: \qquad T=0, \qquad \psi_{\mathrm{e}}=0$$

$$\frac{\partial \psi}{\partial z}=\frac{1}{4}Re_{\mathrm{p}}\, r\left\{3Re\left[(1+\mathrm{i})\frac{\partial \Phi}{\partial r}\frac{\partial^2 \Phi^*}{\partial r^2}\right]+\frac{2}{r}\left|\frac{\partial \Phi}{\partial r}\right|^2\right\} \tag{9.138}$$

$$r=1: \qquad \frac{\partial T}{\partial r}=-Bi(T-T_{\mathrm{a}}), \qquad \psi_{\mathrm{e}}=0$$

$$\varphi = 2Re_{\mathrm{p}}\,\mathrm{Re}\left(\frac{\partial \zeta}{\partial z}\frac{\partial^2 \Phi^*}{\partial z^2}\right)+\frac{Ma}{Pr}\frac{\partial T}{\partial z} \tag{9.139}$$

Here $T_{\mathrm{a}} = \exp(-z^2/\delta^2)$, $E = W^2$ and the following dimensionless parameters are introduced: the vibrational Rayleigh number $Ra_{\mathrm{v}} = \beta\Delta T(b\Omega R)^2/(4\nu\chi)$, the Marangoni number $Ma = \alpha'_{\mathrm{T}}\Delta TR/(\rho\nu\chi)$, the Prandtl number $Pr = \nu/\chi$, the Biot number $Bi = R\epsilon$ and the parameter $\delta = \delta/R$.

Calculations were carried out for the following fixed values: $Pr = 0.02$, $Bi = 2$, $l = 1$, $\delta = 0.5$. The parameters Ma, Ra_{v}, Re_{p}, ω and We were varied. As

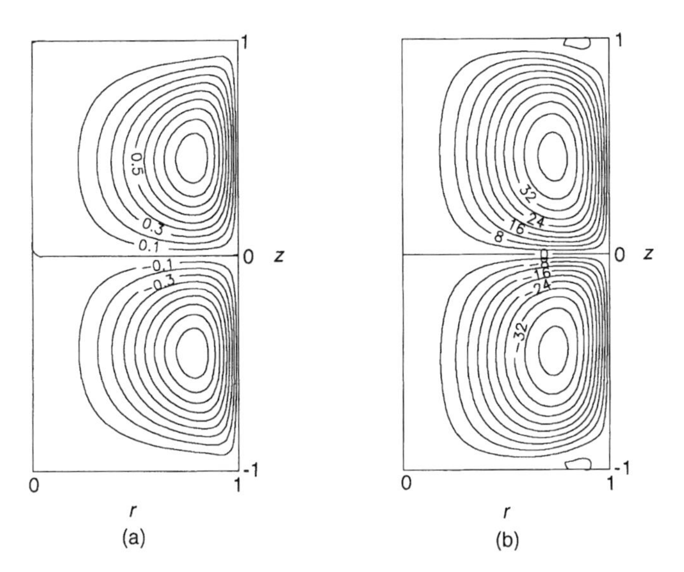

Fig. 169 Structure of the vibrational flow in a heated liquid zone in the absence of the thermocapillary effect ($Ma = 0$, $On = 0.22 \times 10^{-3}$, $We = 7500$, $Ra_{\mathrm{v}}/Re_{\mathrm{p}} = 46.5$): (a) $Ra_{\mathrm{v}} = 18$, (b) $Ra_{\mathrm{v}} = 450$

the calculations show, the flow of a purely thermovibrational origin has a two-vortex structure. Moreover, the direction of motion is the same as that of the flow induced by the surface waves, i.e. the fluid moves from the end walls to the zone center along the free surface.

Let us discuss the structure of the flow generated by the joint action of all the vibrational mechanisms. As the calculations show, at small values of *We*, when

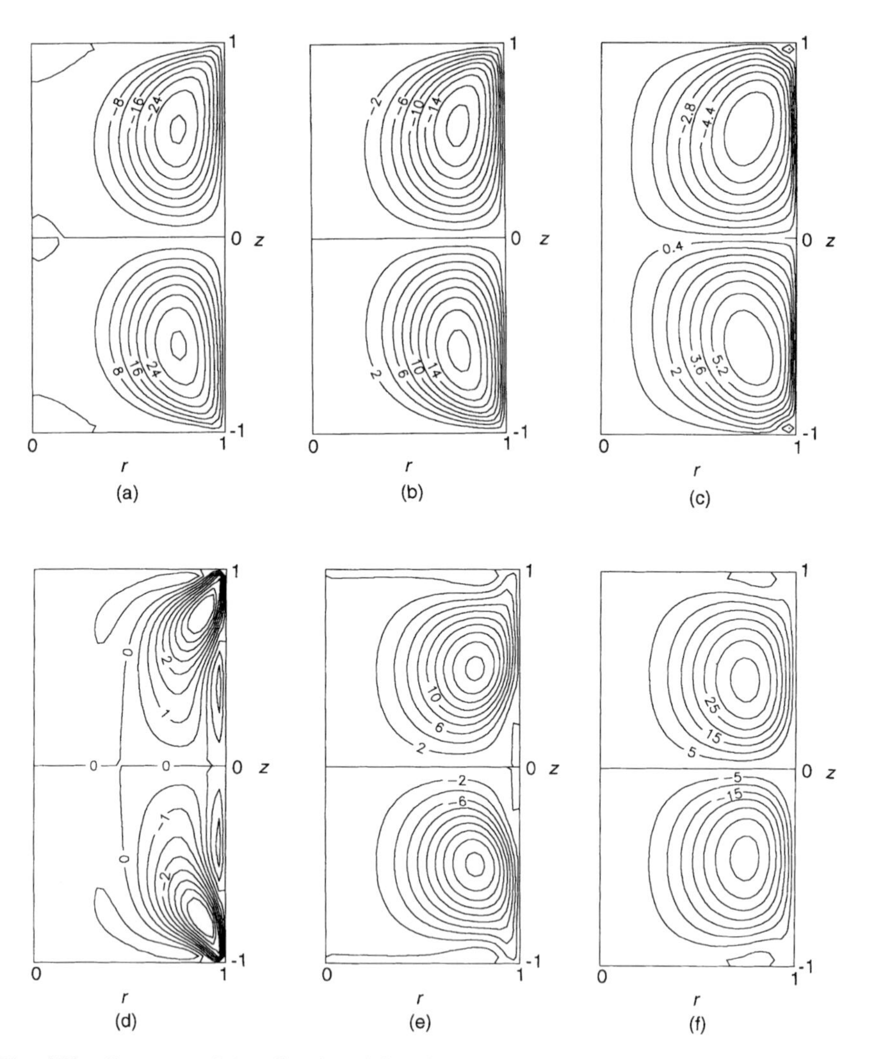

Fig. 170 Structure of the vibrational flow in a heated liquid zone in the presence of the thermocapillary effect ($Ma = 200$, $On = 0.22 \times 10^{-3}$, $We = 7500$, $Ra_v/Re_p = 46.5$): (a) $Ra_v = 18$, (b) $Ra_v = 882$, (c) $Ra_v = 1458$, (d) $Ra_v = 1800$, (e) $Ra_v = 2178$, (f) $Ra_v = 2592$

the relative role of the Schlichting flow is high enough, the structure of the resulting flow is complicated. At higher values of *We* (see Fig. 169) the structure of the flow is close to that of a purely thermovibrational one. This is related to the fact that now the Schlichting flow intensity is relatively small, and the structures of the thermovibrational flow and the flow induced by the surface waves are close. Thus, there is a sufficiently wide range of parameters where the resulting flow has the structure of two vortices with the circulation direction opposite to that of the thermocapillary flow.

In Figs. 170 and 171 the numerical results for convective flows in the presence of the thermocapillary effect are presented. Figs. 170a to f illustrate the change of the resulting flow structure with the increase in the vibrational Rayleigh number. The corresponding calculations were carried out at a fixed ratio of the vibrational Rayleigh number to the pulsational Reynolds number: $Ra_v/Re_p = 46.5$. This means that we analyze the phenomena occurring when the vibration amplitude increases while the frequency remains constant. It can be seen that the flow structure is close to that of the purely thermocapillary one up to $Ra_v \sim 2 \times 10^3$. However, the intensity of the resulting flow abruptly decreases with the increase in the vibration intensity in this range. This is clear from Fig. 171, where the dependences of the total kinetic energy of the average flow on the vibrational Rayleigh number are presented for two different values of the Marangoni number. The kinetic energy is normalized to its value at

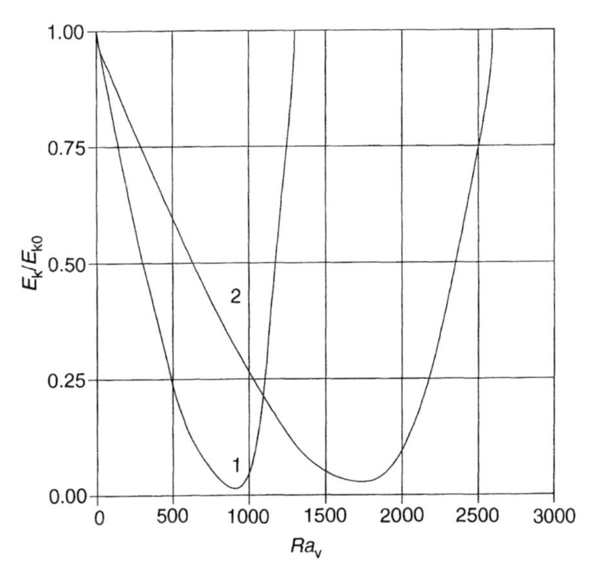

Fig. 171 Total kinetic energy versus vibrational Rayleigh number ($On = 0.22 \times 10^{-3}$, $We = 7500$, $Ra_v/Re_p = 46.5$): (a) $Ma = 100$, (b) $Ma = 200$

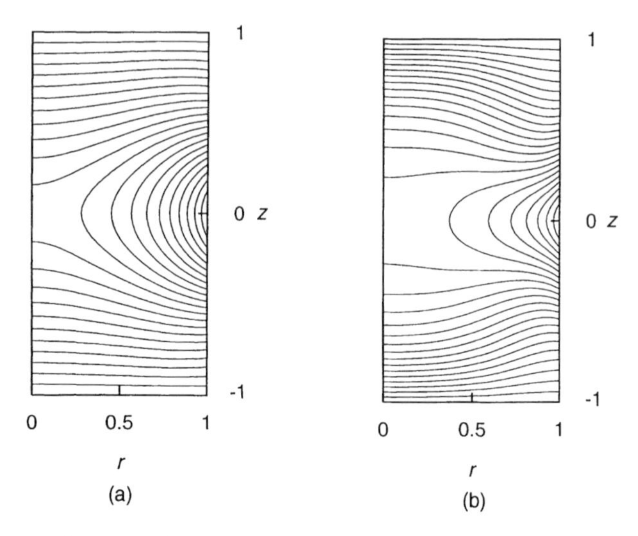

Fig. 172 The effect of vibrational convection on the temperature field ($Ma = 0$, $On = 0.22 \times 10^{-3}$, $We = 7500$, $Ra_v/Re_p = 46.5$): (a) $Ra_v = 18$, (b) $Ra_v = 1800$

$Ra_v = 0$. It can be seen that there is a range of parameters in which a considerable suppression of the thermocapillary flow by vibrations takes place. At yet higher values of Ra_v, vibrations play the dominating role: the structure of the resulting flow is close to that of the flow in the absence of the thermocapillary effect, and the total kinetic energy exceeds E_{k0}.

Vibrations of high enough intensity strongly affect the temperature field. It can be seen from Fig. 172 that the vibrational flow deforms the temperature field in such a way that the surfaces of constant density orient normally to the vibration axis. In technological experiments on growing crystals this may ensure flattening of crystallization fronts. It is well known that a purely thermocapillary flow yields the opposite result [23–25].

4 Convective Flows in a Closed Cavity Subjected to Swing Oscillations

In the preceding sections of this chapter three particular problems were considered corresponding to different classes of non-uniform vibrations. There exists one more class of phenomena characterized by vibration-induced non-uniformities of an isothermal pulsation field. It is related to non-translational oscillations of a rigid vessel completely filled with a fluid. It as known that, translational vibrations of such a container cannot induce any internal average flow. Indeed, in this case the inertial forces are of the gradient type, and the vibration effect reduces just to a pressure renormalization. In the reference frame fixed on the container, the fluid is at rest, and in the laboratory reference frame it moves like a solid body together with the container.

Non-translational vibrations make the inertial forces non-potential. Hence, the fluid cannot remain quiescent relative to the container walls. At high-frequency vibrations the situation looks as follows. In the interior, the pulsation velocity field is described by the equations of inviscid fluid and due to that is potential. However, this flow pattern cannot be global since it is incompatible with the no-slip boundary conditions. In the region adjacent to the walls the effect of viscosity enters the scene, creating a dynamical skin layer. Thus at the cavity walls the no-slip conditions for the complete velocity are recovered.

For the situation under study we adopt the following approach. The problem of the boundary layer near a rigid wall is solved by conventional methods of the boundary layer theory. This solution is then matched to the solution in the interior. As shown by Schlichting, the asymptote of the boundary layer solution for the average velocity at the outer border of the skin layer does not equal zero, i.e. at this boundary the tangential component of the velocity does not vanish, but has a finite value. This value is taken as an effective boundary condition for the tangential component of the average velocity on a rigid wall. It is important that the Schlichting boundary condition obtained in this way does not contain viscosity, though the discussed effect is virtually of a viscous origin.

Therefore, in the case of non-translational vibrations of a closed cavity completely filled with a fluid, the Stokes boundary layer arises near the cavity walls. The flow in this layer is of the vortex type. The average vorticity generated in the skin layer diffuses into the fluid interior, generating there an average flow. The latter organizes in such a way that the fluid moves along the rigid boundaries from the regions of high pulsation energy to areas of a low one. Note that if the cavity has the shape of a body of revolution, the Stokes skin layer is generated as well. However, in this case the energy of pulsations is uniform all over the body surface and, consequently, the average flow does not arise [26]. In this section we consider the situation where the cavity is not a body of revolution and an average flow generation takes place.

Average flows in a closed fluid-filled cavity subject to non-translational vibrations are studied experimentally and theoretically in [27] to [30]. In [27] the average velocity in the near-wall areas is calculated for the case of swing oscillations of a cylinder, with the cross-section being an ellipse or an equilateral triangle. Experiments and a numerical simulation of [28] deal with the average flows in an isothermal fluid-filled cylinder of square cross-section subject to high-frequency swing oscillations about the cylinder axis. It is found that high-frequency swing oscillations of the cavity result in the average flow generation in the bulk of the cavity. The isolines of the pulsation energy and the stream lines of average flow are described in Figs. 173a and b, respectively. It can be seen that the flow in the bulk consists of eight symmetrical vortices. The

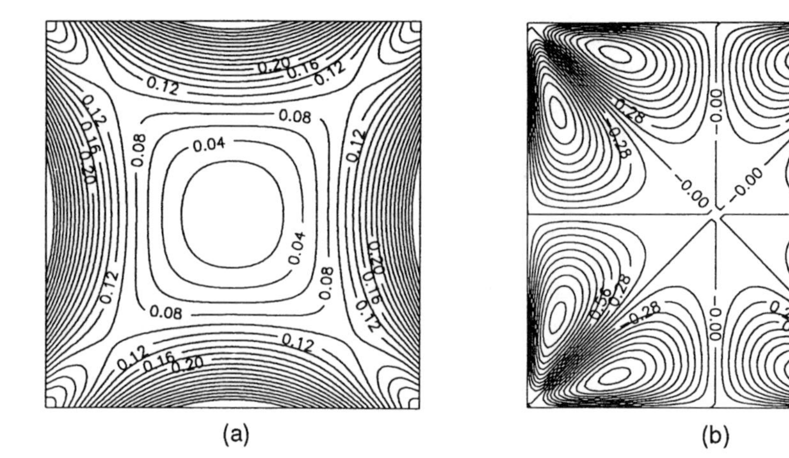

Fig. 173 Pulsation energy (a) and the average flow structure (b) in the case of isothermal fluid ($Re_p = 100$)

fluid moves along the cavity walls from their centers (the areas of maximal pulsation energy) towards the corners (the areas of lower pulsation energy), thus forming the plane jets along the diagonals passing from the square corners to its center.

In [29] convective flows of a non-isothermal fluid in a rectangular cavity subject to low-frequency swing oscillations are studied experimentally and numerically for the case where the swing arm is of the same magnitude as the cavity size. In [30] a coupling of the thermal buoyancy convection and thermovibrational convection in a plane layer is investigated for the limiting case of large-arm swings. In that case, the first-order thermovibrational effect and the Schlichting effect are negligible (see Section 1 in Chapter 8) and the average flow is induced by the second-order effects.

In this section we consider vibrational flows of a non-uniformly heated fluid in a closed cavity subject to high-frequency swing vibrations in the case where the length of the swing arm is comparable with the cavity size. As follows from Section 3 in Chapter 6, in this case the first-order volumic vibrational mechanism of the average flow generation emerges in addition to the isothermal effects investigated in [28]. Interaction and relative roles of different vibrational flow-induction mechanisms are analyzed in weightlessness and under normal gravity.

4.1 Formulation of the Problem

Let us consider the behavior of a non-isothermal fluid in a long cylinder with a square cross-section subject to sinusoidal swing oscillations about the axis

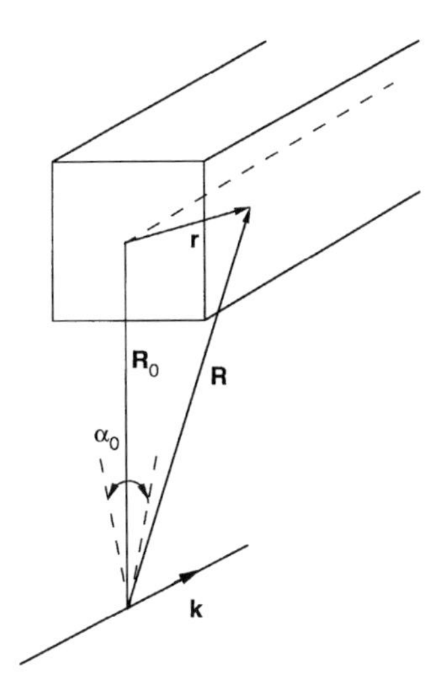

Fig. 174 Geometry of the problem

parallel to the cylinder axis z (Fig. 174). We assume that the vibration frequency Ω is large while their angular amplitude α_0 is small:

$$\left(\frac{\nu}{\Omega}\right)^{1/2} \ll a, \qquad \alpha_0 \ll 1 \tag{9.140}$$

where a is the length of the cavity size.

It is then possible to decompose all the fields into sums of the 'fast' and 'slow' components and use the multiscale method to obtain the governing equations for the average and pulsation components:

$$\mathbf{v} = \mathbf{V} \cos \Omega t + \mathbf{u}, \qquad p = \Phi \sin \Omega t + q \tag{9.141}$$

Here $\mathbf{V}$ and Φ are the amplitudes of the pulsation velocity and pressure, and $\mathbf{u}$ and q are the corresponding average quantities. We assume weakly non-isothermal conditions. In this case the equations for the average and pulsation fields are

$$\frac{\partial \mathbf{u}}{\partial t} + (\mathbf{u}\nabla)\mathbf{u} - \frac{1}{4}\beta V^2 \nabla T = -\frac{1}{\rho}\nabla p + \nu \Delta \mathbf{u} + g\beta T \gamma \tag{9.142}$$

$$\frac{\partial T}{\partial t} + \mathbf{u}\nabla T = \chi \Delta T, \qquad \text{div } \mathbf{u} = 0 \tag{9.143}$$

$$\text{curl } \mathbf{V} = 0, \qquad \text{div } \mathbf{V} = 0 \tag{9.144}$$

(see Section 3 in Chapter 6).

Let us introduce vector $\mathbf{r}_0$, which connects the swing axis with the z axis of the cylinder and is perpendicular to them. We denote $\mathbf{r}$ to be the radius vector of a given point of the boundary with respect to the z axis and $\mathbf{R}$ to be the same but with respect to the swing axis. Evidently, $\mathbf{R} = \mathbf{r}_0 + \mathbf{r}$. Let the angular frequency of swings be time-dependent according to $\omega = \alpha_0 \Omega \cos \Omega t$, where α_0 is the angular amplitude. Then the linear velocity on an arbitrary point on the boundary equals $\mathbf{V}|_\Gamma = \omega \mathbf{k} \times \mathbf{R}$, where $\mathbf{k}$ is the unit vector of the swings axis. Then the boundary condition for the field $\mathbf{V}$ may be rewritten in the form

$$V_n|_\Gamma = \alpha_0 \Omega \mathbf{n}(\mathbf{k} \times \mathbf{R}) \tag{9.145}$$

It is convenient to extract from $\mathbf{V}$ a part that is responsible for a solid rotation. To do that, we present $\mathbf{V}$ as

$$\mathbf{V} = \alpha_0 \Omega(\mathbf{W} + \mathbf{k} \times \mathbf{R}) = \alpha_0 \Omega(\mathbf{W} + \mathbf{A} + \mathbf{k} \times \mathbf{r}), \qquad \mathbf{A} = \mathbf{k} \times \mathbf{r}_0 \tag{9.146}$$

The field $\mathbf{W}$ which has the meaning of the pulsation velocity amplitude in the reference frame of the container is rendered by the solution of the problem

$$\operatorname{curl} \mathbf{W} = -2\mathbf{k}, \qquad \operatorname{div} \mathbf{W} = 0 \tag{9.147}$$

with the boundary conditions

$$W_\mathrm{n}|_\Gamma = 0 \tag{9.148}$$

Introducing the stream function for pulsations by relation

$$\mathbf{W} = \operatorname{curl} \ (\mathbf{k}\,\xi) \tag{9.149}$$

we have

$$\Delta \xi = 2, \qquad \xi|_\Gamma = 0 \tag{9.150}$$

Substituting (9.146) into the thermovibrational force, after simple transformations and omitting the gradient terms we obtain

$$V^2 \nabla T = \alpha_0^2 \Omega^2 \left[W^2 + 2(r_0 - y)\frac{\partial \xi}{\partial y} - 2x\frac{\partial \xi}{\partial x} \right] \nabla T - 2\alpha_0^2 \Omega^2 T(\mathbf{r}_2 + \mathbf{r}_0)$$

Here the y axis is directed along the vector $\mathbf{r}_0$ whereas $\mathbf{r}_2$ is a two-dimensional radius vector, i.e. the projection of $\mathbf{r}$ on the plane normal to the swings axis. In the above formula, the terms proportional to derivatives of ξ represent the Coriolis force and the last term corresponds to the centrifugal force.

Let us restrict ourselves to a consideration of two-dimensional flows for which only the x and y components of the velocity differ from zero and all the variables do not depend on z. In this case it is convenient to introduce the stream function and vorticity of the average flow:

$$\mathbf{u} = \text{curl}\,(\mathbf{k}\,\psi) = \frac{\partial \psi}{\partial y}\,\mathbf{e}_x - \frac{\partial \psi}{\partial x}\,\mathbf{e}_y, \quad \varphi = \mathbf{k}\cdot \text{curl}\;\mathbf{u} = -\Delta\psi \tag{9.151}$$

The average flow equations in terms of ψ and φ can be written as

$$\frac{\partial \varphi}{\partial t} + \frac{\partial \varphi}{\partial x}\frac{\partial \psi}{\partial y} - \frac{\partial \varphi}{\partial y}\frac{\partial \psi}{\partial x} - Pr^{-1}\,Ra_{\text{v}}\left(\frac{\partial D}{\partial x}\frac{\partial T}{\partial y} - \frac{\partial D}{\partial y}\frac{\partial T}{\partial x}\right)$$

$$= \Delta\varphi + Pr^{-1}\frac{\partial T}{\partial x}[Ra - 2\,Ra_{\text{v}}(y + r_0)] + 2\,Pr^{-1}\,Ra_{\text{v}}\,x\frac{\partial T}{\partial y} \tag{9.152}$$

$$\frac{\partial T}{\partial t} + \frac{\partial T}{\partial x}\frac{\partial \psi}{\partial y} - \frac{\partial T}{\partial y}\frac{\partial \psi}{\partial x} = Pr^{-1}\Delta T \tag{9.153}$$

$$\varphi = -\Delta\psi \tag{9.154}$$

Here

$$D = W^2 + 2(r_0 - y)\frac{\partial \xi}{\partial y} - 2x\frac{\partial \xi}{\partial x}$$

With allowance for the Schlichting generation of the average vorticity in the boundary layer, the boundary conditions for the average velocity have the form

$$u_{\text{n}} = 0, \qquad \mathbf{u}_{\text{t}} = -Re_{\text{p}}[(\mathbf{W}_{\text{t}}\nabla_{\text{t}})\mathbf{W}_{\text{t}} + 2\mathbf{W}_{\text{t}}\,\text{div}\,\mathbf{W}_{\text{t}}] \tag{9.155}$$

For the configuration under study, being presented in terms of the stream function ψ, these conditions read as follows: on the horizontal boundaries at $y = \pm\frac{1}{2}$:

$$\psi = 0, \qquad \frac{\partial \psi}{\partial y} = -\frac{3}{4}Re_{\text{p}}\frac{\partial \xi}{\partial y}\frac{\partial^2 \xi}{\partial x\,\partial y} \tag{9.156}$$

and on the vertical boundaries at $x = \pm\frac{1}{2}$:

$$\psi = 0, \qquad \frac{\partial \psi}{\partial x} = \frac{3}{4}Re_{\text{p}}\frac{\partial \xi}{\partial x}\frac{\partial^2 \xi}{\partial x\,\partial y} \tag{9.157}$$

Two variants of thermal boundary conditions were considered. The first case corresponds to the heating from the side walls:

$$\begin{aligned} x = \pm\tfrac{1}{2}&: \qquad T = \pm\tfrac{1}{2} \\ y = \pm\tfrac{1}{2}&: \qquad T = x \end{aligned} \tag{9.158}$$

In the second case, the horizontal boundaries are assumed to have different constant temperatures whereas at the side walls a linear temperature distribution takes place:

$$
\begin{aligned}
x &= \pm\tfrac{1}{2}: \qquad T = -y \\
y &= \pm\tfrac{1}{2}: \qquad T = \mp\tfrac{1}{2}
\end{aligned} \tag{9.159}
$$

Equations (9.152) to (9.154) and boundary conditions (9.155) to (9.159) are written in a dimensionless form. The following quantities are chosen, respectively, as scales for length, stream function, temperature and time: a, ν, θ, a^2/ν. Here θ is the temperature difference. Note that having passed to the dimensionless variables, we retain for them the same notations.

The governing parameters of the problem are the pulsational Reynolds number $Re_{\mathrm{p}} = \alpha_0^2 a^2 \Omega/\nu$, gravitational Rayleigh number $Ra = g\beta\theta a^3/(\nu\chi)$, vibrational Rayleigh number $Ra_{\mathrm{v}} = \alpha_0^2\Omega^2\beta\theta a^4/(4\nu\chi)$ and Prandtl number $Pr = \nu/\chi$; r_0 is the dimensionless arm of the swing oscillations.

4.2 Numerical Results and Discussion

Numerical simulation was carried out by the finite difference method. We present here some of the numerical results obtained a Prandtl number equal 10. A uniform mesh 30×30 was used in most of the calculations.

In Figs. 175 and 176 the stream lines and isotherms of stationary average flows induced by the volumic thermovibrational mechanism in the absence of the Schlichting effect and gravity force ($Ra = 0$, $Re_{\mathrm{p}} = 0$) are presented for $r_0 = 0$ (the swing axis passes through the cavity center) and two different values of the vibrational Rayleigh number $Ra_{\mathrm{v}} = 1.6 \times 10^3$ and 4×10^4

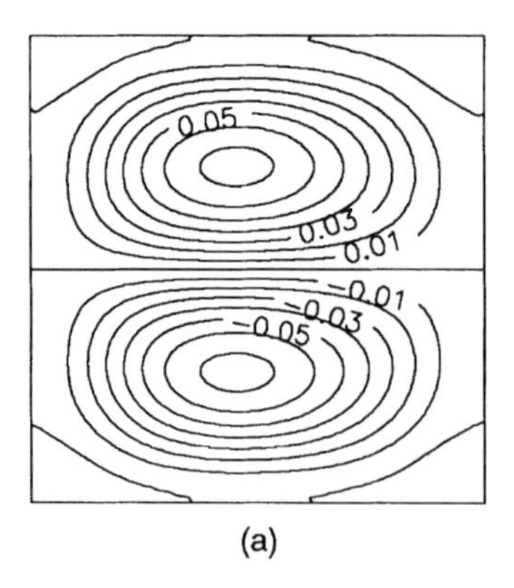

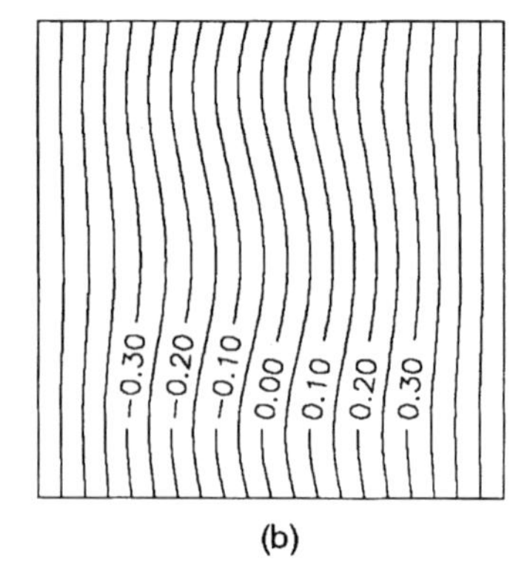

Fig. 175 Flow structure (a) and the temperature field (b) in the case of purely thermovibrational convection ($Ra = 0$, $Re_{\mathrm{p}} = 0$, $Ra_{\mathrm{v}} = 1.6 \times 10^3$, $r_0 = 0$)

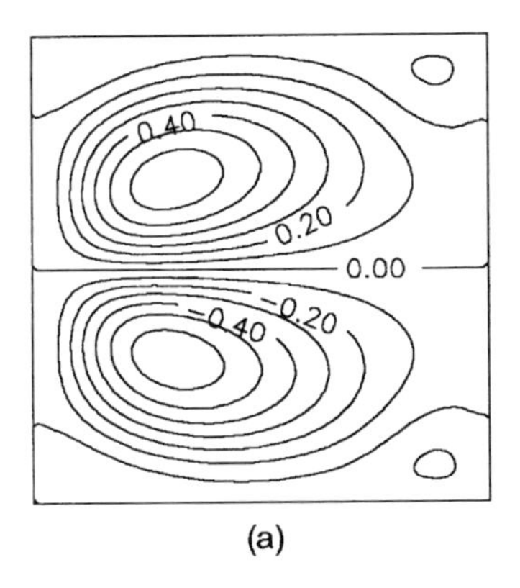

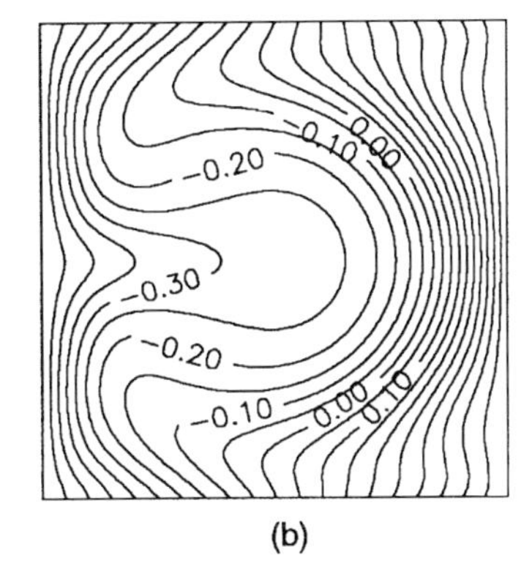

Fig. 176 Flow structure (a) and the temperature field (b) in the case of purely thermovibrational convection ($Ra = 0$, $Re_{\mathrm{p}} = 0$, $Ra_{\mathrm{v}} = 4 \times 10^4$, $r_0 = 0$)

(heating from the side wall). The flow has a two-vortex structure with the boundary between the vortices parallel to the external temperature gradient. The fluid moves from the cold wall to the hot one by the middle part of the cavity and goes back along the walls. This flow structure can easily be understood. Indeed, as one may see from (9.142), the thermovibrational force is directed along the temperature gradient. This force increases with an increase in the pulsation energy density. Since the latter is maximal in the areas located near the mid-points of the walls, the thermovibrational force induces the flow from the wall mid-points to the cavity corners near a hot wall and in the opposite direction near a cold wall.

At small Ra_{v} the vortices are nearly symmetrical with respect to the change of sign of x. As can easily be verified, under the assumption of a creeping flow,

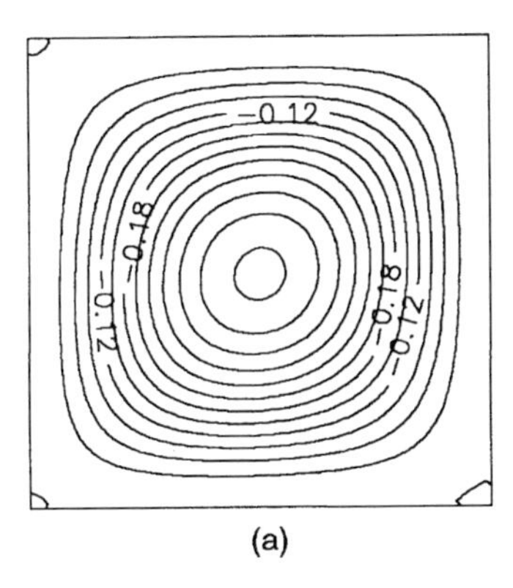

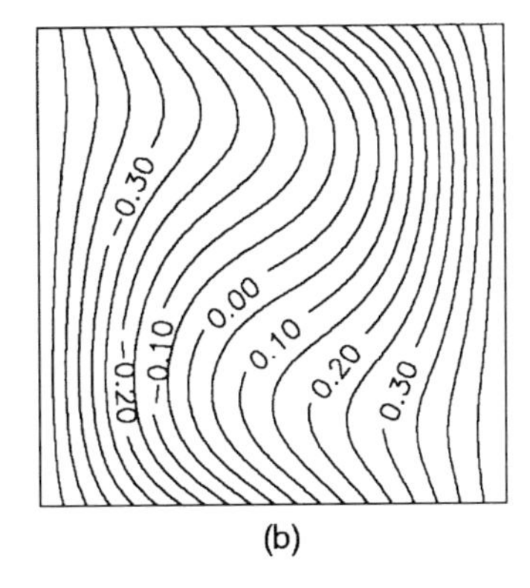

Fig. 177 Flow structure (a) and the temperature field (b) in the case of purely thermovibrational convection ($Ra = 0$, $Re_{\mathrm{p}} = 0$, $Ra_{\mathrm{v}} = 64$, $r_0 = 2$)

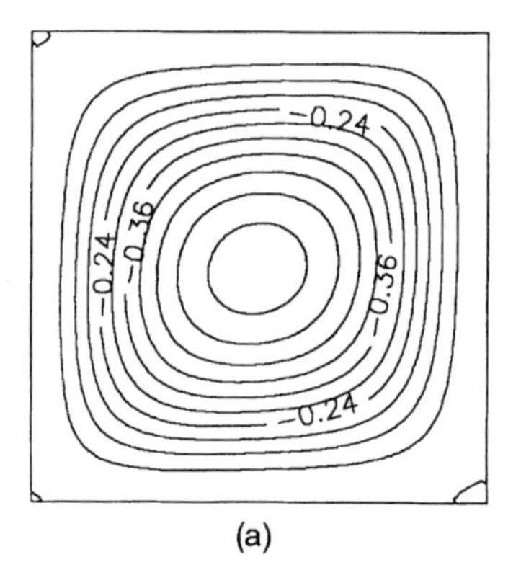

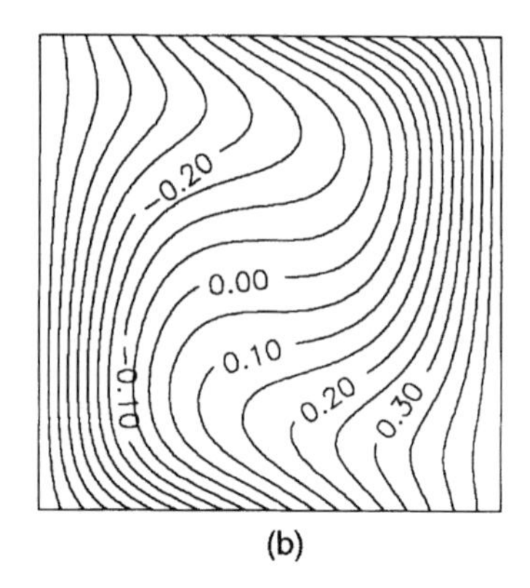

Fig. 178 Flow structure (a) and the temperature field (b) in the case of purely thermovibrational convection ($Ra = 0$, $Re_p = 0$, $Ra_v = 160$, $r_0 = 2$)

this symmetry is rigorous. The flow weakly deforms the temperature field (Fig. 175b). With the increase in Ra_v, the centers of the vortices move to the cold wall (Fig. 176a) and the isotherms become strongly inflected since a temperature boundary layer is formed near the vertical walls (Fig. 176b).

The following figures present the numerical results for a non-zero swing arm. Figures 177 and 178 correspond to $r_0 = 2$, $Re_p = 0$, $Ra_v = 64$ and 160 and heating from the side wall. The flow is one-vortex, its direction reflecting the effect of the centrifugal force. The intensity of flow is markedly higher than in the case $r_0 = 0$.

Figure 179 corresponds to the same values of r_0 and Re_p but the heating is done from below; $Ra_v = 160$. The flow is characterized by a four-vortex

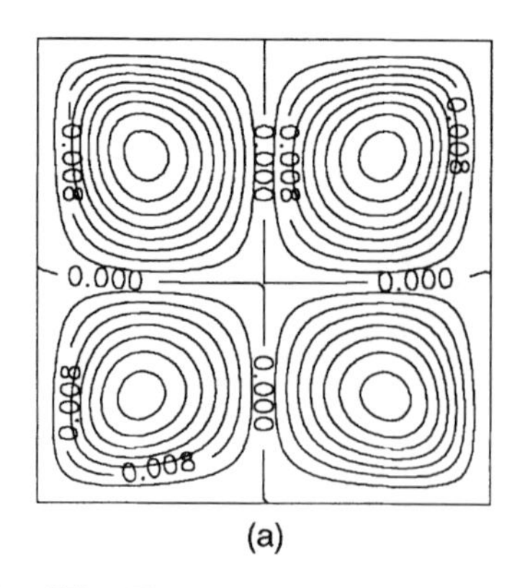

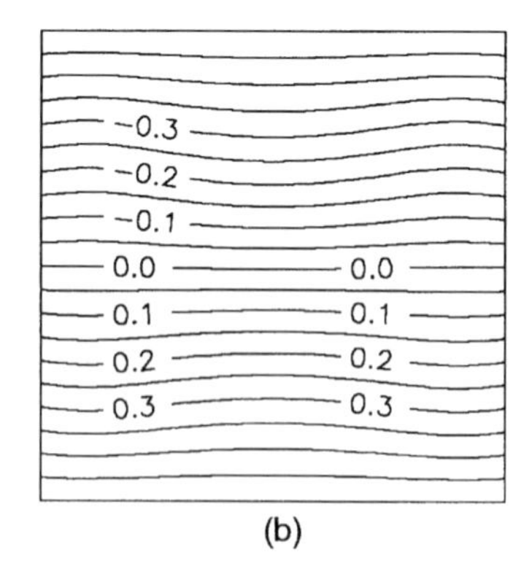

Fig. 179 Flow structure (a) and the temperature field (b) in the case of purely thermovibrational convection for heating from below ($Ra = 0$, $Re_p = 0$, $Ra_v = 160$, $r_0 = 2$)

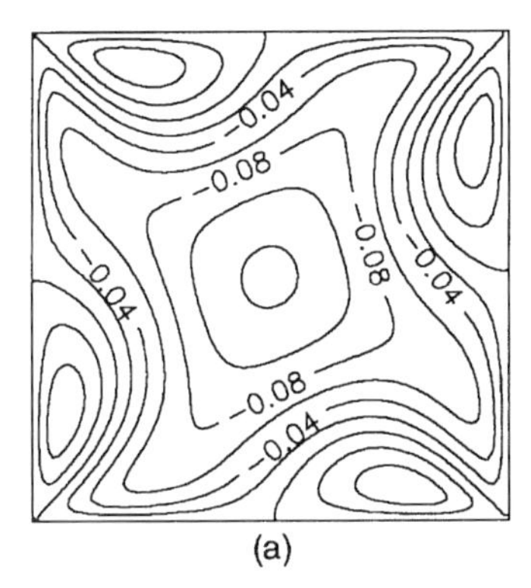

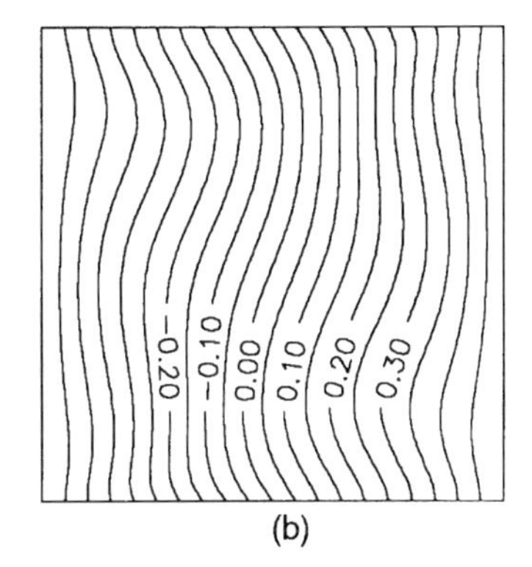

Fig. 180 Flow structure (a) and the temperature field (b) in the case of simultaneous action of the thermovibrational and Schlichting mechanisms ($Ra = 0$, $Re_p = 4.28$, $Ra_v = 16$, $r_0 = 2$)

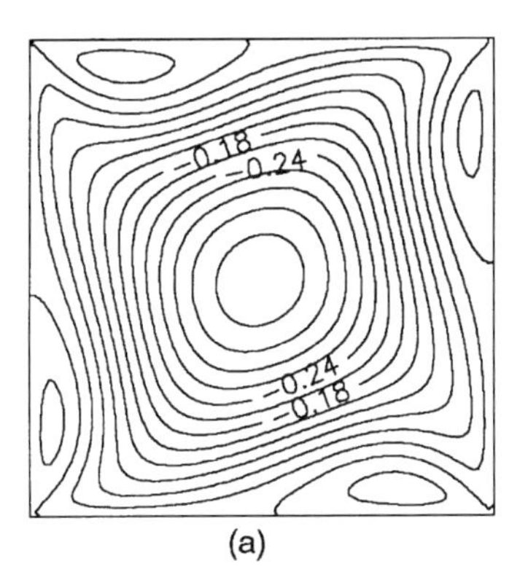

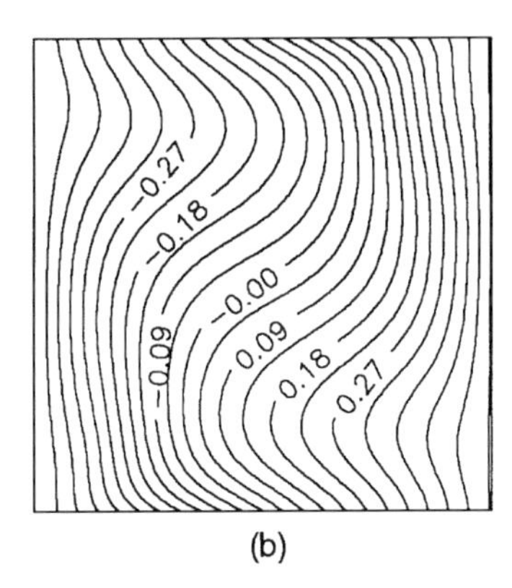

Fig. 181 Flow structure (a) and the temperature field (b) in the case of simultaneous action of the thermovibrational and Schlichting mechanisms ($Ra = 0$, $Re_p = 4.28$, $Ra_v = 64$, $r_0 = 2$)

structure and low intensity, which might have been expected since the temperature gradient and main part of the centrifugal force are antiparallel.

Figures 180 and 181 illustrate the interaction of the thermovibrational and Schlichting mechanisms in the case of heating from the side walls in weightlessness ($r_0 = 2$, $Re_p = 4.28$, $Ra_v = 16$ and 64, respectively). It can be seen that at low values of vibrational Rayleigh number the flow is governed by the isothermal Schlichting mechanism, which induces an eight-vortex average flow (see Fig. 173 for comparison). The flow in the central part of the cavity is one-vortex, which corresponds to the contribution of the centrifugal part of the

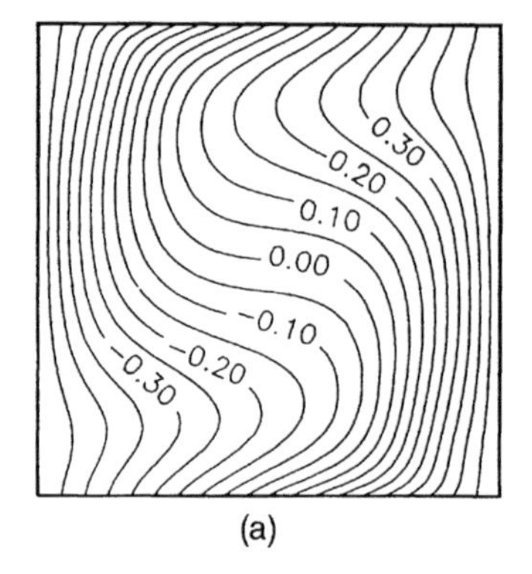

(a)

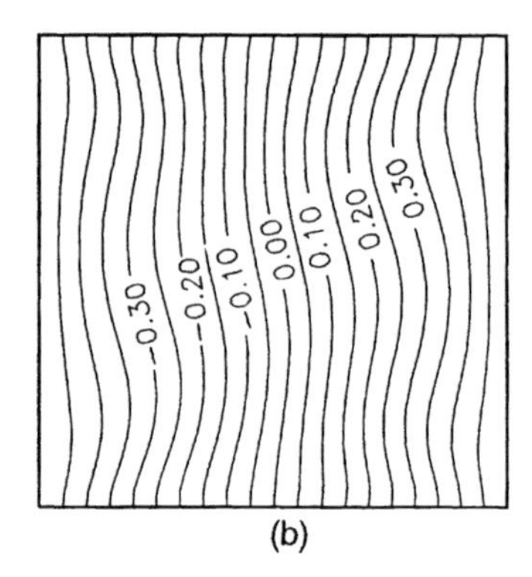

(b)

Fig. 182 Temperature field in the case of simultaneous action of the thermovibrational and thermal buoyancy convection ($Ra = 10^3$, $Re_p = 4.28$, $r_0 = 2$): (a) $Ra_v = 32$, (b) $Ra_v = 160$

thermovibrational force. The interaction between the Schlichting and thermovibrational mechanisms results in coupling of four of the Schlichting vortices with the central, thermovibrational one. The other four Schlichting vortices, which have the opposite direction of circulation, are clearly observed, being located near the cavity walls.

It could be expected that in the parameter range that corresponds to Figs. 180 and 181, the interaction between the vibration flow and thermal buoyancy convection in the gravitational field might result in the suppression of the thermal buoyancy convection by vibrations. In Fig. 182a and b the isotherms of the stationary flow formed by two vibrational mechanisms (Schlichting flow and thermovibrational one) and the thermal buoyancy convection are presented for $r_0 = 2$, $Re_p = 4.28$, $Ra = 10^3$ and two different values of vibrational Rayleigh number: $Ra_v = 32$ and 160. It can be seen that vibrations lead to the significant suppression of the flow which with necessity results in a reduction of the heat transfer: the value of the Nusselt number decreases by a factor of 2.7 and the isotherms become weakly deformed by convection.

References

1. Riley, N. On a sphere oscillating in a viscous fluid, *Q. J. Mech. Appl. Maths*, 1966, **19**, 461–72.
2. Riley, N. Oscillatory viscous flows: review and extension, *J. Inst. Math. Applics*, 1967, **3**, 419–34.
3. Gopinath, A. Steady streaming due to small-amplitude torsional oscillations of a sphere in a viscous fluid, *Q. J. Mech. Appl. Maths*, 1993, **46**, Pt 3, 501–20.

4. Gopinath, A. Steady streaming due to small-amplitude superposed oscillations of a sphere in a viscous fluid, *Q. J. Mech. Appl. Maths*, 1994, **47**, Pt 3, 461–80.
5. Watson, E.J. Slow oscillations of a circular cylinder or sphere in a viscous fluid, *Q. J. Mech. Appl. Maths*, 1992, **45**, Pt 2, 263–75.
6. Bardakov, A.P. and W.E. Nakoryakov. Heat transfer from a sphere in a sound field, *Izv SOAN SSSR*, 1965, Ser. tekh. nauk, **2**(6).
7. Lyubimov, D.V., A.A. Cherepanov, T.P. Lyubimova and B. Roux. The flows induced by a heated oscillating sphere, *Int. J. Heat/Mass Transfer*, 1995, **38**(11), 2089–100.
8. Gershuni, G.Z., E.M. Zhukhovitsky and A.N. Sharifulin. Vibrational convection in cylindrical cavity, *Numer. Meth. Cont. Med. Mech., Novosibirsk*, 1983, **14**(4), 21–33.
9. Chernatynskii, V.I. Numerical study of vibrational convection in a cylindrical layer, in *Convective Flows*, Perm, 1989, pp. 32–7.
10. Brenier, B., P. Bontoux and B. Roux. Comparaison des méthodes Tau-Chebyshev et Galerkin dans l'étude de stabilé des mouvements de connection naturelle. Problems des valers propres parasites, *J. Mec. Th. et Appl.*, 1986, **5**(1), 95–119.
11. Lyubimov, D.V., T.P. Lyubimova, B. Roux and D.N. Volfson. Vibrational flows in the gap between two infinite cylinders, *Eur. J. Mech., B/Fluids*, 1997, **16**, 705–24.
12. Nield, D.A. The thermovibrational Rayleigh-Jeffrey's problem, *J. Fluid Mech.*, 1967, **29**(3), 545–60.
13. Lyubimov, D.V., T.P. Lyubimova and B. Roux. Mechanisms of vibrational control of heat and mass transfer in a liquid bridge. *Int. J. Heat Mass Transfer*, 1997, **40**, 4031–42
14. Kozlov, V.G., D.V. Lyubimov, T.P. Lyubimova and B. Roux. Time-averaged flows induced by vibrations of solid body, in *Int. Workshop on Non-Gravitational Mechanisms of Convection and Heat/Mass Transfer*, Zvenigorod, 1994, p. 28.
15. Gershuni, G.Z., D.V. Lyubimov, T.P. Lyubimova and B. Roux. Coupled thermovibrational and thermocapillary convection in liquid bridge (floating zone system), in *Proc. of the VIIIth Eur. Symp. on Materials and Fluid Sciences in Microgravity*, ESA Publication Division, 1992, pp. 117–22.
16. Birikh, R.V., V.A. Briskman, V.I. Chernatynsky and B. Roux. Control of thermocapillary convection in a liquid bridge by high frequency vibrations, *Microgravity Quart.*, 1993, **3**(1), 23–8.
17. Gershuni, G.Z., D.V. Lyubimov, T.P. Lyubimova and B. Roux. Convective flows in a liquid bridge under the influence of high-frequency vibrations, *Microgravity Quart.*, 1994, **4**(2), 113–21.
18. Lyubimova, T.P., A. Lizee, G.Z. Gershuni, D.V. Lyubimov, G. Chen, M. Wadih and B. Roux. High frequency vibrations influence on heat transfer, *Microgravity Quart.*, 1994, **4**(4), 259–68.
19. Gershuni, G.Z., D.V. Lyubimov, T.P. Lyubimova and B. Roux. Vibrational convection in a heated liquid bridge with a free surface, *C. R. Acad. Sci. Paris*, 1995, **320**, Ser. IIb, 225–30.
20. Alexander, J.I.D. and Y. Zang. The sensitivity of a non-isothermal liquid bridge to residual acceleration, in *Microgravity Fluid Mechanics IUTAM Symposium, Bremen, 1991* (ed. H.J. Rath), Springer-Verlag, 1992, pp. 167–74.
21. Anilcumar, A.V., R.N. Grugel, X.F. Shen and T.G. Wang. Control of thermocapillary convection in a liquid bridge by vibration, *J. Appl. Phys.*, 1993, **73**(9), 4165.
22. Nicolas, J.A., D. Rivas and J.M. Vega. The combined effect of low-frequency mechanical vibrations and thermocapillary convection in liquid bridges, in *Int. Workshop on Non-Gravitational Mechanisms of Convection and Heat/Mass Transfer*, Zvenigorod, Abstracts, 1994, p. 94.

23. Preiser, F., D. Schwabe and A. Sharmann. Steady and oscillatory thermocapillary convection in liquid columns with free cylindrical surface, *J. Fluid Mech.*, 1983, **126**, 545–67.
24. Kuhlman, H. Small amplitude thermocapillary flow and surface deformations in a liquid bridge, *Phys. Fluids A*, 1988, **4**, 672–7.
25. Chen, G. and B. Roux. An analytical study of thermocapillary flow and surface deformations in floating zones, *Microgravity Quart.*, 1991, **1**(2), 73–80.
26. Lyubimov, D.V. Convective flows under the influence of high frequency vibrations, *Eur. J. Mech. B/Fluids*, 1995, **14**(4), 439–58.
27. Povitskii, A.S. and L.Ya. Lyubin. *Foundations of Dynamics and Heat/Mass Transfer in Fluids and Gases in Weightlessness*, Machinostroenie, Moscow, 1972.
28. Ivanova, A., V. Kozlov, D. Lyubimov and T. Lyubimova. Convective states in a fluid subjected to static gravity and non-translational oscillations, in *IX Eur. Symposium on Gravity Dependent Phenomena in Physical Sciences*, Berlin, Abstracts, 1995, pp. 67–8.
29 Bogatyrev, G.P., M.K. Ermakov, A.I. Ivanov, S.A. Nikitin, D.S. Pavlovskii, V.I. Polezhaev, G.V. Putin and S.F. Savin. Experimental and theoretical investigation of thermal convection for ground-based model of convective sensor, *Izv. RAN, Mekhanika Zhidkosti i Gaza*, 1994, **5**, 67–75.
30. Kozlov, V.G. On vibrational thermal convection in a cavity subject to high frequency vibrational swings, *Izv. AN SSSR, Mekhanika Zhidkosti i Gaza*, 1988, **3**, 138–44.

Index